Thanks for your
support & tolerance of
these projects.

ACS SYMPOSIUM SERIES **788**

Omega-3 Fatty Acids

Chemistry, Nutrition, and Health Effects

Fereidoon Shahidi, EDITOR
Memorial University of Newfoundland

John W. Finley, EDITOR
Kraft Foods

American Chemical Society, Washington, DC

Library of Congress Cataloging-in-Publication Data

Omega-3 fatty acids : chemistry, nutrition, and health effects / Fereidoon Shahidi, John W. Finley, editors.

    p. cm.—(ACS symposium series ; 788)

    Includes bibliographical references and index.

    ISBN 0–8412–3688–7

    1. Omega-3 fatty acids—Congresses.

    I. Shahidi, Fereidoon, 1951- II. Finley, John W. Finley, 1942- III. Series.

QP752.O44 O435   2001
612.3′97—dc21   00–68978

The paper used in this publication meets the minimum requirements of American National Standard for Information Sciences—Permanence of Paper for Printed Library Materials, ANSI Z39.48–1984.

Copyright © 2001 American Chemical Society

Distributed by Oxford University Press

All Rights Reserved. Reprographic copying beyond that permitted by Sections 107 or 108 of the U.S. Copyright Act is allowed for internal use only, provided that a per-chapter fee of $20.50 plus $0.75 per page is paid to the Copyright Clearance Center, Inc., 222 Rosewood Drive, Danvers, MA 01923, USA. Republication or reproduction for sale of pages in this book is permitted only under license from ACS. Direct these and other permission requests to ACS Copyright Office, Publications Division, 1155 16th St., N.W., Washington, DC 20036.

The citation of trade names and/or names of manufacturers in this publication is not to be construed as an endorsement or as approval by ACS of the commercial products or services referenced herein; nor should the mere reference herein to any drawing, specification, chemical process, or other data be regarded as a license or as a conveyance of any right or permission to the holder, reader, or any other person or corporation, to manufacture, reproduce, use, or sell any patented invention or copyrighted work that may in any way be related thereto. Registered names, trademarks, etc., used in this publication, even without specific indication thereof, are not to be considered unprotected by law.

PRINTED IN THE UNITED STATES OF AMERICA

Foreword

The ACS Symposium Series was first published in 1974 to provide a mechanism for publishing symposia quickly in book form. The purpose of the series is to publish timely, comprehensive books developed from ACS sponsored symposia based on current scientific research. Occasionally, books are developed from symposia sponsored by other organizations when the topic is of keen interest to the chemistry audience.

Before agreeing to publish a book, the proposed table of contents is reviewed for appropriate and comprehensive coverage and for interest to the audience. Some papers may be excluded to better focus the book; others may be added to provide comprehensiveness. When appropriate, overview or introductory chapters are added. Drafts of chapters are peer-reviewed prior to final acceptance or rejection, and manuscripts are prepared in camera-ready format.

As a rule, only original research papers and original review papers are included in the volumes. Verbatim reproductions of previously published papers are not accepted.

ACS Books Department

Contents

Preface..xi

Introduction

1. The Chemistry, Processing, and Health Benefits of Highly 2
   Unsaturated Fatty Acids: An Overview
   John W. Finley and Fereidoon Shahidi

Polyunsaturated Fatty Acids in Health and Nutrition

2. Long-Chain Fatty Acids in Health and Nutrition...................................14
   Ian S. Newton

3. Electrophysiologic and Functional Effects
   of Polyunsaturated Fatty Acids on Excitable Tissues:
   Heart and Brain..28
   A. Leaf, J. X. Kang, Y-F. Xiao, G. E. Billman, and R. A. Voskuyl

4. The Omega-6 versus Omega-3 Fatty Acid Modulation
   of Lipoprotein Metabolism...37
   K. C. Hayes

5. Docosahexaenoic Acid in Human Health...54
   Bruce J. Holub

6. Dietary Polyunsaturated Fat in Cardiovascular Disease:
   Boon or Bane?..66
   T. R. Watkins and M. L. Bierenbaum

7. The Effect of Alpha-Linolenic Acid on Retinal Function
   in Mammals..79
   Andrew J. Sinclair and Lavinia Abedin

Production of Polyunsaturated Fatty Acids and Special Nutraceutical Products

8. The Large-Scale Production and Use of a Single-Cell Oil Highly Enriched in Docosahexaenoic Acid..................92
   David J. Kyle

9. Production of Docosahexaenoic Acid from Microalgae..................108
   Sam Zeller, William Barclay, and Ruben Abril

10. Blue-Green Alga Aphanizomenon flos-aquae as a Source of Dietary Polyunsaturated Fatty Acids and a Hypocholesterolemic Agent in Rats..................125
    Christian Drapeau, Rafail I. Kushak, Elizabeth M. Van Cott, and Harland H. Winter

11. Seal Blubber Oil and Its Nutraceutical Products..................142
    Fereidoon Shahidi and Udaya N. Wanasundara

12. Structured Lipids Containing Omega-3 Highly Unsaturated Fatty Acids..................151
    Casimir C. Akoh

13. Modified Oils Containing Highly Unsaturated Fatty Acids and Their Stability..................162
    S. P. J. Namal Senanayake and Fereidoon Shahidi

Aroma Effects, Stabilization, and Analytical Procedures

14. Effect of Lipase Hydrolysis on Lipid Peroxidation in Fish Oil Emulsion..................176
    Junji Terao and Takashi Nagai

15. Farmed Atlantic Salmon as Dietary Sources of Long-Chain Omega-3 Fatty Acids: Effect of High-Energy (High-Fat) Feeds on This Functional Food..................191
    Robert G. Ackman, Xueliang Xu, and Catherine A. McLeod

16. Highly Unsaturated Fatty Acids as Precursors of Fish Aroma..................208
    Toshiaki Ohshima, Janthira Kaewsrithong, Hiroyuki Utsunomiya, Hideki Ushio, and Chiaki Koizumi

17. Identification of Potent Odorants in Seal Blubber Oil by Direct Thermal Desorption–Gas Chromatography–Olfactometry...221
    Keith R. Cadwallader and Fereidoon Shahidi

18. Errors in the Identification by Gas–Liquid Chromatography of Conjugated Linoleic Acids in Seafoods...235
    Robert G. Ackman

19. Impact of Emulsifiers on the Oxidative Stability of Lipid Dispersions High in Omega-3 Fatty Acids.............243
    Eric A. Decker, D. Julian McClements, Jennifer R. Mancuso, Larry Tong, Longyuan Mei, Shigefumi Sasaki, Sam G. Zeller, and James H. Flatt

20. Application of Natural Antioxidants in Stabilizing Polyunsaturated Fatty Acids in Model Systems and Foods.................258
    Leon C. Boyd

21. The Effect of Additives on the Rancidity of Fish Oils..........................280
    R. J. Hamilton, G. B. Simpson, and C. Kalu

Indexes

Author Index..301

Subject Index...303

Preface

Interest in the omega-3 fatty acids as health-promoting nutrients has expanded dramatically in recent years. A rapidly growing literature illustrates the benefits of polyunsaturated fatty acids (PUFA), in general, and long-chain PUFA (LC PUFA) of the omega-3 family, in particular, in alleviating cardiovascular disease, atherosclerosis, autoimmune disorder, diabetes, and other diseases. In addition, essentiality of LC PUFA, especially docosahexaenoic acid (DHA), in the development of brain gray matter and retina in the fetus has been well documented. Presence of DHA and docosapentaenoic acid (DPA) and a much smaller amount of eicosapetaenoic acid (EPA) in human milk is associated with better performance of breast-fed infants when compared to those on infant formula. Therefore, pregnant and lactating women are encouraged to ensure their dietary intake of appropriate amounts of omega-3 fatty acids.

Evolutionary assessments suggest that most Western populations consume far less omega-3 fatty acids than historically or presently considered nutritionally desirable. High-oil fish are the best sources of omega-3 fatty acids, but consumption of fish is too low to meet the requirements. Efforts to supplement foods with omega-3 fatty acids have been slow because of off-flavors associated with the oxidation products of PUFA due to the highly labile nature of raw materials. Thus, scientists and technologists are presented with the difficult challenge of delivering highly unsaturated fatty acid foods that are appealing and do not have off-flavors.

The stability of omega-3 fatty acids is dictated not only by the number of double bonds present in the molecules but also by the endogenous and exogenous antioxidants employed as well as the system in which they are incorporated. Microencapsulation of products as a means of delivering these highly labile materials to foods has thus been perfected and introduction of PUFA to bread, cereal, and dairy-based products and spreads has slowly been taking place. Furthermore, production of omega-3 capsules as well as concentrates and structured lipids containing omega-3 fatty acids has led to a large number of over-the-counter supplements and nutraceuticals.

It is the purpose of this book to present a comprehensive assessment of the current state of the chemistry, nutrition, and health aspects of omega-3 fatty acids and to address stability issues and the potential for their delivery in functional foods. We are indebted to participating authors who provided authoritative views and results of their latest investigations on different aspects of omega-3 fatty acids.

FEREIDOON SHAHIDI
Department of Biochemistry
Memorial University of Newfoundland
St. John's, Newfoundland A1B 3X9, Canada

JOHN W. FINLEY
Kraft Foods
801 Waukegan Road
Glenview, IL 60025

Introduction

Chapter 1

The Chemistry, Processing, and Health Benefits of Highly Unsaturated Fatty Acids: An Overview

John W. Finley[1] and Fereidoon Shahidi[2,3]

[1]Kraft Foods, 801 Waukegan Road, Glenview, IL 60025
[2]Department of Biochemistry, Memorial University of Newfoundland, St. John's, Newfoundland A1B 3X9, Canada
[3]Visiting professor, Department of Chemistry, National University of Singapore, 3 Science Drive 3, Singapore 117543

Long-chain polyunsaturated fatty acids (LCPUFA) are primarily referred to eicosapentaenoic acid (EPA), docosapentaenoic acid (DPA) and docosahexaenoic acid (DHA) in the omega-3 series and arachidonic acid (AA) in the omega-6 series. The beneficial health effects of omega-3 fatty acids are related to their protection against cardiovascular disease, autoimmune disorders, diabetes, arthritis and arrhythmia. DHA has also been recognized as an essential nutrient in the brain and retina. However, the highly unsaturated nature of oils containing LCPUFA brings about oxidation and rancidity problems. Thus, use of antioxidants and novel processing techniques is necessary in order to allow their food applications. Omega-3 concentrates may also be produced and used as over-the-counter drugs or as nutraceuticals.

Polyunsaturated fatty acids (PUFA) provide unique health benefits to consumers but also present the scientists and technologist with a difficult challenge in delivering the highly unsaturated fatty acids (HUFA) foods that are appealing and do not have off flavors associated with the oxidation products of these labile materials*(1)*. Interest in omega-3 fatty acids as health promoting nutrients has expanded dramatically in the last several years*(2,6)*. There is a rapidly growing body of literature illustrating cardiovascular and other health benefits of HUFA. Evolutionary assessments suggest that most western populations are consuming far less omega-3 fatty acids than historically and much less than now appears to be nutritionally desirable*(7,8)*. High oil fish are the best sources of omega-3 fatty acids. The consumption of fish is too low to meet the requirements. Efforts to supplement foods with omega-3 fatty acids have been slowed because of off flavors associated with the oils. The omega-3 rich fish oils are extremely labile to oxidation, thus requiring control of oxidation and off flavor development *(9-11)*.

It is the purpose of this publication to present a comprehensive assessment of the current state of the nutritional effects, the stability issues and the potential for delivery of highly unsaturated fatty acids of the omega-3 family in foods.

Chemistry of Highly Unsaturated Fatty Acids

Highly unsaturated fatty acids include any fatty acid with multiple unsaturated olefinic groups in the molecule. For the purpose of this book the emphasis on the polyunsaturates containing omega-3 fatty acids will be on those typically found in marine (e.g. Chapter 12) and algal oils (Chapters 9-11). The chemical structures and common nomenclature for the highly unsaturated fatty acids are shown in Table I.

The degree of unsaturation of these fatty acids makes them extremely vulnerable to oxidative degradation. The oxidation of the fatty acids results in off flavors that can range from desirable, to unpleasant, to offensive. In the chapter by Ohshima et al (Chapter 17) the route by which unsaturated fatty acids lead to "fishy" flavors is discussed. The origin of the off flavors are breakdown products of hydroperoxides formed from the highly unsaturated lipids in the fish (e.g. 12). The variations in aromas are related to the starting fatty acids in the fish species. Terao and Nagai (Chapter 15) show data suggesting that in some systems free polyunsaturated fatty acids provide protection to the triacylglycerols containing highly unsaturated fatty acids. Meanwhile, Senanyake and Shahidi (Chapter 14) prepared structured lipids containing both omega-3 fatty acids and gamma-linolenic acid and discussed stability characteristics of the resultant structured lipids.

Several approaches are described in this volume to protect HUFAs from oxidation. One indirect approach presented by Senanayake and Shahidi (Chapter 14) is to interesterify the HUFAs into a stable oil such as borage oil. Another approach (Chapter 20) is to partially hydrolyze the unsaturated oil and combine it with emulsifiers. It is interesting to note that cationic emulsifiers were significantly more effective in inhibiting oxidation than either neutral or anionic emulsifiers. In addition the inclusion of chelating agents to scavenge metal ions, particularly those of iron were shown to be beneficial. Hamilton et al (Chapter 22) discuss the use of binary and ternary combinations of tocopherols, lecithin, ascorbyl palmitate and cholestane-3,5-diene from corn oil to delay the oxidation of fish oils.

Akoh (Chapter 13) presents an interesting approach to developing a healthy ingredient by interesterifying HUFAs with medium chain lipids resulting in fairly stable oils which provide the health benefits of the HUFAs with the caloric advantage of the medium chain fatty acids. The reader is left with several alternatives that offer varying degrees of protection of the highly unsaturated fatty acids. In the real world of food products it is likely that all of these approaches need to be considered and combinations may offer potential. Currently there are few formulated food products containing high levels of HUFAs that provide long shelf life and good flavor. In these, stabilization of the oil (Chapters 21 and 22) is essential. Effect of oxidation on flavor of products (Chapters 17 and 18) and influence of hydrolysis on stability of unsaturates in emulsion systems (Chapter 15) provides incentives for devising strategies to obtain shelf-stable products. Obviously, fish, as such, may still be used as an important source of omega-3 fatty acids and the advent of aquaculture has played an important role in this area (Chapter 16). However, analytical techniques and appropriate peak assignment of specific fatty acids, such as conjugated linoleic acid (CLA) may not be treated simplistically (Chapter 19).

Evolution of Human Diet

As humans have evolved form a hunter gatherer to the current Western diet, several components in the diet have changed significantly. In prehistoric times omega-3 fatty acids were approximately equal to the omega-6 fatty acids in the diet. In modern time the amount is less than 10% of the fat as omega-3 fatty acids in the diet. Simopoulos (Chapter 2) reviews how the changes in diet have evolved with emphasis on the polyunsaturated fatty acids. She goes on to discuss how these changes may impact the incidence of certain forms of cardiovascular disease.

Nutritional Benefits of Highly Unsaturated Fatty Acids

Highly unsaturated fatty acids have been documented to have positive effects in reducing the risk of certain forms of cardiovascular disease, inflammatory diseases, maternal and infant nutrition. Newton (Chapter 3) presents an excellent review of the potential health benefits of long chain unsaturated fatty acids in the diet. Sinclair and Abedin (Chapter 8) report that the ability of α-linoleic acid to provide for retinal biosynthesis was only about 10% as effective as docosohexanenoic acid. Hayes (Chapter 5) demonstrated that the main effect of omega-3 fatty acids in blood lipid composition is that moderate levels depress plasma triacylglycerols and postprandial lipemia. This suppression may be due to depressed output of VLDL and chylomicrons. Meanwhile, benefits of docosahexaenoic acid (DHA) are discussed by Holub (Chapter 6) and the effect of PUFA on cardiovascular disease (Chapter 7) and their electrophysiological as functional effects are addressed by Leaf et al. (Chapter 4) and issues related to modulation of lipoprotein metabolism are further discussed and clarified by Hayes (Chapter 5).

Nomenclature for Highly Unsaturated Fatty Acids

The nomenclature for the unsaturated fatty acids does not always make it easy to relate to the structure. Table I brings together nomenclature and structure to help the reader relate to different descriptors of the fatty acids used in this book and elsewhere.

Marine Oil Processing

Crude fish oil is a byproduct of the fish meal industry *(13)*. The type of fish caught for the meal industry are white anchovy, black anchovy, sardine, mackerel (Chile and Peru), capelin, blue whiting, herring, menhaden, sandeel (Iceland, Norway), and sprat (Denmark).

Typical commercial products receiving this process are:

- Shark Liver oil
- Salmon oil
- Cod Liver oil
- Tuna oil
- Menhaden oil
- Anchovy oil

The extraction of the oil from the fish is done through a grinding process. Whole fish or heads and tails are being used for fish oil and meal production. The crude materials pass through a rotating grinder where the oil is freed and pressed out. At this stage, the crude oil contains dispersed matters, water, impurities, ketones and aldehydes which must be removed by further processing before the oil is suitable for consumption. The crude oil can be used for the tanning industry, margarine oil (hydrogenated), supplement industry and for aquaculture. Typically the crude oil is stored in drums with nitrogen flushing and sold to a refiner. Crude oil is often stored for up to 3 years.

Table I. Nomenclature of Polyunsaturated Fatty Acids (PUFA)

Common Name	Chemical Name	Shorthand Notation	Chemical Formula
Linoleic (LA)	cis, cis-9,12-Octadecadienoic	18:2 n-6	$CH_3(CH_2)_3(CH_2CH=CH)_2(CH_2)_7COOH$
α-Linolenic (LNA)	all cis-9, 12, 15-Octadecatrienoic	18:3 n-3	$CH_3(CH_2CH=CH)_3(CH_2)_7COOH$
γ-Linolenic (GLA)	all cis-6, 9, 12-Octadecatrienoic	18:3 n-6	$CH_3(CH_2)_3(CH_2CH=CH)_3(CH_2)_4COOH$
Stearidonic	all cis-6,9,12,15-Octadecatertaenoi	18:4 n-3	$CH_3(CH_2CH=CH)_4(CH_2)_4COOH$
Dihomo-γ-linolenic	all cis-8,11,14-eicosatrienoic	20:3 n-6	$CH_3(CH_2)_3(CH_2CH=CH)_3(CH_2)_4COOH$
Meads	all cis-5,8,11-eicosatrienoic	20:3 n-9	$CH_3(CH_2)_6(CH_2CH=CH)_3(CH_2)_2COOH$
Arahidonic (AA)	all cis-5,8,11,15-Eicosatetraenoic	20:4 n-6	$CH_3(CH_2)_6(CH_2CH=CH)_4(CH_2)_3COOH$
Eicosapentaenoic (EPA)	all cis-5,8,11,14, 17-eicosapentaenoic	20:5 n-3	$CH_3(CH_2CH=CH)_5(CH_2)_3COOH$
Docosapentaenoic (DPA)	all cis-7,10,13,16, 19-docosapentaenoic	22:5 n-3	$CH_3(CH_2CH=CH)_5(CH_2)_5COOH$
Docosahexaenoic (DHA)	all cis-4,7,10,13, 16,19-docosahexaenoic	24:6 n-3	$CH_3(CH_2CH=CH)_6(CH_2)_2COOH$

Some of the highest levels of EPA/DHA fish oil come from the anchovy and the sardine, which contain 21% EPA - 9% DHA for the anchovy and 16% EPA - 16% DHA for the sardine. These fish are caught in waters off the coast of Peru and Chile.

The first process used in the refining of crude fish oil is an alkaline refining step. Food grade dilute sodium hydroxide is added to the crude oil to convert the free fatty acids to soap. The FFA soaps then precipitate and are removed with the water in a separator. The oil is then filtered through 5 micron filters which removes any remaining proteins and other particulates.

Bleaching is a simple process where the oil is mixed in a tank with bleaching clay and the mixture is heated to 50-60°C under vacuum. The clay adsorbs the color components of the oil leaving a clear oil. In this process, the hydroperoxides left in the oil are decomposed to carbonyl compounds by the bleaching clay.

The deodorization process used for most marine oils is molecular distillation. The molecular distillation process removes volatile contaminants such as pesticides, remaining free fatty acids, water, ketones and aldehydes.

The deodorization process generally is a 2 stage process.

- The first stage in the molecular distillation is conducted in a wiped film still which consists of two concentric cylinders.
- The inside cylinder contains oil and wipers, which are rotating up to 200 rpm.
- The fast rotating wipers disperse the oil into a thin layer against the wall of the cylinder.
- The top of the cylinder is connected to a vacuum. The vacuum and mid heat remove volatile components form the thin film of oil on the wall. The vacuum also keeps the oxygen levels low so the oxidative damage is minimized despite the heat applied in the process.
- At this stage, the oil and heavy contaminants are collected at the bottom.
- In the second stage of deodorization the oil is heated the outer portion of the still and a condenser in the middle of the "inside cylinder" is chilled to trap volatiles.
- The outside cylinder will be heated up to 250°C which strips more volatiles from the oil and they will be condensed on the chilled inner portion.
- The vapors will condense on the condenser and drop to the bottom of the cylinder for removal through the central drain of the condenser.
- The liquid oil moves down the wall of the inside cylinder and exits where it is cooled through a heat exchanger.

Enriched Omega-3 Oils

For some applications, particularly supplements enrichment of the EPA and DHA in fish oil is desirable (*14*). The enrichment process also reduces the levels of palmitic acid normally present in fish oil. The palmitic acid is known to be

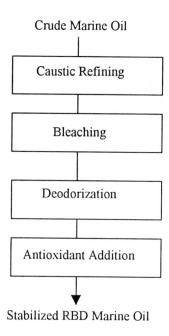

Figure 1. Production of refined, bleached and deodorized (RBD) marine oils.

hypercholesterolemic. Thus reduction of palmitic acid levels greatly enhances the desirability of the final product. EPA/DHA concentrates are prepared starting with high quality oils described above. There are different methodologies for production of omega-3 concentrates. These oils may be in the free fatty acid, alkyl ester or acylglycerol forms. Generally, the first step is to convert all of the fatty acids in the fish oil to esters. Ethyl esters are the preferred form. The process of interesterifcaton to ethyl esters consists of adding ethanol and NaOH to the oil, heated to 80°C.

Distillation or urea complexation is then used to separate the fatty acid esters. Generally the distillation process will be done twice. The first distillation will yield EPA/DHA concentrations of around 25% EPA and 18%DHA. The second distillation will further concentrate the esters to 30% EPA and 20% DHA.

The ethyl esters may then be converted to acylglycerols with or without the addition of other fatty acid sources.
- The re-esterification process is accomplished by adding glycerol, ethanol (to create a blend), and a catalyst, usually sodium methoxide.
- After this reaction, most of the molecule is back to a triacylglycerol (in addition mono- and diacylglycerols may be present).

Bleaching is a process where the esters are mixed in a tank with bleaching clay and the mixture is heated to 50-60°C under vacuum. The clay binds the pigments and some of the volatiles, leaving a clear, bright yellow color.

Finally, if desired, deodorization of the oil is achieved as described earlier.

Encapsulation for Supplements. Either triacylglycerols or ethyl esters can be used for capsules. While esters have traditionally been produced, the triacylglycerol concentrates are becoming more popular because of their perceived true-to-nature status. Microencapsulation might also be considered, however, an encapsulated oil might affect the shelf-life and odor characteristics of products. This issue has received considerable attention in the existing literature.

Summary

A comprehensive review of the current state of knowledge on the chemistry of highly unsaturated fatty acids is provided in this book. The authors have reviewed the health benefits and alternatives to processing and stabilization of the highly unsaturated oils. Clearly there is opportunity in the future to find new creative ways to stabilize these highly beneficial oils for delivery in food systems. To gain consumer acceptance the foods supplemented with these products must taste as good or better than current products and be delivered to the consumer at a modest cost. Supplements offer an alternative approach, but cost and compliance are continuing issues with their use.

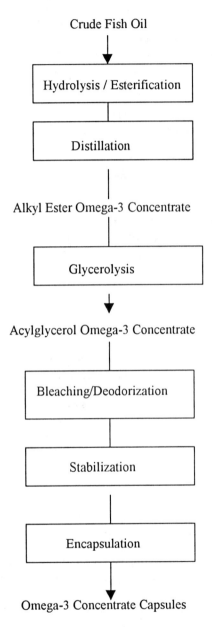

Figure 2. Production of omega-3 concentrates from marine oils.

References

1. Shahidi, F.; Cadwallader, K.R., Eds. Flavor and Lipid Chemistry of Seafoods. ACS Symposium Series 674, Washington, DC, 1997.
2. Bang, H.O.; Bang, J. *Lancet* **1956**, *1*, 381-383.
3. Dyerberg, J.; Bang, H.O.; Hjorne, N. *Am. J. Clin. Nutr.* **1975**, *28*, 958-966.
4. Bang, H.O.; Dyerberg, J.; Hjorne, N. *Acta Med. Scand.* **1976**, *200*, 69-73.
5. Dyerberg, J.; Bang, H.O.; Stofferson, E. *Lancet* **1978**, *2*, 117-119.
6. Dyerberg, J.; Bang, H.O. *Lancet* **1979**, *2*, 433-435.
7. Simopoulos, A.P. *Am. J. Clin. Nutr.* **1991**, *54*, 438-463.
8. Eaton, S.B.; Konner, M. *N. Engl. J. Med.* **1985**, *312*, 283-289.
9. Ke, P.J.; Ackman, R.G.; Linke, B.A. *J. Am. Oil Chem. Soc.* **1975**, *52*, 349-353.
10. Shahidi, F.; Synowiecki, J.; Amarowicz, R.; Wanasundara, U.N. In *Lipids in Food Flavors*; Ho, C.-T.; Hartmen, T.G., Eds. ACS Symposium Series 558. American Chemical Society: Washington, D.C., 1994, pp. 233-243.
11. Shahidi, F.; Wanasundara, U.N.; Brunet, N. *Food Res. Int.* **1994**, *27*, 555-562.
12. Shahidi, F.; Wanasundara, U.N.; He, Y.; Shukla, V.K.S. In *Flavor and Lipid Chemistry of Seafoods*. Shahidi, F. and Cadwallader, K.R. Eds., ACS Symposium Series 674, American Chemical Society: Washington, D.C. 1997, pp. 186-197.
13. Bimbo, A.P. In *Seafood in Health and Nutrition - Transformation in Fisheries and Aquaculture: Global Perspectives.* Shahidi, F. Ed. ScienceTech Publishing Co.: St. John's, Canada, 2000, pp. 45-68.
14. Shahidi, F.; Wanasundara, U.N. *Trends Food Sci. Technol.* **1998**, *9*, 230-240.

Polyunsaturated Fatty Acids in Health and Nutrition

Chapter 2

Long-Chain Fatty Acids in Health and Nutrition

Ian S. Newton

Business Development, Human Nutrition Department, Roche Vitamins Inc., 45 Waterview Boulevard, Parsippany, NJ 07054

The health benefits of omega-3 long chain polyunsaturated fatty acids (LC PUFA) are well established for cardiovascular disease, maternal and infant nutrition and certain inflammatory diseases. New research is leading scientists to study neurological disorders that may have their etiology in fatty acid deficiencies. Many organizations and governments now recognizing the dietary need for LC fatty acids are making intake recommendations. Until the development of special refined highly unsaturated omega-3 oils the possibilities of fortifying foods with these ingredients for health was not possible. Special refining techniques, antioxidants and application developments have combined making addition to foods possible. Infant and maternal formulas are well established but functional foods with combinations of beneficial ingredients are entering the mainstream. Rising healthcare costs, and aging populations are propelling the development of foods for health conscious consumers.

Evolution of the Human Diet

In prehistoric times early man had a diet of roughly equal portions of omega-3 and omega-6 fatty acids ([1], Figure 1). The initial studies by Dyerberg and Bang

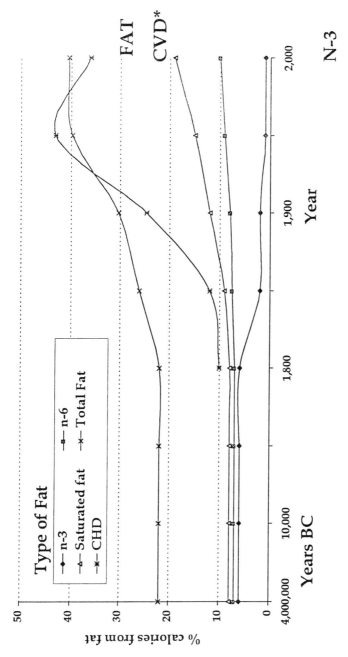

Figure 1. Relationship between fat intake and incidence of coronary heart disease (CHD) as percentage of total mortality.

(2) have led to additional significant research producing sound evidence on the relationship between omega-3 PUFA and CVD (Table I). Data on the benefits of incorporating long chain polyunsaturated fatty acids (LC PUFA) into the diet have been reported for many years. Most experts agree we should consume a ratio of between 5:1-10:1 of omega-6 to omega-3 in our diet (3,4). This balance is recommended by nutrition authorities to be more ideal. It may also explain why the modern diseases such as cardiovascular disease, arthritis, skin disorders, asthma and possibly cancers have increased in the last two centuries (1). Our genetic make-up simply cannot respond fast enough to this about-turn in diet. In the past, for example, we used to eat considerably more fruit and fresh leafy vegetables. Our diet has gradually come to include more meat and more calories. Moreover, the meat and fat from artificially fattened animals has a completely different composition of fatty acids than meat from animals in the wild. The high-energy protein feed given to commercially farmed animals contains a higher proportion of linoleic acid which leads to a shift in the fatty acid profile of farmed animals' meat and fat. This coupled with the widespread use of vegetable oils rich in linoleic acid, such as sunflower, corn and soy bean oil, have resulted in a modern diet containing a large excess of omega-6 fatty acids. PUFA's from the omega-3 family, on the other hand, are often under-represented in the modern western diet.

Critically for infants the current reliance upon infant formula which has increased in the latter half of this century places them at particular risk. In contrast to human milk, North American formula is devoid of the essential omega-3 PUFA docosahexaenoic acid (DHA) (5).

Table I. Myocardial infarction and dietary fat intake in Eskimos and Danes.

	Eskimos	*Danes*
Myocardial Infarction	3	40
Energy from fat (%)	39	42
n-6 PUFA (g/d)	5	10
n-3 PUFA (g/d)	14	3
n-3/n-6	2.8	0.3
Cholesterol (mg)	790	420

From ref. 2.

Saturated and Unsaturated Fatty Acids

Fatty acids are classified as saturated, monounsaturated or polyunsaturated, depending on the number of double bonds. Saturated fatty acids are predominantly found in animal foods: meats and dairy products such as cheese, milk, butter and eggs and fats from beef and pork (6). They are also abundant in the tropical oils, coconut, palm and palm kernel and vegetable shortening. The term "unsaturated" means the hydrocarbon chain contains at least one double bond. Unsaturated fatty acids fall into two categories, monounsaturated (MUFA) and polyunsaturated (PUFA). Monounsaturated fatty acids contain only one double bond and are synthesized within the human body.

Polyunsaturated fatty acids contain two or more double bonds

Linoleic acid and alpha-linolenic acid are precursors or "parent" compounds of omega-6 and omega-3 long chain PUFA respectively. Linoleic acid can be metabolized into gamma-linolenic acid (GLA), dihomo-gamma linolenic acid (DGLA) and arachidonic acid (AA). Alpha-linolenic acid can be metabolized into eicosapentaenoic acid (EPA) and docosahexaenoic acid (DHA). EPA and DHA can also be obtained directly from the diet from fish oils of both marine and freshwater sources (7).

Essential Fatty Acids

Linoleic acid and alpha-linolenic acid are considered to be essential fatty acids (EFA) for human health because humans cannot synthesize them and must obtain them from dietary sources. Similarly, DHA and AA may be considered conditionally essential fatty acids under certain circumstances, particularly during infancy when the body's capacity to convert alpha-linolenic acid and linoleic acid to their higher homologues is limited (8-10).

PUFA Metabolism

The metabolic process that converts linoleic acid to GLA and AA and alpha-linolenic acid to EPA and DHA involves elongation of the carbon chain through the addition of carbon atoms and desaturation of the molecule through the addition of double bonds. This requires a series of special desaturation and elongation enzymes. They regulate the body's delicate chemistry, maintain important hormone-like substances such as prostaglandins, thromboxanes and leukotrienes at the required levels, and play a key role in preventing certain diseases and keeping us generally healthy.

Roles in Human Health

Both classes of PUFA are important for normal biological function and are involved in a variety of physiological processes. Most clinical studies on the protective and therapeutic effects of omega-3 PUFA have used preformed EPA and DHA in the form of fish oil. The potential health benefits of fish oil include reduced risk of coronary vascular disease, hypertension and atherosclerosis as well as inflammatory and autoimmune disorders (*11-13*).

Coronary Vascular Disease

Omega-3 fatty acids have been extensively studied for their effect upon coronary vascular disease (CVD). Although the exact mechanism remains unclear, research suggests that omega-3 PUFA in fish oil may prevent CVD by lowering serum triacylglycerols, reducing the occurrence of arrythmia and acting as antiatherogenetic and antithrombotic agents. Kromhout reported that by increasing fish intake following a myocardial infarction (MI) that time of survival was increased. After four years into the trial, the differences between the two groups became apparent. At the termination of the trial MI rates were 18.5% for non-fish eaters versus 7.2% for fish eaters. Fish intake averaged 42 gram per day (Figure 2). The Zutphen Study, by Kromhout, and the Western Electric Study by Shekelle, both indicate that increasing fish intake in a stepwise fashion reduces risk of cardiovascular disease. At the highest levels of intake, CVD risk was reduced in the Zutphen Study by 60% and in the Wetsern Electric Study, CHD declined from 20% to 12% of deaths. (*14-16*, Figure 3).

Studies investigating the effects of omega-3 PUFA on blood lipid and lipoprotein levels have consistently demonstrated that omega-3 PUFA supplementation lowers blood triglyceride and Very Low Density Lipoprotein (VLDL) concentrations in a dose dependent fashion.

The triacylglycerols lowering effect is seen both in patients with high triacylglycerols and those with normal levels. The magnitude of this effect is large; decreases of 50% or more are frequently observed. Five studies reported consistent lowering of triacylglycerols depending on dose and time of intervention. In general lower doses can be as effective as high doses over an extended period (*17-21*, Table II).

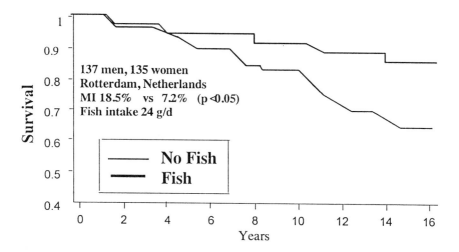

Figure 2.   Dependence of fish intake and mortality (or survival) from coronary heart disease (CHD) indicating long term positive effects. Adapted from ref. 15.

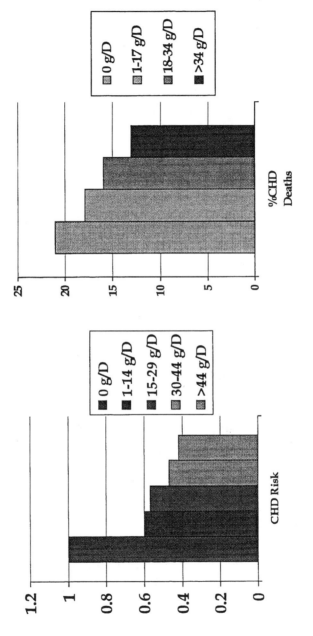

Figure 3. Dependence of coronary heart disease (CHD) risk and death on the consumption of long chain polyunsaturated fatty acids. From Zutphen Study (15) and Western Elect Study (16).

Table II. The effect of omega-3 long chain polyunsaturated fatty acids (LC PUFA) on postprandial lipemia.

ω3 LC-PUFA (g)	Weeks	Δ Fasting TG	Δ Postprandial TG	Reference
28.0	3	-43.7	-44.1	17
7.0	3.5	-42.7	-54.0	18
2.7	6	-24.8	-43.5	19
2.3	15	-26.5	-31.6	20
0.8	16	-21.2	-31.8	21

TAG, triacylglycerol.

Cardiac arrhythmia is believed to be one of the major causes of sudden death in patients with coronary heart disease. Animal experiments have shown that omega-3 PUFA can reduce susceptibility to heart arrthymia (22).

Observational and other studies indicate that a small amount of omega-3 PUFA, equivalent to one fatty fish meal per week was associated with a 50% reduction in risk of cardiac arrest. Siscovick (23) reported a retrospective study on dietary intake of fish and sudden MI. Increasing intake of fatty fish to provide 32-455 mg of EPA plus DHA per day resulted in a stepwise reduction in MI. At the highest level MI was reduced by 70% (Figure 4).

Recent data from a large 11,000 patient well controlled trial in Italy (GISSI) show that following an MI, patients receiving one gram of long chain omega-3 fatty acids as an ethyl ester had a reduction in sudden fatal death of 40%, which surpassed reductions in mortality from various drug treatments. All patients continued on their normal "Mediterranean type" diets and had in addition to the omega-3 fatty acids various drug therapies and thus were already maximally protected on cardiac drugs of various types (24).

Restenosis

Omega-3 fatty acids have been shown to be helpful in preventing restenosis—the reclogging of arteries—in patients undergoing angioplasty. Restenosis commonly occurs in 30 to 45% of the dilated lesions approximately 6 months after the procedure (25). In approximately half of the reported studies, a benefit has been reported when omega-3 fatty acid supplementation was provided (26).

Inflammatory and Autoimmune Disorders

Many experimental studies have provided evidence that omega-3 PUFA may modify and provide modest therapeutic benefits by reducing inflammatory and autoimmune disorders including rheumatoid arthritis, psoriasis, and ulcerative colitis. Omega-3 PUFA are anti inflammatory and decrease monocyte prostaglandins PGI2 and PGI3. Reduction is also noted for neutrophil leucotrienes B4 and B5. Effects have been seen at dosage ranges from 0.5-6gm fish oil per day. Common doses are about 3g/day. The benefit from omega-3 PUFA is seen as a decrease in long term use of non steroidal anti inflammatory drugs (NSAID) with few if any side effects. Results are modest but consistent and include morning stiffness, joint pain, swollen joints, grip strength (*27*).

Infant Development

During intrauterine life DHA and AA are incorporated into the phospholipid membranes of the retina and brain and continue to accumulate during the first two years of life after birth. Therefore adequate intake of DHA and AA is critical during pregnancy, lactation and infancy for proper development of these tissues. Deposition of DHA is especially critical through the third trimester to eighteen months of age. Furthermore it is important to maintain infant plasma levels of LC PUFA since embryonic enzyme systems are not well developed and unable to convert ALA to DHA. In prematurity even human milk can not supply enough DHA to maintain plasma levels at birth. Formula unfortified with DHA places the infant at risk for deficiency of DHA based on plasma levels, Hornstra 1995 (*28*).

The unborn child needs an adequate supply of docosahexaenoic acid (DHA) if the gray matter in its brain and the tissue and cell membranes of the retina are to develop fully and properly. Lucas (*29*) reported that perhaps intellectual development is compromised when comparing babies fed human milk versus DHA unfortified infant formula. Premature babies are affected particularly badly by DHA deficiency because they miss the vital phase before birth during which DHA is supplied and they are not able to synthesize enough of this PUFA to ensure the normal development of the brain and retina.

Recommended Intakes

In the United Kingdom, the Department of Health recommends an omega-3 PUFA intake of a minimum of 0.2 percent of energy (*30*). In addition, the Task Force of the British Nutrition Foundation proposes a daily omega-3 PUFA intake ranging from 0.5 to 2.5 percent of energy in the form of alpha-linolenic acid, which corresponds to 1-6 grams linolenic acid for men and 1-5 grams for women. This translates to 1.25 grams of long chain omega-3 EPA/DHA polunsaturated fatty acid per day.

No specific Recommended Dietary Allowances (RDA's) for either omega-6 or omega-3 PUFA have been established in the United States but the possibility exists for establishing RDAs for these fatty acids in the near future (*31*).

Currently the dietary intake of omega-3 fatty acids in the USA is estimated at 150-200mg per day similar to other western industrialized countries (*32-43*). This would indicate a significant "dietary gap" from generally regarded prudent intakes. A review of various studies indicates current intakes are around 100-200 mgs. per day of EPA plus DHA. This compares to recommendations varying from 400-1250mgs per day (Figure 5).

Special Refining and Quality Assurance Means Bland Fish Oils

An interesting new development for the fats and oils industry is the use of "novel" refining techniques that are able to produce marine oils which can be added to a range of foods without affecting the flavor profile of the product. Previously, fish oils have been used only in a "hardened" or hydrogenated form to prevent the occurrence of fishy off-tastes and smells. Now, fish oils and dry powders, with microencapsulated oils are available for food fortification use. Only oils that have been specifically refined, protected and packed are suitable.

Production of these special types of products is not simple. Normal vegetable oil processing techniques are employed in the production of refined fish oils. However in highly refined oils from Roche Vitamins Inc. special proprietary deodorizing techniques, and the use of special antioxidants ensure oil preparations suitable for wide spread food use even in delicate foods (Figure 6). The quality of the incoming raw oils, the individual manufacturing stages and the resulting refined products from the omega-3 and omega-6 families are all part of the program to produce odor free oils and powders. The products are tested to evaluate the color, appearance, organoleptic profiles and the elimination of any unwanted impurities. The novel refined oil is protected with a powerful antioxidant system to prevent further oxidation to peroxides or aldehydes which may cause off-flavors. The shelf stability of the finished ingredients is good with 6-12 months stability in original unopened containers.

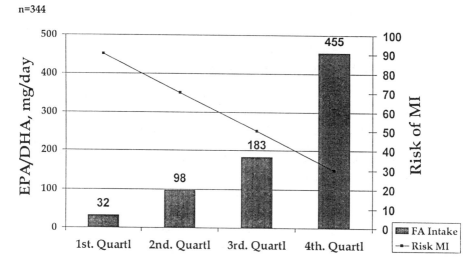

Figure 4.   Dietary intake of fatty acids and risk of myocardial infarction (MI). Adapted from ref. 23.

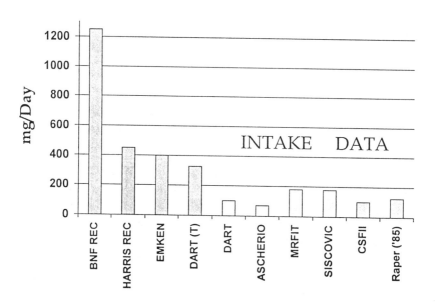

Figure 5.   Intake of omega-3 long-chain polyunsaturated fatty acids (LC PUFA) recommended as by EPA-DHA per day.

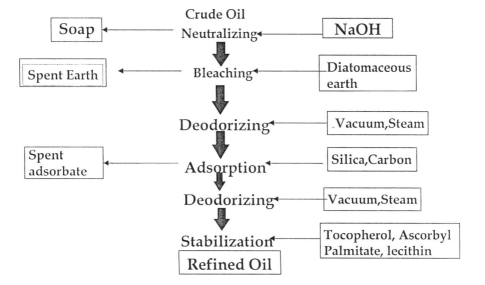

Figure 6. Production process for ROPUFA omega-3 oils.

References

1. Leaf, A.; Weber, P. C. *Am. J. Clin. Nutr.* **1987**;*45*(supple), 1048-1053.
2. Dyerberg, J.; Bang, H.O.; Hjorne, H. *Acta. Med. Scand.* **1976**, *200*, 69-73.
3. Salem, N. Jr.; Ward, G. R. *Rev. Nutr. Diet.* **1993**, *72*,128 -147.
4. Hibbeln, J.R.; Salem, N. Jr. *Am. J. Clin. Nutr.* **1995**, *62*, 1-9.
5. Connor, W.E.; Neuringer, M.; Reisbick, S. *Nutr. Rev.* **1992**, *50* (4), 21-29.
6. *Agricultural Research Service*, U.S. Department of Agriculture. Handbook Number 8, Washington, DC, 1976.
7. Nettleton, J.A. *J. Am. Diet. Assoc.* **1991**, *91*, 331-337.
8. Budowski, P. *World Rev. Nutr. Diet.* **1988**, *57*, 214-74.
9. Clandinin, M.T.; Chappel, J.E.; Leong, S.; Heim, T.; Swyer, P.R. *Early Hum. Dev.* **1980**, *4*, 131-138.
10. Endres, S.; De Caterina, R.; Schmidt, E. B.; Kristensen, S. D. *Eur. J. Clin. Invest.* **1995**, *25*, 629-638.
11. Weber, P.C.; Leaf, A. *World Rev. Nutr. Diet.* **1991**, *66*, 218-232.
12. Leaf, A. *Circulation* **1990**, *82*, 624-62.
13. Schmidt, E.B.; Dyerberg, J. *Drugs* **1994**, *47*, 405-424.
14. Conner, W.E. Evaluation of publicly available evidence regarding certain nutrient-disease relationships: 7. Omega-3 fatty acids and heart disease. Bethesda, MD: Fed. Am. Soc. Ex.Biol., 1991. FDA Contract No. 223-88-2124.
15. Kromhout, D.; Bosschieter, E.B.; Coulander Cde L. *New Engl. J.Med.* **1985**, *312*, 1205-1209.
16. Shekelle, R.B.; Missell, L.V.; Paul, O.; Shyrock, A.M.; Stamler, J. *N. Engl. J. Med.* **1985**, *313*, 820.
17. Harris, W.S.; Conner, W.E.; Alam, N.; Illingworth, D.R. *J. Lipid Res.* **1988**, *29*, 1451-1460.
18. Weintraub, M.S.; Zechner, R.; Brown, A.; Eisenberg, S.; Breslow, J.L. *J.Clin.Invest.* **1988**, *82*, 1884-1893.
19. Williams, C.M.; Moore, F.; Morgan, L.; Wright, J. *Brit. J. Nutr.* **1992**, *68*, 655-666.
20. Agren, J.J.; Hanninen, O.; Julkunen, A.; Fogelholm, L.; Schwab, U.; Pynnonen, O.; Uusitupa, M. *Eur. J. Clin. Nutr.* **1996**, *50*, 765-771.
21. Roche, H.M.; Gibney, M.J. *Eur. J.Clin. Nutr.* **1996**, *50*, 617-624.
22. Leaf, A. *Prosta. Leuko. Essen. Fatty Acid* **1995**, *52*, 197-198.
23. Siscovick, D.S.; Raghunathan, T.E.; King, I.; Weinmann, S.; Wieklund, K.G. *J. Am. Med. Assoc.* **1995**, *274*, 1363-1367.
24. GISSI- Prevenzione Investigators. *Lancet* **1999**, *354*, 447-455.
25. Gapinski, J.P.; VanRuiswyk, J.V.; Heudebert, G.R.; Schectman, G.S. *Arch. Intern. Med.* **1993**, *153*, 1595-1601.

26. Dehmer, G.J.; Pompa, J.J.; Van Den Berg, E.K.; Eichorn, E.J.; Prewitt, J.B.; Campbell, W.B.; Jennings, L.; Willerson, J.T.; Scmitz, J.M. *New Engl. J. Med.* **1988**, *319*, 733-40.
27. Fortin, P.R. *J. Clin. Epidemiol.* **1995**, *48*, 1379-1390.
28. Hornstra, G.; Al, M.D.M.; van Houwelingen, A.C. *Eur. J. Obstet. Gynecol. Reprod. Biol.* **1995**, *61*, 57-62.
29. Lucas A.; Morley, R.; Cole, T.J.; Lister, G.; Leeson-Payne, C. *Lancet* **1992**, *339*, 261-264.
30. Department of Health 1991 Dietary Reference Values for Food Energy and Nutrients for the United Kingdom. Committee on Medical Aspects of Food Policy. Report on Health and Social Subjects, 41, HMSO, London.
31. Food and Nutrition Board, National Research Council, National Academy of Sciences. Recommended Dietary Allowances, 10th Edition. Washington DC, 1989.
32. Raper, N.R.; Cronin, F. J.; Exler, J. *J. Am. Coll. Nutr.* **1992**, *11*, 304-308.
33. Ascherio, A. *New Engl. J. Med.* **1995**, *332*, 977-982.
34. Dolecek, T.A. *P.S.E.B.M.* **1992**, *200*, 177-182.

Chapter 3

Electrophysiologic and Functional Effects of Polyunsaturated Fatty Acids on Excitable Tissues: Heart and Brain

A. Leaf[1], J. X. Kang[1], Y-F. Xiao[2], G. E. Billman[3], and R. A. Voskuyl[4]

[1]Department of Medicine, Massachusetts General Hospital East and Harvard Medical School, 149 Thirteenth Street–1494001, Charlestown, MA 02129–6020
[2]Cardiology Research, Beth Israel Deaconess Medical Center, 330 Brookline Avenue, Boston, MA 02215
[3]Department of Physiology, The Ohio State University, Columbus, OH 45210
[4]Department of Physiology, Leiden University Medical Center and the Stichting Epilepsie Instellingen Nederland, P.O. Box 21, 2100 AA Heemstede, The Netherlands

The n-3 polyunsaturated fatty acids (PUFAs) have been shown to be antiarrhythmic in animals and probably in humans. The PUFAs stabilize the electrical activity of isolated cardiac myocytes by inhibiting sarcolemmal ion channels, so that a stronger electrical stimulus is required to elicit an action potential and the relative refractory period is markedly prolonged. Inhibition of voltage dependent sodium currents which initiate action potentials in excitable tissues, and of the L-type calcium currents, which initiate release of sarcoplasmic calcium stores, that increase cytosolic free calcium concentrations and activate the contractile proteins in myocytes, appear at present to be the probable major antiarrhytmic mechanism of the PUFAs.

Following earlier suggestions (*1, 2*) that unsaturated fatty acids may have antiarrhytmic effects, McLennan et al. (*3, 4*) reported, that feeding rats a diet in which the fat content was largely saturated or monounsaturated resulted in a high incidence of irreversible ventricular fibrillation when their coronary arteries were subsequently experimentally heated. When vegetable oils were the major source of the dietary fat, there was a reduction in artliythmic mortality by some 70%.

With tuna fish oil, however, McLennan (4) reported irreversible ventricular arrhythmias to be completely prevented with or without reflow to the ischemic myocardium. The essential findings were confirmed in marmosets (5). These striking observations led us to pursue the possible mechanism(s) for such an antiarrhythmic action of the fish oil.

Beneficial Effects of Polyunsaturated Fatty Acids

To confirm previous findings, we first studied a canine model of sudden cardiac death. A surgically induced myocardial infarction was produced by ligating the left main coronary artery and an inflatable cuff was placed around the left circumflex artery of dogs as experimental animals. The dogs were allowed about a month to recover froi-n the surgery and their myocardial infarction during which they were trained to run on a treadmill. The animals were then screened for susceptibility to fatal ventricular arrhythmias (VF) when their left circumflex artery was occluded while they were running on a treadmill. Some 60% of animals were found susceptible and these were the dogs studied. Once an animal is "susceptible" it remains susceptible on further exercise-ischemia trials. In 10 of the 13 such dogs intravenous infusion of an emulsion of a concentrate of the fish oil free fatty acids (PUFAs) just prior to the exercise-ischemia test prevented the fatal VF ($p<0.005$) (6). In the control exercise-ischemia tests one week prior to the test with the infusion of the (PUFAs) and one week following that test all animals developed ischemia-induced VF requiring prompt defibrillation. In additional studies we have found that pure eicosapentaenoic acid C20:5n-3, EPA) of docosahexaenoic acid (C22:6n-3, DHA), or α-linolenic acid (C18:3n-3, LNA) were apparently equally antiarrhythmic in this dog preparation (7). We purposely infused the n-3 fatty acids rather than fed the dogs fish oil to be certain exactly what ingredient of the fish oil prevented the fatal VF. Indietary studies invariably several things must change, which may confound the study, but when the free fatty acids were infused intravenously just prior to producing the ischemia and the fatal VF is prevented, then one can feel confident that the effect results from what has just been infused.

Having thus confirmed directly the findings of the earlier workers, the mechanism by which the n-3 polyunsaturated fatty acids (PUFAs) produced their antiarrhythmic effect could be studied. To have a simple, available model to visualize the production of arrhythmias and possible prevention of the arrhythmias by the PUFA, we studied cultured neonatal rat cardiac myocytes (8). Hearts were quickly removed from one to two day old decapitated rat pups. The cardiac cells were separated with trypsin digestion and the cells were plated on microscope cover slips. By the second day of culture one could see clumps of growing myocytes of a few to several hundred cells adherent to the cover slips. Each group of cells was contracting spontaneously, synchronously, and rhythmically. With a microscope, a video camera, and an edge monitor we could focus on a single myocyte in a clump of cells to see and record the rate and amplitude of contractions. With this *in vitro* model we produced arrhythmias with a number of chemicals known to produce fatal VH in humans: elevated extracellular [Ca^{2+}], toxic levels of the cardiac glycoside ouabain (8), excessive β-adrenergic agonist isoproterenol (9), lysophosphatidyl choline, acyl carnitine and even the the calcium inophore A23187 (10). With each agent a tachyrhythmia was induced. If the PUFAs were added to the fluid perfusing the isolated myocytes before the arrhythmogenic toxins were administered, they would in every instance prevent the expected arrhythmia. If the arrhythmia was first induced by the toxin and the PUFAs added to the superfusate in the continued presence of the toxin within a few minutes the arrhythmia would be terminated and the cells would commence beating again regularly. Then in the continued presence of the toxin, the PUFAs could be extracted from the cells with delipidated bovine serum albumin and the arrhythmia would promptly resume (8). These results indicated that it was only the free PUFAs partitioning into the membrane phospholipids that prevented the arrhythmias. If the fatty acid had been covalently bound to any constituent in the membrane, we would not have been able to extract it from the emebrane. When the ethyl ester or the triacylglycerol of the PUFAs were tested, they were not antiarrhythmic in this model; the free carboxylic acid group was essential for this antiarrhythmic action.

Among PUFAs tested for antiarrhythmic activity (8), both the n-3 and n-6 PUFAs were antiarrhythmic, but arachidonic acid (C20:n-6,AA) was anomalous. Cyclooxygenase metabolites of AA (except protacyclin, PGI_2) caused arrhythmias whereas cyclooxygenase metabolites of n-3 EPA did not (11). Thus, the n-3 PUFAs should be tested in clinical trials as antiarrhythmic agents.

The structural requirements for an antiarrhythmic compound that acted in the manner as PUFAs were a long acyl or hydrocarbon chain with two or more C=C unsaturated bonds and a free carboxyl group at one end. With this guideline, all-*trans*-retinoic acid was also found to be specifically antiarrhythmic (12).

The antiarrhythmic action of the PUFAs resulted from their effects on the electrophysiology of cardiac myocytes (13). They caused slight hyperpolarization of the resting or diastolic membrane potential and the threshold voltage for the opening of the Na^+ channel became more positive. This resulted in an increased depolarizing stimulus of about 50% required to induce an action potential. In addition, the refractory period, phase 4 of the cardiac cycle, was prolonged by some 3-fold. These two effects on every myocyte in the heart accounted for the increased electrical stability and resistance of the heart to lethal arrhythmias.

This electrical stabilizing effect of the n-3 PUFAs on every cardiomyocyte could be readily demonstrated *in vitro* (10). Figure 1 shows the tracing of the rate

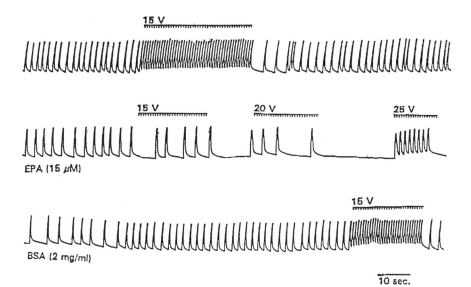

Figure 1. An in vitro demonstration of the electrical stabilizing effect of the n-3 polyunsaturated fatty acids on isolated cultured neonatal rat hearts. The rate and the amplitude of contractions of a single cardiac myocyte in a clump of cells on a microscope cover slip has been shown. The three strips were recorded without interruption from the same myocyte.

and amplitude of contractions of a single cardiomyocyte in a clump of cells growing on a microscope coverslip. The cell was visualized with an inverted microscope. A video camera photographed the contractions which were traced and recorded with an edge monitor. When two platinum electrodes were placed across the microscope coverslip in a perfusion chamber and connected to an external voltage source, the regular beating rate could easily be doubled by stimulating the myocyte by an external field of 15 volts (Figure 1). When the external voltage source was turned off the myocyte regained its prior beating rate. When the same cell was exposed to n-3 EPA (15µM) added to the superfusate, the beating rate began to slow down – a highly reproducible effect of the PUFAs on the neonatal rat cardiomyocytes – and now the myocyte paid no attention to the stimuli from the external voltage source at 15 or 20 volts. However, the external stimuli delivered at 25 volts succeeded in eliciting myocyte contractions, but only in response to every other electrical stimulus. When delipidated bovine serum albumin (2 mg/ml) was added to the superfusate of the same coverslip to extract the free fatty acids from the cardiomyocytes, the beating rate returned to its control frequency and the myocytes responded to the external electrical stimuli delivered at 15 volts, just as they had initially (Figure 1). When one considers that this electrical stabilization is a direct effect of the PUFAs on every cardiac myocyte, both atrial and ventricular, in the absence of neural or humoral effects, one can sense what a potent antiarrhythmic action these n-3 PUFAs may exert.

The antiarrhythmic effects in turn result from the action of PUFAs to modulate the conductance of ion channels in the plasma membranes of heart cells. By whole cell voltage clamp measurements we have found that the PUFAs inhibit the voltage dependent Na^+ (I_{Na}), [14], K^+ (the transient outward current, I_{to7} and the delayed rectifier current, I_K, but not the inward rectifying current, I_{K1}) (Xiao et al, unpublished results), as well as the L-type Ca^{2+} current, $I_{Ca,L3}$ (15). In the case of the Na^+ channels we have shown that only the antiarrhythmic PUFAs noncompetitivwely displace ^3H-batrachotoxinin-20-α-benzoate bound to the sodium channel pore protein (16). This is similar to the finding that PUFAs noncompetitively displaced ^3H-nitrendipine (a specific L-type calcium channel antagonist) from its binding site at the external pore of the calcium channel protein (17). Because the displacement in each case, though specific, was noncompetitive, the urgent question of whether the PUFAs bind specifically and primarily to ion channel proteins directly or interact primarily with the phospholipid of the cell membranes to allosterically change the conformation of transmembrane protein channels could not be resolved as of yet.

In the cultured neonatal rat heart cells the PUFAs were shown to shift the steady state potential for inactivation of the Na^+ channels to more hyperpolarized potentials (14). A recent study of the effects of the PUFAs on the Na^+ currents in α-subunits of the human myocardial Na^+ channel transiently expressed in stable human embryonic kidney cells (18) showed no effect on the activated opening of the Na⁺ channel. The only effect seen was a large shift in the steady state inactivation potential to more negative potentials, i.e., the hyperpolarization of the membrane potential necessary to close the Na⁺ channel and bring the channel to a resting, but activatable state. Attaining a closed, resting, but activatable state is required before the channel is again susceptible to initiation of a new action potential. This makes the I_{Na} voltage-dependent in the presence of the PUFAs, as we have found (14, 18). Myocytes must apparently maintain their normal resting membrane potentials in order to avoid becoming nonfunctional as a result of the voltage-dependent inhibitory actions of the PUFAs on I_{Na}.

These effects, primarily on inhibition of the Na⁺ and Ca²⁺ currents, may account for the potent antiarrhythmic effects of PUFAs examined. With occlusion of a coronary artery myocytes in the core of the ischemic tissue rapidly depolarize and die. The myocytes in the border region of the ischemic zone, however, become partially depolarized due to the reduction of the Na,K-ATPase pump activity with increase in the interstitial K⁺ concentration in the ischemic zone. These cells become "hyperexcitable" and subject to premature action potentials and arrhythmias. Their resting membrane potential is more positive and closer to the threshold for the gating of the I_{Na} so that any further small depolarizing current, such as a current of injury, may elicit a premature action potential and initiate an arrhythmia. Because they have a reduced (more positive) resting membrane potential, it is just these ischemic "hyperexcitable" myocytes that are eliminated quickly from further mischief by the PUFAs. The necessity for a physiologically unattainable more negative resting potential in order to revert the channel to a closed but activatable resting state makes the channel unresponsive and eliminates it as an arrhythmogenic risk. Furthermore, partially depolarized cardiac myocytes can slip spontaneously and quickly from a resting state directlyl into the inactivated, unresponsive state without eliciting an action potential. This effect of the PUFAs on the sodium current, is important in preventing the ischemia-induced fatal arrhythmias associated with acute myocardial infarctions and ischemia.

It should, however, be noted that not all fatal arrhythmias result from dysfunction of sodium channels. The potent inhibitory action of the PUFAs on L-type Ca²⁺ currents, $I_{Ca,L}$ (15), complements this action on I_{Na+}. Significant inhibition of $L_{Ca,L}$ is observed at 10 nM Ca²⁺ in the medium bathing the cultured rat myocytes. This effect prevented triggered arrhythmias induced by overload of cytosolic Ca²⁺ and increased cytosolic Ca²⁺ fluctuations. Most arrhythmogenic cardiotoxins may induce fatal VT or VF by triggered after-potentials from excessive cytosolic Ca²⁺ fluctuations, which are prevented by the PUFAs, e.g., cardiac glycosides, lysophosphatidyl choline, excessive catecholeamines and thromboxane A_2, among others.

Although, at present we cannot be certain that PUFAs prevent lethal arrhythmias in patients, however, two secondary prevention trials, which unexpectedly showed prevention of ischemia-induced sudden cardiac death, are encouraging. One dietary study (*19*) was a prospective, randomized, single-blinded, secondary prevention trial which compared the effect of a Mediterranean α-linolenic acid-rich diet to the usual post-infarct prudent diet. The subjects on the more fat restricted experimental diet receiving α-linolenic acid (C:18n-3) showed a remarkable reduction in mortality and morbidity of some 70%, including prevention of sudden death. The other study (*20*) was also a randomized, prospective, secondary prevention trial in which advice to eat oily fish two or three times weekly was compared with no such advice. This study did not record arrhythmic deaths. It, however, found no reduction in new events, but a 29% reduction in mortality suggested a reduction in sudden deaths which comprise 50% to 60% of the acute mortality from heart attacks (*21*). In both studies the survival curves showed a very early beneficial separation of the experimental versus the control groups, quite unlike the two years required in the cholesterol lowering trial before the lower mortality was significant (*22*). The beneficial effects associated with significant reductions of plasma cholesterol levels was not evident in any of these studies. A recent case control study reported an inverse relationship between fish consumption and sudden cardiac death, suggesting an antiarrhythmic effect from ingestion of fish (*23*). An epidemiologic review based on the Physicians' Health Study reported a 52% reduction in the risk of sudden cardiac death in subjects who ate at least one fish meal per week (*24*).

Perhaps a bonus for pursuing the mechanisms by which n-3 PUFAs prevent fatal cardiac arrhythmias was our resulting expectation that these fatty acids would also have important effects on the brain, as well. Once it was known that these fatty acids were affecting the ion currents in one excitable tissue, the heart, we could predict that they would have effects on other excitable tissues, musche and brain – and in fact they did. All excitable tissues utilize the same ionic currents for cell signaling and the transmembrane protein ion channels are highly homologous. Whole cell voltage-clamp studies on hippocampal CA1 neurons showed that the same PUFAs, which are antiarrhythmic, inhibit the sodium and calcium currents in the brain in a manner quite similar to their effects in the heart (*25*). To determine a possible functional consequence of this electrical effect on the brain, we tested the possible anticonvulsant action of the PUFAs in rats using the cortical stimulation model and found that the PUFAs infused over 30 min via a tail vein in prepared rats promptly, but modestly, raised the electrical threshold for seizure activity significantly for most of a day (26). Whether this will translate into a useful medication for epilepsy, we do not know yet. However, it indicates a clear action of these interesting fatty acids on brain cells as well as on muscle cells.

Conclusion

It is apparent that there exists a basic control of cardiac and neural function by common dietary fatty acids which has been lartely overlooked. The n-3 PUFAs have been part of the human diet for some 2-4 million years during which our genes have adapted to our diet as hunter-gatherers (29) and they are safe. With some 250,000 sudden cardiac deaths annually, lartely due to ventricular fibrillation, in the USA alone (21) there may be a potential large public health benefit from the practical application of this recent understanding. Also with reports that the n-3 PUFAs are producing beneficial effects on depression (27), bipolar and other behavioral disorders (28), the knowledge that these fatty acids have direct physical effects on the fundamental property of the nervous system, namely its electrical activity, should encourage further exploration of potential beneficial effects on brain activities both normal and pathological. It seems likely that we are just scratching the surface of the potential health effects of these interesting dietary polyunsaturated fatty acids.

Acknowledgements

Studies from the authors' laboratories have been supported in part by research grants DK38165 from NIDDK of the National Institutes of Health of the U.S. Public Health Service (AL), the American Heart Association, Ohio Affliate (GEB), American Heart Association Grant-in-Aid (Y-F X) and Christelijke Vereniging voor de Verpleging van Lijders aan Epilepsie (RAV).

References

1. Gudbjarnason, S.; Hallgrimsson, J. *Acta Med. Scand.* **1975**, suppl. *587*:17-26.
2. Murnaghan, M.F. *Br. J. Pharmacol.* **1981**, *73*, 909-915.
3. McLennan, P.L; Abeywardena, M.Y.; Charnock, J.S. *Am. Heart J.* **1988**, *116*, 709-717.
4. McLennan, P.L. *Am. J. Clin. Nutr.* **1993**, *57*, 207-212.
5. McLennan, P.L.; Bridle, T.M.; Abeywardena, M.Y.; Charnock, J.S. *Am. Heart J.* **1992**, *123*, 1555-1561.
6. Billman, G.E.; Kang, J.X.; Leaf, A. *Lipids* **1997**, *32*, 1161-1168.
7. Billman, G.E.; Kang, J.X.; Leaf, A. *Circ.* **1999**, *99*, 2452-2457.
8. Kang, J.X.; Leaf, A. *Proc. Natl. Acad. Sci. USA* **1994**, *91*, 9886-9890.
9. Kang, J.X.; Leaf, A. *Biochem. Biophy. Res. Comm.* **1995**, *208*, 629-636.

10. Kang, J.X.; Leaf, A. *Eur. J. Pharmacol.* **1996**, *297*, 97-106.
11. Li, Y.; Kang, J.X.; Leaf, A. *Prostaglandins* **1997**, *54*, 511-530.
12. Kang, J.X.; Leaf, A. *J. Cardiovasc. Pharmacol.* **1995**, *26*, 943-948.
13. Kang, J.X.; Xiao, Y.F.; Leaf, A. *Proc. Natl. Acad. Sci. USA* **1995**, *92*, 3997-4001.
14. Xiao, Y.F.; Kang, J.X.; Morgan, J.P.; Leaf, A. *Proc. Natl. Acad. Sci. USA* **1995**, *92*, 11000-11004.
15. Xiao, Y.F.; Gomez, A.M.; Morgan, J.P.; Lederer, W.J.; Leaf, A. *Proc. Nat. Acad. Sci. USA* **1997**, *94*, 4182-4187.
16. Kang, J.X.; Leaf, A. *Proc. Natl. Acad. Med. Sci. USA* **1996**, *93*, 3542-3546.
17. Hallaq, H.; Smith, T.W.; Leaf, A. *Proc. Natl. Acad. Med. Sci. USA* **1992**, *89*, 1760-1764.
18. Xiao, Y.F.; Wright, S.N.; Wang, G.K.; Morgan, J.P.; Leaf, A. *Proc. Natl. Acad. Med. Sci. USA* **1998**, *95*, 2680-2685.
19. deLogeril, M.; Renaud, S.; Mamelle, N.; Salen, P.; Martin, J.L.; Monjaud, I.; Guidollet, J.; Touboul, P.; Delaye, J. *Lancet* **1994**, *143*, 1454-1459.
20. Burr, M.; Gilbert, J.F.; Holliday, R.M.; Elwood, P.C.; Fehily, A.M.; Rogers, S.; Sweetnam, P.M.; Deadman, N.M. *Lancet* **1989**, *334*, 757-761.
21. American Heart Association. *Heart and Stroke Facts: 1995 Statistical Supplement,* Am. Heart Assn, Dallas TX.
22. Lipids Research Clinics Program The Lipid Research Clinics Coronary Primary Prevention Trial Results: I. Reduction in Incidence of coronary heart disease. *J. Am. Med. Assoc.* **1984**, *251*, 351-364.
23. Siscovick, D.S.; Raghunathan, T.E.; King, I.; Weinmann, S.; Wicklund, K.G.; Albright, J.; Ovbjerg, V.; Arbogast, P.; Smith, H.; Kushi, LH. *J. Am. Med. Assoc.* **1995**, *274*, 1363-1367.
24. Albert, C.M.; Hennekens, C.H.; O'Donnel, C.J.; Ajani, U.A.; Carey, V.J.; Willett, W.C.; Ruskin, J.N.; Manson, J.E. *J. Am. Med. Assoc.* **1998**, *279*, 23-28.
25. Vreugdenhil, M.; Breuhl, C.; Voskuyl, R.A.; Kang, J.X.; Leaf, A.; Wadman, W.J. *Proc. Natl. Acad. Med. Sci. USA,* **1996**, *93*, 12559-12563.
26. Voskuyl, R.A.; Vreugdenhil, M.; Kang, J.X.; Leaf, A. *Euro. J. Pharmacol.* **1998**, *31*, 145-152.
27. Hibbeln, J.R. *Lancet* **1998**, *351*, 1210-1213.
28. Stoll, A.L.; Severus, E.; Freeman, M.P.; Rueter, S.; Zboyan, H.A.; Diamond, E.; Gress, K.K.; Marangell, L.B. *Arch. Gen. Psychiatry,* 2000. In Press.
29. Leaf, A.; Weber, P.C. *Am. J. Clin. Nutr.* **1987**, *45*, 1048-1053

Chapter 4

The Omega-6 versus Omega-3 Fatty Acid Modulation of Lipoprotein Metabolism

K. C. Hayes

Foster Biomedical Research Laboratory, Brandeis University, Waltham, MA 02454

Both 18:2n-6 and 18:3n-3 fatty acids contribute to lowering total cholesterol (TC) and low density lipoprotein cholesterol (LDL-C) if an individual is below their polyunsaturated fatty acid (PUFA) threshold requirement. Generally speaking, the appropriate intake of 18:2n-6 at or near the 18:2 requirement facilitates lipoprotein metabolism by reducing LDL and increasing high density lipoprotein (HDL), if either lipoprotein fraction was not being metabolized at maximal efficiency at the time of intervention. At high intakes of 18:2n-6 (>200% energy) HDL-C also becomes depressed as very low density lipoprotein (VLDL) output decreases and HDL-C removal increases via enhanced activity of hepatic triacylglycerol lipase (HTGL) and hepatic scavenger receptor-B1 (SR-B1) receptors. The main effect of n-3 high unsaturated fatty acids (HUFAs) at moderately high levels is to depress plasma triacylglycerols (TG) and postprandial lipemia via depressed output of VLDL and chylomicrons, respectively. In humans, n3-HUFAs typically increase LDL-C slightly, whereas HDL-C remains largely unchanged.

    The cholesterol-lowering effect of dietary polyunsaturated fatty acids has been appreciated for many years (*1*). The mechanisms for reducing total cholesterol as constituted by the various lipoproteins, are not fully understood, but the process undoubtedly has many facets affecting both production and clearance of all three lipoprotein classes.
    To better understand the complex relationship between dietary fatty acids and lipoprotein metabolism, particularly related to polyunsaturated fatty acid (PUFA) intake, it is helpful to review the overall impact of the various classes of fatty acids

on coronary heart disease (CHD) risk. In large part this CHD risk derives from the association between total plasma cholesterol, particularly low density lipoprotein (LDL) and high density lipoprotein (HDL), and the atherogenic process.

PUFAs reduce CHD

The largest set of epidemiological data that illustrates this point derives from the Nurses Health Study where 90,000 nurses were followed for 14 years (2). Figures 1 and 2 summarize the relationship between classes of fatty acids consumed and the relative risk for CHD over these years. The first figure describes total CHD risk contributed by each fatty acid class incorporated in a typical US diet having 35% energy as fat. The total risk for this typical fat was calculated to be +68% more than a hypothetical diet based on equivalent carbohydrate intake, ie. the simple substitution of the average US fat for carbohydrate. Both trans and saturated (SATs) fats were found to increase risk, while monounsaturated (MONOs) and polyunsaturated (POLYs) fats reduced it. Figure 2 is even more revealing because it describes the equivalent risk contributed by 1 % energy from each fat class. It is remarkable that trans fat, gram for gram, exerts almost 15x the risk of an equivalent 1% energy from SATs. MONOs appear to reduce risk about half as effectively as POLYs and equal to the collective increase from all types of SATs, primarily because MONOs lack 12:0+14:0 SFAs and generally contain some PUFAs.

Similarly, before examining PUFAs individually, it is instructive to consider the atherogenic potential of dietary fats having various degrees of fatty acid saturation, with emphasis on the protective effect of PUFAs. This is conveniently accomplished by examining lipoprotein profiles and associated arterial cholesterol deposition. In most animal models the fatty acid effect is often complicated by excessive intake of dietary cholesterol that masks the true fatty acid effect. However, a recent study utilized gene-modified mice to circumvent the dietary cholesterol issue by directly comparing fats of different saturation for their effect on atherosclerosis (3). Dietary cholesterol was not required to artificially boost total cholesterol (TC) and LDL-C because these mice lacked LDL receptor activity, and the human apob 100 gene was inserted in their livers. These two gene modifications readily accentuated the fatty acid impact on LDL production, which exerts the major fatty acid influence on the circulating LDL level (4). Recall that LDL-C concentration and size represent the main lipoprotein risk factor for atherosclerosis and that LDL concentration in plasma reflects the balance between production and clearance. Clearance was not a variable here because LDL receptors were absent, impairing LDL removal in all mice.

Figure 3 summarizes the results from these mice and reveals the superior potential of fish oil and safflower oil to reduce both LDL-C and total plasma cholesterol, as well as aortic cholesterol deposition, relative to olive oil and palm

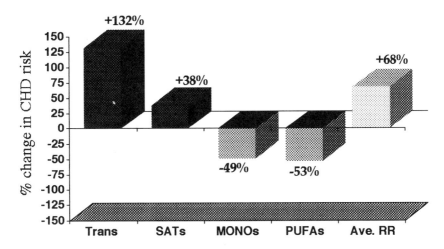

Figure 1. The relative risk for CHD derived from the typical American fat intake compared to a carbohydrate diet. The total contribution from the four classes of dietary fats assumes 35% calories from fat (from ref. 2).

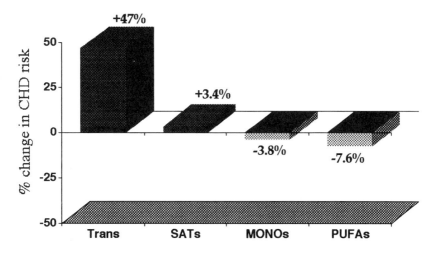

Figure 2. The relative CHD risk for each 1% calories contributed by each class of fat as compared to equal calories from carbohydrate (from ref. 2).

oil. The last two oils were equivalent and superior to partially-hydrogenated soybean oil containing trans fatty acids, which was significantly worse than all other oils for both categories of lipoprotein response and atherosclerosis. Unfortunately, mice under these conditions have exhausted their HDL pool, so the overall pattern of lipoprotein effects is not exactly relevant to humans. However, note the marked decrease in very LDL-cholesterol (VLDL-C) (decreased triacylglycerol output) attributable to n-3 highly unsaturated fatty acids (HUFAs) from fish oil, and to a lesser extent to n-6 PUFAs in safflower oil. By contrast, a major effect of trans fatty acids appeared to be a marked increase in VLDL output or decreased total triacylglycerol (TG) clearance, factors also implicated in the gerbil response to trans fatty acids (*5*).

Fatty Acids, Total Cholesterol, and the Lipoprotein Profile

Before we conclude that n-3 fatty acids exert different or similar lipoprotein effects as n-6 fatty acids, it is helpful to consider an overview of fats (and their fatty acids) on plasma cholesterol and the lipoprotein response. Figure 4 provides a general summary of the relative potency of dietary fatty acids on TC as derived from human and animal studies where fatty acid intake was carefully documented (*6, 7*). In humans the bulk of the dietary fat effect is on LDL-C, with lesser mass changes in VLDL-C and HDL-C. Note that the response to 18:2 can vary appreciably, with added increments of 1% energy having almost no effect to as much as -4mg/dl lowering of TC. The reason for this variation in the response to 18:2 reflects the fact that the PUFA consumed determines the "setpoint" for the lipoprotein profile, which, in turn, influences how any given individual will respond to dietary fat (*6-9*).

A practical example of the effect of different lipoprotein setpoints on the response to 18:2 relative to other fatty acids in Benedictine nuns is provided by Baudet et al. (*10*) as depicted in Figure 5. This study was unique because it compared fats of different saturation in both normolipemic and hyperlipemic individuals. The main fatty acid comparisons (in these diets providing 30% energy from fat) emphasize exchanges between 20%energy as 18:2 (sunflower oil), 12% energy as 16: 0 (palm olein), 18% as 18: 1 (peanut oil), and 20% as 12:0+14:0+16:0 (milk fat). Recall that milkfat not only contains excess SFAs, but it also is severely lacking in 18: 2 (only 1% from PUFA).

The response to the above fats is informative because the normolipemic women revealed no significant differences in their TC response among 18:2, 16:0, and 18: 1 -rich fats, with only the 14:0-rich+1 8:2-poor milk fat raising TC and LDL-C appreciably. By contrast, the hypercholesterolemic nuns revealed a significant decline in TC from baseline during 18:2 intake, while the TC responses to 16:0 and 18:1 were equally reduced and higher than 18:2, but still significantly below the 14:0-rich+1 8:2-poor milk fat, which showed no improvement from

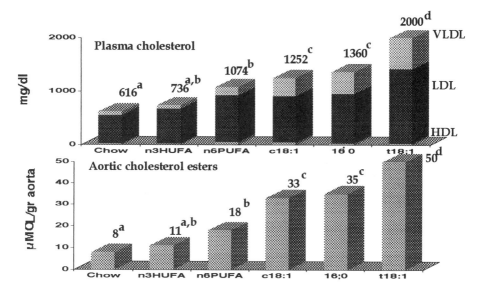

Figure 3. The plasma and aorta cholesterol response to dietary fat in gene-modified mice. The key fatty acid contribution for fish oil, safflower oil, canola oil, palm oil, and partially hydrogenated soybean oil is emphasized. Means with different superscripts are significant (from ref. 3).

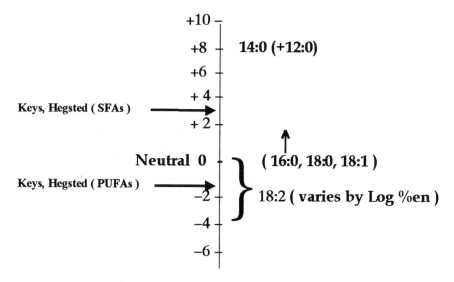

Figure 4. The putative cholesterolemic response (in mg/dl) to each 1% energy from individual dietary fatty acids during low-cholesterol diets. The historical, collective lumping of fatty acid classes by Keys and Hegsted is indicated as +2.7 for SFAs and -1.6 for PUFAs. Our version emphasizes the separation in SFAs (keying on 14:0 in certain saturated fats) and the variable response to 18:2. These fatty acid relationships can be estimated for any given dietary fat composition with the equation: TC=8x14.0 % energy 36 x log 18:2% energy, the log function indicating that the 18:2 response is nonlinear. The arrow above 16:0 reflects the fact that it can become cholesterol-raising during high cholesterol intake or in persons with an elevated lipoprotein profile.

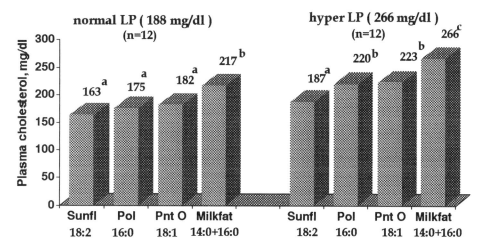

Figure 5. Plasma cholesterol affected differently by fats (fatty acids) in Benedictine nuns experiencing normal or elevated lipoprotein profiles (setpoints) at the time of intervention (from ref. 10).

baseline. This implies that individuals are sensitive or not to fatty acids in the diet depending on their requirement for 18:2 at the time of the challenge. Thus, the women with a normal lipoprotein profile (normal setpoint) decreased their baseline TC only 1 mg/dl for each 1% energy derived from 18:2, whereas the subjects with a hyper setpoint decreased TC about –4 mg/dl for each 1% energy from 18:2. In both settings (where diet cholesterol was low) 16:0 and 18:1 were neutral and equal. Because the palm olein and peanut oil also provided more 18:2 than the traditional diet consumed by these Benedictine nuns, TC tended to be lower than baseline in both these groups even though 16:0 or 18:1 were dominant, albeit neutral. Thus, it is much more revealing to focus initially on PUFA intake than either the SFAs or MUFAs, but ultimately all three fatty acid classes must be considered together to predict the TC response to fat.

Other human studies have subsequently revealed the same result, ie. that tile setpoint or lipoprotein profile at the time of intervention greatly influences the ultimate response to dietary fat saturation and its overall fatty acid profile (6, 9,11,12). Figure 6 summarizes the above concept in its simplest form. Here, the rise in LDL-C induced by dietary fat in humans is described as a function of two or more dietary fatty acids, with the intake of 18:2 being most critical because it dictates the "threshold" of responsiveness to the other fatty acids present in the diet. The idea is that any individual or population has a dietary requirement (threshold) for 18:2 that is determined by the "stress" on lipoprotein metabolism at the time of intervention. Some people need more 18:2, some less, to meet their specific needs. When below personal threshold for 18:2, one is extremely variable to raising TC (LDL-C) during the consumption of certain saturated fatty acids (SFAs), even (MUFAs), when these other fatty acids are included in the diet. This sensitivity to other fatty acids is especially true for 14:0-rich fats, in part because they contribute almost no 18:2 of their own, thereby increasing the risk that the individual will fall below threshold. On the other hand, the model suggests that once above the 18:2 requirement, other fatty acids exert minimal impact on plasma LDL (and TC). This, in essence, is what the Baudet data (Figure 5) teaches, and it represents the critical message of this review, ie. because dietary fatty acids are so interrelated, one cannot discuss the effect of a single class, eg. PUFAs, on the lipoprotein response without simultaneously considering the other fatty acids present, eg. SFAs and MUFAs, because other fatty acids indirectly influence the amount of 18:2 needed to maintain normal lipoprotein metabolism.

A recent report of 25 year mortality data from The Seven Countries Study highlights several of the points just made (13). Using dietary fatty acid profiles of reconstituted diets based on those consumed during the original period of induction into the study, the authors assessed the 25yr CHD percent mortality as a function of fat (fatty acid) intake. Using their published data, CHD mortality was plotted as a function of the dietary PUFAs/12:0+14:0 fatty acid ratio (Fig. 7). Similar to the model in Fig. 6, this fatty acid relationship revealed a nonlinear curve that explains 50% of the variation in CHD mortality between countries.

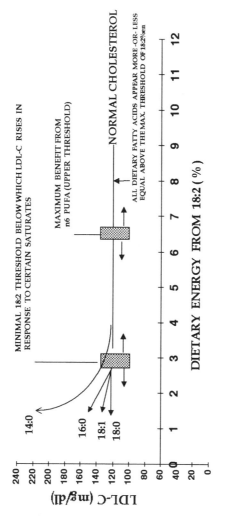

Figure 6. Putative relationship between the dietary 18:2 threshold and LDL-C in humans.

Can this scenario of fatty acid relation be applied to specific clinical studies in humans? In fact, the balance between dietary fatty acids becomes critical when attempting to generate the lowest TC and LDL/HDL ratio. One illustrative experiment, where dietary fatty acid intake was exquisitely controlled (*14*), applied AHA and NCEP dietary guidelines to assess the effectiveness of lowering TC and the LDL/HDL ratio by removing dietary SFAs. In this case the focus was exclusively on SFAs, as the objective was to decrease total fat from 35 to 25% energy by stepwise removal of SFAs while maintaining the % energy from MUFAs and PUFAs at constant intakes. As seen in Figure 8, this decrease in SFAs eliminated most of the 12:0+14:0-fat and decreased TC, LDL-C, and HDL-C proportionately in a stepwise fashion such that an overall decrease in lipoprotein cholesterol of about 10% was achieved as the dietary polyunsaturates to saturates (P/S) ratio steadily increased from 0.4 to 1.2. Although the decline in LDL-C was laudable, it would have been better if HDL-C had been sustained as LDL-C declined. The question is what role, if any, did PUFAs play in this response?

In partial response to the question, it is noteworthy that a more favorable result, ie. decreasing the LDL/HDL ratio along with TC, was achieved by Schwandt et al. (*15*) without altering total fat intake. In essence, these investigators found that simply rebalancing the fatty acid classes while maintaining fat intake at 37% energy, lowered LDL-C by 14% while sustaining HDL-C (Figure 9). Thus, an saturates: monounsaturates: polyunsaturates (S:M:P) imbalance in % energy of 18:13:6, with a P/S ratio of 0.3 was corrected to 12:12:12, and a P/S of 1.0, with great success. In fact, the decreases in both TC and LDL-C were comparable to subjects studied by Ginsberg et al. (*14*), but HDL-C was sustained. A possible explanation for the negative effect on HDL in the Ginsberg study may be that a diet too high in PUFAs relative to SFAs (especially if SFAs are low) enhances LDL activity and reduces LDL production (decreasing LDL-C), even as it increased scavenger receptor B1 (SR-B1) and hepatic HDL binding activity (decreasing HDL-C) (*16*). The data of Schwandt et al. (*15*) suggest that a strategic balance exists between fatty acids whereby the LDLdecline can be maximized by 18:2, while SFAs can minimize hepatic SR-B 1receptor and triacylglycerol lipase activities to sustain plasma HDL.

Sundram et al. (*17*) further examined the concept of fatty acid balance and observed that a 10: 13:8 ratio in dietary S:M:P at 30% energy from fat generated the best LDL/HDL ratio (Figure 10). This balancing of dietary fatty acids did not further lower TC or LDL-C, but did increase HDL-C by 15%. The failure to reduce TC may reflect the normal lipoprotein profile at the outset in these subjects with TC of 175 mg/dl and LDL-C of 100mg/dl.

With these observations in mind, we used cebus monkeys to explore the possibility of fine-tuning the dietary fatty acid balance to enhance the lipoprotein profile. Mimicking the Ginsberg design (14), dietary SFAs were removed in a similar stepwise fashion from the average American diet (35% energy, 500 mg/day cholesterol) to a Step II Diet (25% energy, 200 mg/day cholesterol) (Figure 11).

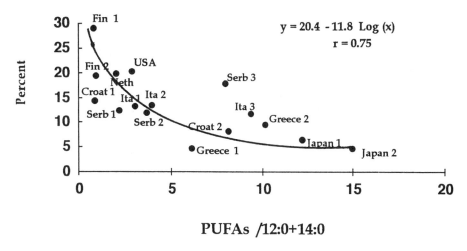

Figure 7. Percent CHD mortality data from the Seven Countries Study and relationship with the fatty acid ratio of 18:2/12:0+14:0 in the original diets (from ref. 13).

Fig.8

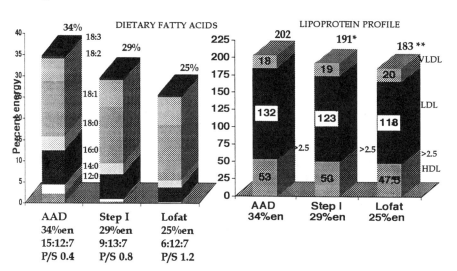

Figure 8. Stepwise removal of saturated fatty acids from 35% energy total fat (S:M:P of 15:12:7) to 25% energy (S:M:P of 6:12:7) decreases TC by lowering both LDL and HDL fractions so that the LDL/HDL ratio (@2.5) remains unchanged between diet periods (from ref. 14).

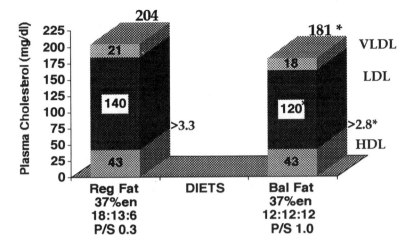

Figure 9. Rebalancing the S:M:P ratio at 37% energy lowers TC as effectively as the approach in Fig.8, but the LDL/HDL ratio is substantially improved because the dietary P/S ratio is not distorted in favor of PUFAs (from ref. 15).

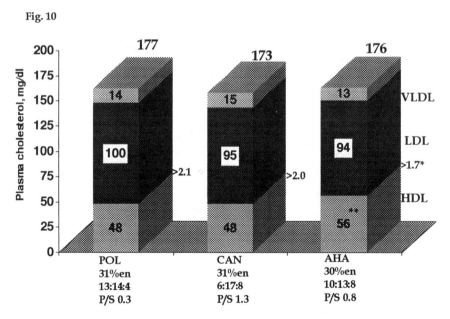

Figure 10. Rebalancing the dietary S:M:P ratio close to 10:10:10 irnproves the LDL/HDL ratio in normolipemic subjects, even though TC remains unchanged (from ref. 17).

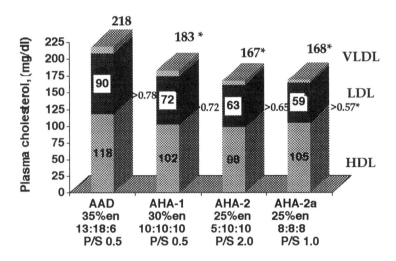

Figure 11. Stepwise reduction of dietary SFAs in cebus monkeys and improvement in TC. Rebalancing the S:M:P ratio of the Step II diet to 8:8:8 improved the LDL/HDL ratio (from ref. 23)

The relative decreases in TC, LDL-C, and HDL-C in these monkeys were remarkably similar to the human data. However, after completing the 6 wk Step 11 Diet period with an imbalance in the S:M:P ratio, the fatty acid intake was rebalanced by blending fats to produce an 8:8:8 ratio in S:M:P. The result produced no change in TC, but the LDL/HDL ratio improved because LDL slightly decreased as HDL rose 7%, not unlike the human results of Sundram et al. (*17*).

These empirical results suggest that the "metabolic tension" between SFAs and PUFAs is an important determinant of the LDL/HDL ratio. Furthermore, in addition to the total fat intake, the absolute intakes of 18:2 and SFAs have a critical bearing on the total plasma cholesterol, such that reducing only SFAs increases the P/S ratio of the diet and decreases TC, LDL-C as well as HDL-C if the fatty acid balance becomes overly distorted.

The n-6 versus n-3 Balance and Lipoproteins

From the above emphasis on the primary role of PUFAs in lipoprotein metabolism, the question arises as to the relative importance of n-6 vs n-3 fatty acids in this process. Within the normal range of intake for these two PUFA types, there appears to be no apparent impact of the n-3 fatty acids on the dominant role played by the more abundant 18:2n-6. For example, exchange of half the 7% energy from PUFAs between 18:2n-6 and 18:3n-3 had no effect on human lipoprotems (*18*). Thus, under normal circumstances 18:2n-6 dictates the PUFA effect on lipoprotein metabolism.

This conclusion also was demonstrated by Sanders et al. (*19*), who fed human subjects diets in which 2% energy was exchanged between n-6 and n-3 HUFAs within a total pool of 6% energy from all polyunsaturates. The exchange did not alter TC or the LDL/HDL ratio (Figure 12).

A summary of the literature reports describing studies where fish oil on 18:3 n-3 replaced other unsaturated fat (mostly olive oil) was generated by Harris (*20*) who included 36 crossover studies for their impact on lipoproteins (Table I). As generally agreed, the data indicate that 18:3n-3 was similar to 18:2n-6, whereas fish oil n-3 HUFAs exert their main effect in humans by lowering triacylglycerols (TC) without having much effect on TC, LDL, or HDL. The TG lowering reflects the ability of HUFAs to depress hepatic TG secretion as VLDL, interfering with TG and VLDL formation in hepatocytes by disrupting apo-B assembly (*21*). In theory, the decrease in VLDL output should ultimately decrease LDL formation. The reason that LDL-C does not typically decrease with n-3 HUFA consumption would appear to be related to the measured depression in LDL receptor activity induced by these fatty acids when consumed in atypical amounts, eg. 10 g per day (21,22). Thus, the decrease in LDL formation, coupled with a decrease in LDL clearance, combine to leave LDL-C relatively unchanged, or even slightly elevated.

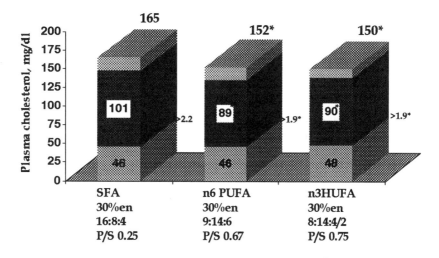

Figure 12. Exchanging 2% energy from 18:2n6 with n3 HUFAs (EPA & DHA) and lack of effect on TC or the LDL/HDL ratio in humans (from ref. 26)

Table I. Human Crossover Studies of Fish Oil vs. Control Oil, Each Relative to Baseline Lipid Values

Criteria	Baseline TG	
	<180 mg/dl	>180 mg/dl
Data sets	16	20
Subjects	281	254
TOTAL CHOLESTEROL		
Control oil	-1	-1
Fish oil	0	-2
LDL-C		
Control oil	-2	-1
Fish oil	+2	+12
HDL-C		
Control oil	+1	+4
Fish oil	+4	+5
TGs		
Control oil	0	+2
Fish oil	-25	-32

Adapted from ref. 20.

When total PUFA intake is limited and close to the "threshold requirement for 18:2", appreciable substitution with 18:3n-3 may even have a negative effect on HDL-C, especially if HDL is the predominant lipoprotein. This result was encountered in cebus monkeys receiving only 4% energy as PUFA in a diet with 30% energy as fat rich in 12:0+14:0 SFAs(ie. stressing the lipoprotein setpoint). When one-third of the PUFA was shifted to 18:3n-3 from flaxseed oil, TC decreased 10%, with the total decrease derived from the HDL-C pool. This had a major detrimental effect on the LDL/HDL ratio, increasing it 10% (Figure 13) (*23*). This observation would support the notion that n-3 fatty acids can depress VLDL output to result in decreased HDL formation associated with a reduced pool of VLDL for catabolism.

Although the focus of this review has been to emphasize the nature of the lipoprotein changes associated with PUFA intake, where n-6 PUFA normally reigns supreme, the major impact of n-3 HUFAs on atherogenic risk undoubtedly reflects their ability to stabilize cardiac rhythm and reduce thrombosis, while enhancing fibrinolysis (*24*). By way of example, it was recently shown in a large clinical intervention trial with patients recently recovering from myocardial infarction that the cardioprotective effects of n-3 HUFAs can substantially improve their ability to avoid subsequent adverse CHD events (*25*).

References

1. Hegsted, D. M.;McGandy, R. B.;Myers, M. L. Stare, F. J. *Am. J. Clin. Nutr.* **1965**, *17*, 281-295.
2. Hu, F. B.; Stampfer, M. J.;Manson, J. E.;Rimm, E.;Colditz, G. A.;Rosner, B. A.; Hennekens, C. H.Willett, W. C. *N. Engl. J. Med.* **1997**, *337*, 1491-1499.
3. Rudel, L. L.;Kelley, K.;Sawyer, J. K.;Shah, R.Wilson, M. D. *Arteriosclerosis Thromb. Vasc. Biol.* **1998**, *18*, pp. 1818-27.
4. Hajri, T.; Khosla, P.; Pronczuk, A. Hayes, K. C. *J. Nutr.* **1998**, *128*, 477-84.
5. Dictenberg, J.B.; Pronczuk, A. Hayes, K.C. *J. Nutr. Biochem.* **1995**, *6*, 353-361.
6. Hayes, K. C. Khosla, P. *FASEB J* **1992**, *6*, 2600-2607.
7. Pronczuk, A.; Khosla, P.; Hayes, K. C. *FASEB J* **1994**, *8*, 1191-1200.
8. Hayes, K. C.; Pronczuk, A.; Khosla, P. *J. Nutr. Biochem.* **1995**, *6*, 188-194.
9. Hayes, K.C. *Can. J. Cardiol.* **1995**, *11*, 39G-46G.
10. Baudet, M. F.; Dachet, C.; Lasserre, M.; Esteva, O. Jacotot, B. *J. Lipid Res.* **1984**, *25*, 456-468.
11. Mattson, F.H.; Grundy, S. M. *J. Lipid Res.* **1985**, *26*, 194-202.
12. Cuesta, C.; Rodenas, S.; Merinero, M. C.; Rodriguez-Gil, S.; Sanchez-Muniz, F. J. *Europ. J. Clin. Nutr.* **1998**, *52*, 675-683.
13. Kromhout, D.; Menotti, A.; Bloemberg, B. *Preventive Medicine* **1995**, *24*, pp. 308-315.

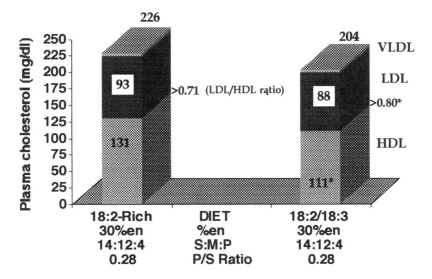

Figure 13. Replacing one-third of 4% energy from 18:2n-6 with 18:3n-3 and its detrimental impact on HDL-C in cebus monkeys (from ref. 23)

14. Ginsberg, H. N.; Kris-Etherton, P.; Dennis, B.; Elmer, P. J.; Ershow, A.; Lefevre, M.; Pearson, T.; Roheim, P.; Ramakrishnan, R.; Reed, R.; Stewart, K.; Stewart, P.; Phillips, K. Anderson, N. *Arteroscierosis Thrombosis Vascular Biol.* **1998**, *18*, 441-449.
15. Schwandt, P.; Janetschek, P.; Weisweiler, P. *Atherosclerosis* **1982**, *44*, 9-17.
16. Spady, D. K.; Kearney, D. M.; Hobbs, H. H. *J. Lipid Res.* **1999**, *40*, 1384-94.
17. Sundram, K.; Hayes, K. C.; Siru, O. H. *J. Nutr. Biochem.* **1995**, 179-187.
18. Pang, D.; Allman-Farnelli, A.; Wong, T.; Barnes, R.; Kingham, K. M. *Brit. J. Nutr.* **1998**, *80*, 163-167.
19. Sanders, K.; Johnson, L.; O'Dea, K.; Sinclair, A. *J. Lipids* **1994**, *29*, 129-138.
20. Harris, W. S. *Am. J. Clin. Nutr.* **1997**, *65*, 1645S-1654S.
21. Wilkinson, J.; Higgins, J.A.; Fitzsimmons, C.; Bowyer, D.E. *Arteriosclerosis Thrombosis Vascular Biol.* **1998**, *18*, 1490-1497.
22. Lindsey, S.; Pronczuk, A.; Hayes, K. C. *J. Lipid Res.* **1992**, *33*, 647-658.
23. Pronczuk, A.; Hayes, K. C. *FASEB* **1998**, *12*, A484.
24. Leaf, A.; Kang, J. X.; Xiao, Y. F.; Billman, G. E.; Voskuyl, R. A. *Prostaglandins Leukotrienes and Essential Fatty Acids* **1999**, *60*, 307-312.
25. Annonymous. *Lancet* **1999**, *354*, 447-455.
26. Sanders, T.A.; Oakley, F.R.; Miller, G.J.; Mitropoulos, K.A.; Crooke, D.; Oliver, M.F. *Arterio. Thromb. Vasc. Biol.* **1997**, *17*, 3449-60.

Chapter 5

Docosahexaenoic Acid in Human Health

Bruce J. Holub

Department of Human Biology and Nutritional Sciences, University of Guelph, Guelph, Ontario N1G 2W1, Canada

Docosahexaenoic acid (DHA, 22:6n-3) is now recognized as a physiologically-essential nutrient in the brain and retina for neuronal functioning and visual acuity, respectively. DHA is present in aquatic organisms (including fish/fish oils, aquatic mammals, algae, etc.) but is absent from common plant sources and edible vegetable oils. The limited metabolic convertibility of dietary alpha-linolenic acid (LNA, 18:3n-3) to DHA via desaturation plus elongation reactions in the liver and other tissues has led to consideration of preformed dietary DHA as an important nutrient in order to ensure optimal levels for maximal biological performance including protective effects on various chronic disorders. Epidemiological and intervention studies have indicated the potential for dietary DHA to protect against cardiovascular disease (morbidity and mortality) as well as favorably attenuating associated risk factors including plasma triglyceride-lowering. Evidence supporting the anti-arrhythmic potential of DHA based on experimental studies has led to considerations that the apparent cardioprotective effects of fish/fish oil consumption may be due in part to the enrichment of cardiac tissue in DHA subsequent to its dietary consumption. Vegan vegetarians are at particular risk with respect to having depressed physiological levels of DHA as well as sub-optimal concentrations of DHA in breast milk of lactating women. Depressed levels of DHA have become apparent from various measurements on patients with various neurological/behavioral disorders although controlled intervention trials with preformed DHA (lacking EPA) have been very limited in number to date.

Controlled intervention studies with supplemental DHA have indicated benefit in hyperaggressive behavior as induced via mental stress in university students. Controlled clinical trials using supplemental DHA (free of EPA) are required and many are now currently underway to directly assess the potential of DHA itself as distinguished from fish/fish oils containing DHA plus EPA in various studies related to human health indices and the attenuation of chronic disorders and associated risk factors. The employment of nutraceutical preparations of DHA and their inclusion in a wide assortment of functional foods are expected to dramatically increase in numbers and availability in the marketplace within the coming decade in view of mounting evidence for the health benefits associated with increased consumption of dietary DHA in various forms.

Sources and Metabolism of DHA

The long chain omega-3 (n-3) fatty acids of human health interest as found in fish/fish oils plus diverse aquatic organisms include docosahexaenoic acid (DHA, 22:6n-3) in addition to eicosapentaenoic acid (EPA, 20:5n-3). These n-3 polyunsaturated fatty acids (n-3 PUFA) are found almost exclusively in aquatic organisms (fish, mammals, others) where they exist in varying amounts and ratios (*1*). While fish and fish oils contain varying levels and mixtures of DHA plus EPA, it is DHA which accumulates at high levels in neuronal and retinal tissue where it appears to play a physiologically essential role in functioning. DHA is found primarily in esterified form associated with cellular membrane lipid components (phospholipids) as well as within neutral (storage) lipid, primarily in the form of triglyceride. DHA is lacking from all common plant food sources including the commercial vegetable oils although microalgae and other sources of DHA are now available for nutritional and food applications. In view of its very low melting point (approaching -50°C), DHA is highly fluid at both room and body temperatures and is considered to play an important role in biochemical adaptation to colder environmental temperatures in aquatic ectothermic organisms. Although common plant sources and vegetable oils lack DHA, some (for e.g., canola oil, soybean oil, flaxseed/flax oil) do contain significant amounts of the omega-3 fatty acid known as alpha-linolenic acid (LNA, 18:3n-3) which, in mammalian organisms (including the human body), can be metabolically converted to DHA via desaturation plus elongation reactions as found in the liver and elsewhere (see Figure 1). However, the conversion efficiency of LNA to DHA is very limited in human adults with conversion efficiencies estimated to be approximately 4% (*2*) and even more limited in infants where the conversion efficiency of LNA to DHA appears to be less than 1% (*3*). Thus, the most direct

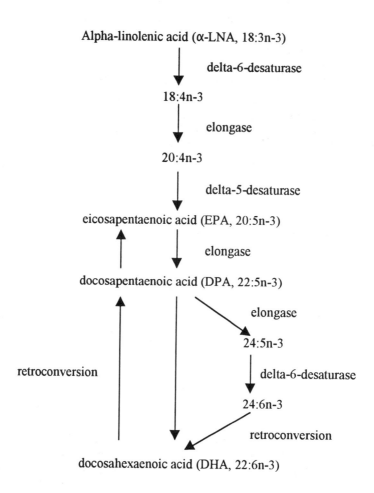

Figure 1. Desaturation, Elongation, and Retroconversion of Omega-3 Fatty Acids.

means of maintaining optimal/high levels of DHA in the human body is by direct consumption of preformed dietary DHA. DHA in its fatty cellular fatty acyl-CoA form is readily esterified to membrane phospholipid (predominantly in the 2-position) as well as undergoing β-oxidation and other metabolic transformations (including selected oxygenation reactions). As indicated in Figure 1, DHA can also be metabolically retroconverted to EPA via peroxisomal-mediated β-oxidation with estimates of approximately 10% efficiency in the retroconversion of DHA to EPA based on human feeding trials employing algae-derived DHA free of EPA (*3, 4*).

DHA as a Physiologically-Important Nutrient for Neuronal and Visual Functioning

Mounting evidence from published research indicates that DHA is a physiologically-essential nutrient for early human development and functioning (*6*) including neuronal performance and visual acuity (*7, 8*). DHA is markedly concentrated in brain and retinal membrane phospholipid (particularly phosphatidylethanolamine and phosphatidylserine) of human infants, non-human primates, and in various other animals. This accumulation of cellular DHA is particularly pronounced during late fetal and early neonatal life. The depletion of DHA levels to sub-optimal concentrations in the brain and retina, due to insufficient dietary intakes of n-3 polyunsaturated fatty acids (PUFA) as precursor LNA and/or DHA, has been found to result in impaired learning ability and a loss of visual acuity, respectively (*9-11*).

While much of the published research on the importance of maintaining physiological levels of DHA for optimal neurological development and functioning has been performed in infants (pre-term, full-term, and breast-fed vs. formula-fed), the importance and health implications of maintaining an adequate DHA status in human adults is also becoming recognized. The intake of dietary DHA during pre-conception and pregnancy has a close relationship to the level and availability of circulating DHA in the mother and to the developing fetus. The levels of DHA present in breast milk and available to the infant are dependent upon DHA intakes during lactation by the mother. Many groups have recommended that DHA should represent at least 0.4% of the total fatty acids in breast milk; approximately one half or more of North American women have breast milk levels which do not meet this minimal criteria. Furthermore, the significantly lower levels of DHA in the breast milk of vegan/vegetarian women when compared to omnivorous controls (*12*) further supports the very limited convertibility of dietary LNA (found in plant food sources) to DHA. It has been estimated that the average DHA intake amongst pregnant and lactating women in North America is approximately 80mg/day on average; furthermore, intakes of 200 - 300mg/day are considered necessary for maintaining the minimal level of 0.4% in breast milk. The presence of DHA in human breast milk, but not in infant formulas, may possibly underlie the structural development of the nervous system as reflected in subsequent intelligence quotients (higher IQ scores) in children at approximately 8 years of age (*13*). An association between birth weight and cognitive function as determined in adult life (20 years later) has been reported (*14*). It is becoming apparent that DHA is a physiologically-essential nutrient for cognitive performance from infancy into adulthood. It has been reported (*15*) that a relation exists between dietary factors (including dietary PUFA) and cognitive function in a cohort of elderly men (aged 69 to 89 years). Interestingly, the mean daily intake of fish and DHA, but not EPA, was significantly lower in those with impaired cognitive functioning relative to normal subjects.

A very recent study (*16*) has indicated that an early dietary supply of DHA is a major dietary determinant of improved performance on the Mental Development Index in term infants. Others have concluded that breast-fed infants achieve a higher rate of brain and whole body docosahexaenoate accumulation than formula-fed infants not consuming dietary docosahexaenoate (*17*). The latter group has concluded that dietary DHA should likely be provided during at least the first 6 months of life.

DHA in Cardiovascular Health and Disease Prevention/Management

Epidemiological studies have indicated that the increased consumption of fish containing n-3 fatty acids (DHA plus EPA) can protect against death from cardiovascular disease (*18*) including the risk of primary cardiac arrest (*19*) and sudden cardiac death (*20*). Results from the Multiple Risk Factor Intervention Trial (MRFIT) have indicated that increasing the consumption of combined DHA/EPA up to approximately 700 mg/day lowered both cardiovascular and all-cause mortality (*21*). It should be pointed out that in the various fish trials to date, a distinction between EPA plus DHA has not been possible since both n-3 PUFA are consumed concomitantly. In a case-control study, Simon et al. (1995) (*22*) examined the relation between serum fatty acids as measured by gas-liquid chromatography and coronary heart disease (CHD) in men with and without CHD. Interestingly, the levels of DHA in serum phospholipid were inversely correlated with CHD risk in a multivariate model that controlled for the effects of the HDL-cholesterol to LDL-cholesterol ratio. These findings are consistent with other evidence indicating that DHA consumption in the form of fish/fish oils and physiological DHA status are inversely correlated with CHD. Recent intervention trials, (*23, 24*) indicating the potential for fish oils containing DHA plus EPA to retard the progression of cardiovascular disease and to reduce secondary mortality from myocardial infarctions and sudden cardiac death has not allowed a specific assessment of these beneficial clinical effects to DHA directly. DHA has been found to exhibit anti-platelet aggregatory potential in some studies (25), triglyceride-lowering in the majority of human studies using supplemental DHA (devoid or containing minimal amounts of EPA), and important anti-arrhythmic effects in experimental studies (26). These various effects of DHA may underlie and contribute to the overall cardioprotective effects of fish consumption and the constituent DHA.

The recent availability of DHA-enriched preparations (depleted of EPA) has allowed for human intervention trials which directly assess the potential of dietary DHA (independent of dietary EPA) to influence selected risk factors for cardiovascular disease (CVD) and overall physiological DHA status. A number of published trials using supplemental DHA have indicated a significant triglyceride-lowering potential. This is of interest since even moderately increased triglyceride levels in the circulation (fasting) have been implicated in an increased risk of myocardial infarctions in both men and women. One of the early double-blind studies (4) investigated the influence of dietary supplementation with an algae source of DHA (devoid of EPA) on selected risk factors for CVD, and on the DHA status of serum and platelet phosphopliplid, in vegetarian subjects given 1.6 gm/day over a six week period. A significant lowering of the fasting triglyceride levels (by approximately 20%) was seen in the DHA-supplemented groups along with reduction (by 16%) in the total cholesterol:HDL-cholesterol ratio. Measurements of DHA levels in serum phospholipid, a recognized biomarker for DHA status in the body and DHA intakes, showed a relative rise of 246% in serum phospholipid (from 2.4 to 8.3 weight % of total fatty acids) and 225% in platelet phospholipid (from 1.2 to 3.9 weight %). The lipid/lipoprotein changes observed were consistent with an overall favorable effect of dietary DHA on these recognized risk factors for CVD. The authors also demonstrated that the levels of DHA in the circulation of vegetarian subjects, which have often been reported to be well below the levels found in omnivorous humans, can be elevated to levels which surpass that for omnivores via supplementation of vegetarians using an algae source of DHA. The level of supplementary DHA employed (1600 mg/day) is well above current/low dietary intakes of approximately 80 mg (12); however, these supplemental intake levels represent approximately 2% of the total fat intake and approximately 0.6% of the total energy intake.

Other workers (27) have used highly purified preparations of DHA and EPA in human studies and have reported similar serum triglyceride-lowering effects of each but divergent effects on serum fatty acids. Ethyl ester preparations of DHA and EPA (at intake levels of approximately 3.6-3.8 gm/day) over a seven week period provided a decrease in serum triglyceride levels by 26% in the DHA group which was accompanied by a 4% rise in HDL-cholesterol and a 4% decrease in the total cholesterol:HDL-cholesterol ratio. DHA levels in serum phospholipid rose by 69% with DHA supplementation. These and other results indicate that dietary DHA can contribute favorable effects on selected lipid/lipoprotein parameters associated with an increased risk of cardiovascular disease. Interestingly, these authors (27) also found a significant decrease in heart rate with DHA supplementation but not with EPA (and suggest that DHA may be responsible for the postulated anti-arrhythmic effect of n-3 PUFA. The apparent anti-arrhythmic effect of DHA (26) in cardiac tissue (in free fatty acid form) may be responsible for the apparent reduction in sudden cardiac death in post-myocardial infarction subjects given supplementary fish oil concentrates (24). However, future work will have to employ clinically-significant endpoints such as the progression of cardiovascular disease and atherosclerosis (measured by angiography and other parameters) as well as its effects on sudden cardiac death to determine whether the previous findings with fish oil concentrates are due to DHA and/or EPA. DHA supplementation (0.75 or 1.5 gm/day) of human subjects has been found to elevate circulating non-esterified fatty acid levels to concentrations that exhibit anti-thrombotic and anti-arrhythmic potential *in vitro* (28).

DHA and Neurological/Behavioral Disorders

A number of studies have been preformed to compare the physiological levels of DHA in patients with various neurological disorders as compared to control subjects. In many of these studies, depressed levels of DHA have been found in the patient groups although controlled intervention studies with DHA supplementation (devoid of EPA) are currently underway to determine if any significant clinical benefit can be realized from restoration of the DHA levels independent of EPA. Neurological disorders associated with reduced levels of DHA have been reported in schizophrenia, Alzheimer's disease, depression, various peroxisomal disorders, as well as retinitis pigmentosa plus others (12). Behavioral disorders associated with decreased levels of DHA include Attention Deficit Hyperactivity Disorder (ADHD) and hyperactivity. Schizophrenic patients in some studies have shown to have a substantial depletion of DHA (as well as AA) in red blood cells when compared to a healthy control group, which may possibly be due to increased lipid peroxidation in these patients (29). Alzheimer brains have shown significantly lower levels of DHA in the phosphatidylethanolamine and phosphatidylcholine fractions in frontal grey matter as well as in white matter and the hypocampus (30). It has been postulated (31) that adequate intakes and status of DHA may reduce the development of depression. There is evidence from research studies suggesting that decreased n-3 fatty acid consumption correlates with increasing rates of depression and a positive correlation between depression and coronary artery disease (31). Peroxisomal disorders are a group of genetic disorders characterized by extremely low levels or absence of peroxisomes, especially in the liver and kidney. These disorders include Zellweger syndrome, neonatal adrenaleukodystrophy, and Infantile Refsum's Disease. Children with such disorders often live from a few months (Zellweger syndrome) to their teenage years. This disease exhibits profound alterations in PUFA status with DHA levels being much lower as compared to control individuals in the erythrocytes and plasma (*32*). The hereditary retinal degenerative disorder, retinitis pigmentosa, is associated with reduced night vision and loss of peripheral visual field. It is associated with decreased plasma phospholipid DHA levels as well as decreased DHA levels in the circulating red blood cells (*33*). Sub-normal levels of DHA have been reported in individuals with behavioral disorders such as ADHD and hyperactivity; for example, it has been reported (*34*) that DHA levels were approximately 13% lower than controls in plasma polar lipids and 26% lower in erythrocyte lipid. A greater number of behavioral problems, temper tantrums, and sleep problems have been reported in subjects with lower omega-3 fatty acid concentrations (*35*). Very few intervention/supplementation trials involving DHA (devoid of EPA) have been performed in neurological patients although there have been reports of some beneficial effects in selected individuals (*36*). Further rigorous double-blind placebo-controlled studies using various levels of DHA supplementation (devoid of EPA) over varying time periods with extensive clinical monitoring are required to determine if normalizing physiological DHA levels can improve any of the aforementioned clinical conditions involving neurological/behavioral disorders.

In a placebo-controlled double-blind study, it was observed that DHA intake could favorably attenuate aggression enhancement at times of mental stress (*37*). The administration of DHA-rich oil capsules (1.5-1.8 gm DHA/day) for 3 months significantly dampened mental stress-induced extra-aggressive behavior as determined via psychological testing. In subsequent research, DHA supplementation was found to reduce plasma norepinephrine levels in medical students during final exams as well as increasing the plasma ratio epinephrine to norepinephrine (*38*). These latter authors concluded that these effects of DHA may be applied to people under long-lasting psychological stress to prevent stress-related diseases.

DHA and Other Pathophysiological Conditions

Little information is available, due to the absence of long-term controlled clinical trials using supplemental DHA (devoid of EPA), to determine if dietary DHA itself can exhibit potentially favorable effects on other clinical conditions (including cancer). DHA has been found to inhibit carcinogenesis (including mammary and liver cancer) in animal models (*39, 40*) and metastasis, such as the inhibitory effect of DHA on colon carcinoma metastasis to the lung (*41*). A comparison of the fatty acid status of breast adipose tissue in breast cancer patients with that from patients of benign breast disease has been reported (*42*). This prospective study analyzed and compared the fatty acid composition of lipids from the breast tissue of women with breast cancer as compared to those with benign breast disease. The percentage of DHA in the phospholipids from breast tissue of post-menopausal women with breast cancer was significantly lower (by 55% overall) as compared with the corresponding women having benign breast disease. The dietary intake of DHA was also lower in the former groups relative to the latter. These results indicate the possibility that increased intakes of n-3 PUFA (particularly DHA) may have a protective effect against breast cancer in post-menopausal women. These authors propose that the apparent protective effect of DHA against breast may be mediated by the inhibitory effect of prostaglandin synthesis in the tumor tissue and modulation of the host immune mechanisms such as interleukin and tumor necrosis factor production as well as lymphocyte activation.

Since the lipid composition of the sperm membrane has a significant effect upon the functional characteristics of spermatozoa, and considering the high levels of DHA in human ejaculate and spermatozoa, analyses have been performed on the overall fatty acid and DHA status specifically of spermatozoa including the fatty acid compositions of phospholipids from sperm heads and tails in both normal and abnormal semen samples (43). DHA levels were found to be significantly lower in oligozoospermic samples and in asthenozoospermic samples as compared with normozoospermic samples. A very recent report (44) indicated that the DHA levels in the phospholipid of both seminal plasma and spermatozoa derived from asthenozoospermic males were significantly lower that those derived from normozoospermic individuals despite both groups having non-differing levels of DHA in their circulation (based on serum phospholipid analysis). These results indicate that the lower concentration of DHA in the spermatozoa of asthenozoospermic males are not due to dietary differences but potentially due to defects in uptake and/or metabolic parameters involving DHA. It remains to be established whether long-term supplementation with DHA can restore the sub-normal DHA levels in the spermatozoa of the asthenozoospermic males with beneficial effects on clinical parameters (sperm motility, etc.).

In other studies using DHA supplementation in healthy men, it has been reported that DHA ingestion inhibits natural killer cell activity and the production of inflammatory mediators (45). These results indicated that inhibitory effects of DHA on the immune cell functions varied with the cell type and that the inhibitory effects are not mediated via increased production of prostaglandin E_2 and leukotriene B_4. Furthermore, DHA consumption was found not to inhibit many of the lymphocyte functions which had been reported to be inhibited by fish oil consumption when the total fat intake is low and held constant (46).

Conclusions

DHA is a physiologically-essential nutrient in the brain and retina for optimal neuronal functioning and visual performance, respectively, via its enrichment in the 2-position of membrane phospholipid. It also has considerable functional importance in other tissues and cells including the potential to offer protection against cardiovascular disease, associated risk factors (including triglyceride-lowering), as well as morbidity and mortality via anti-arrhythmic and other physiological effects. Current consumption levels of DHA in most countries of the world per capita are well below intakes which are considered to be optimal for human health including the prevention and/or attenuation of various chronic disorders. The development of various DHA concentrates from conventional (aquatic) and alternative sources and their inclusion in various forms (oils, powders, micro-encapsulated preparations, emulsions, etc.) as nutraceutical ingredients in a wide variety of functional foods is currently undergoing a dramatic upsurge in activity (47). Over the coming decade, a dramatic increase in the availability and distribution of functional foods enriched in DHA is anticipated which should significantly improve the overall DHA status of the global population. Hopefully, this surge in the overall physiological DHA status will be accompanied by enhancements in human health and a greater capacity for the prevention and attenuation of cardiovascular and other chronic disorders.

References

1. Thomas, L.M.; Holub, B.J. In *Technological Advances in Improved and Alternative Sources of Lipids*; Kamel, B.S. and Kakuda, Y., Eds.; Blackie Academic & Professional: London, **1994**; pp 16-49.
2. Emken, E.A.; Adlof, R.O.; Gulley, R.M. *Biochimica et Biophysica Acta.* **1994**, *1213*, 277-288.
3. Salem, N. Jr; Wegher, B.; Mena, P.; Uauy, R. *Proc. Natl. Acad. Sci.* **1996**, *93*, 49-54.
4. Conquer, J.A.; Holub, B.J. *J. Nutr.* **1996**, *126*, 3032-3039.
5. Conquer, J.A.; Holub, B.J. *Lipids* **1997**, *32*, 341-345.
6. Horrocks, L.A.; Yeo, Y, K. *Pharmacol. Res.* **1999**, *40*, 211-225.
7. British Nutrition Foundation. Task force on unsaturated fatty acids, a report of the British Nutrition Foundation. Chapmann & Hall. London. **1992**.
8. Nettleton, J.A. *J. Am. Diet. Assoc.* **1993**, *93*, 58-64
9. Connor, W.E.; Neuringer, M. and Reisbeck, S. *Nutr. Rev.* **1992**, *50*, 21-29.
10. Makrides, M.; Neumann, M.; Simmer, K.; Peter, J.; Gibson, R. *Lancet* **1995**, *345*, 1463-1468
11. Werkman, S. Carlson, S.E. *Lipids* **1996**, *31*, 91-97.
12. Conquer, J.A.; Holub, B.J. *Veg. Nutr.* **1997**, *1/2*, 42-49.
13. Lucas, A.; Morley, R.; Cole, T.J.; Lister, G.; Leeson-Payne, C. *Lancet* **1992**, *339*, 261-264.

14. Sorensen, H.T.; Sabroe, S.; Olsen, J.; Rothman, K.J.; Gillman, M.W.; Fischer, P. *BMJ* **1997**, *315*, 401-3.
15. Kalmijn, S.; Faskens, E.J.M.; Launer; Kromhout, D. *Am. J. Epid.* **1997**, *145*, 33-41.
16. Birch, E.E.; Farfield, S.; Hoffman, D.R.; Uauy, R.; Birch, D.G. *Dev. Med. Child. Neurol.* **2000**, *42*, 174-181.
17. Cunnane, S.C.; Francescutti, V.; Brenna, J.T.; Crawford MA. *Lipids* **2000**, *35*, 105-111.
18. Simopoulos, A.T. *Can. J. Physiol. Pharmacol.* **1997**, *75*, 234-239.
19. Siscovick, D.S.; Raghunathan, T.E.; King, I.; Weinmann, S.; Wicklund, K.G.; Albright, J.; Bovbjerg, V., Arbogast, P.; Smith, H.; Kushi, L.H.; Cobb, L.A.; Copass, M.K.; Pasty, B.M.; Lemaitre, R.; Retzlaff, B., Childs, M. Knopp, R.H. *JAMA* **1995**, *274*, 1363-1367.
20. Albert, C.M.; Hennekens, C.H.; O'Donnell CJ, Ajani, U.A.; Carey, V.J.; Willett. W.C.; Ruskin, J.N. Manson, J.E. *JAMA* **1998**, *279*, 23-29.
21. Dolecek, T.A. Proc. *Soc. Exp. Biol. Med.* **1992**, *200*, 177-182.
22. Simon, J.A.; Hodgkins, M.L.; Browner, W.S.; Neuhaus, J.M.; Bernert, J.T.; Hulley, S.B. *Am. J. Epid.* **1995**, *142*, 469-476.
23. Von Schacky, C.; Angerer, P.; Kothny, W.; Theisen, K.; Mudra, H.; *Ann. Intern. Med.* **1999**, *130*, 554-562.
24. Gruppo Italiano per lo Studio della Sopravvivenza nell'Infarto miocardico. *Lancet* **1999**, *354*, 447-455.
25. Gaudette, D.C.; Holub B.J. *J. Nutr. Biochem.* **1991**, *2*, 116-121.
26. Kang, J.X.; Leaf, A. *Lipids* **1996**, *31*, 541-544.
27. Grimsgaard, S.; Bonaa, K.; Hansen, J.; Nordoy, A. *Am. J. Clin. Nutr.* **1997** *66*, 649-659.
28. Conquer, J.A.; Holub, B.J. *J. Lipid Res.* **1998**, *39*, 286-292.
29. Laugharne, J.D.E.; Mellor, J.E.; Peet, M. *Lipids* **1996**, *31*, 163S-165S.
30. Söderburg M.; Edlund, C.; Kristensson, K.; Dallner, G. *Lipids* **1991**, *26*, 1363-1367.
31. Hibbeln, J.R.; Salem, N. *Am. J. Clin. Nutr.* **1995**, *62*, 1-9.
32. Martinez, M.; Mougan, I.; Roig, M.; Ballabriga, A. *Lipids* **1994**, *29*, 273-280.
33. Holman, R.T.; Bibus, D.M.; Jeffrey, G.H.; Smethurst, P; Croft, J.W. *Lipids* **1994**, *29*, 61-65.
34. Stevens. L.J.; Zentall, S.S.; Deck, J.L.; Abate, M.L.; Watkins, B.A.; Lipp, S.R.; Burgess, J.R. *Am.J.Clin.Nutr.* **1995**, *62*, 761-768.
35. Stevens. L.J. Zentall, S.S., Abate, M.L. Kuckek, T., Burgess, J.R. *Physiol. Behavior* **1997**, *59*, 915-920.
36. Martinez, M. *Lipids* **1996**, *31*, S145-S52.
37. Hamazaki, T., Sawazaki, S.; Itomura, M.; Asaoka, E.; Nagao, Y.; Nishimura, N.; Yazawa, K.; Kuwamori, T.; Kobayashi, M. *J. Clin. Invest.* **1996**, *97*, 1129-1133.

38. Sawazaki, S.; Hamazaki, T., Yazawa, K.; Kobayashi, M. *J. Nutr. Sci. Vitaminol.* **1999**, *45*, 655-665.
39. Noguchi, M.; Minami, M. et al. B*r. J. Cancer.* **1997**, 75, 348-353.
40. Calviello, G.; Palozza, P.; Piccioni, E.; Maggiano, N.; Frattucci, A.; Franceschelli, P.; Bartoli, G.M. *Int. J. Cancer* **1998**, *75*, 669-705.
41. Iigo, M.; Nakagawa, T.; Iwahori, Y.; Asamoto, M.; Yazawa, K.; Araki, E.; Tsuda, H. *Br. J. Cancer* **1997**, *75*, 650-655.
42. Zhu, A.R.; Agren, J.; Mannisto, S.; Pietinen, P.; Eskelinen, M.; Syrjanen K; Uusitupa, M. *Nutr. Cancer* **1995**, *24*, 151-160.
43. Zalata, A.A.; Christophe, A.B.; Depuydt, C.E.; Schoonjans, F; Combaire, F.H. *Mol. Hum Reprod.* **1998**, *4*, 111-118.
44. Conquer, J.A.; Martin, J.B.; Tummon, I.; Watson, L.; Tekpetey, F. *Lipids* **1999**, *34*, 793-799.
45. Kelly, D.S.; Taylor, P.C.; Nelson, G.L.; Schmidt, P.C.; Ferretti, A.; Erickson, K.L.; Yu, R.; Chandra, R.K.; Mackey, B.E. *Lipids* **1999**, *34*, 317-324.
46. Kelley, D.S.; Taylor, P.C.; Nelson, G.J.; Mackey, B.E. *Lipids* **1998**, *33*, 559-566.
47. Shahidi, F.; Wanasundara, U.N. *Trends Food Sci. Tech.* **1998,** *9*, 230-240.

Chapter 6

Dietary Polyunsaturated Fat in Cardiovascular Disease: Boon or Bane?

T. R. Watkins and M. L. Bierenbaum

K. L. Jordan Heart Research Foundation, 48 Plymouth Street, Montclair, NJ 07042

Cardiovascular disease (CVD) still accounts for one out of two deaths in the United States. Among changes in life style recommended by the U. S. Government a change in the amount and kind of dietary fat has received more emphasis than any other. An habitual shift from saturated to polyunsaturated fat leads to decreased serum cholesterol levels, a well known risk factor. Such a change leads also to decreased concentrations of low density lipoprotein cholesterol (LDL) and high density lipoprotein cholesterol (HDL) lipid fractions. In general, the resulting altered lipid profile also modulates the platelet response to agonists, such as arachidonic acid and thrombin, as well as prolonged bleeding times. These kinds of changes lead to an improved cardiovascular risk profile. However, at the same time, enriching the diet with polyunsaturated fat (as ene proportion increased) also leads to elevated serum peroxides and depleted antioxidant reserves. These observations will be illustrated with data from a rat model and human supplementation studies with α-tocopherol. In order to promote the public health and well being, recommendations to increase the proportion of calories derived from polyunsaturated fatty acids ought to be accompanied with a warning to increase the intake of antioxidants, such as tocopherol.

Does chemistry in the body mirror chemistry in the test tube? Atwater (*1*) built a whole body calorimeter to test the First Law of Thermodynamics in the human body. Energy was conserved in the whole body as in the bomb calorimeter. Considering fat behavior, saturated fatty acids prefer ordered states at ambient temperatures. Such crystalline arrays can be disrupted by the presence of unsaturation sites. Ghosh et al. (*2-4*) demonstrated that a two-dimensional lipid array in a Langmuir balance so disrupted expanded significantly. She studied various fatty acyl configurations and noted that both mono- and polyunsaturated ones effectively expanded the film. The behavior of such films has been illustrated also for purified phospholipid in water by Smaby et al. (*5*). Jain and Wagner (*6*) and Ogston (*7*) have reported that the transition temperature of a film of pure distearoyl-phosphatidylcholine (di-18:0 PC), 58° C, decreases to 3° C with the introduction of a single unsaturation site, in the case of 1-stearoyl-2-oleoyl-phosphatidylcholine (1-18:0, 2-18:1-PC). Would this phenomenon and subsequent behavior also prevail in a three-dimensional space, such as a phospholipid array in a biological membrane?

To test the notion that unsaturation sites also disrupt order in biological membranes, we constructed membranes highly enriched in unsaturated fatty acids. This we did (*8*) by culturing an auxotrophic mutant of *S. cerevisiae* (KD115), which has an absolute growth requirement for ethylenic bonds in the medium. Levels of unsaturated fatty acid enrichment of 55-75 % (by weight) were commonly achieved. A diverse assortment of pure fatty acids was evaluated, including some with acetylenic bonds, cultivating the cells under aerobic conditions, such that mitochondrial development was enhanced (*9*). By introducing a fatty acid with an unsaturation site at modest levels (5 mole %), the physical melt state of the membrane lipids decreased. These were measured with a lipophilic nitroxide spin probe which partitioned into the membrane bilayer. In the same cultures, we also measured respiratory ability of the cells with a Clark electrode. In either case, when the results were plotted in Arrhenius fashion (1/°K), the phase transition temperature shifted to lower temperatures with increased ration unsaturation. The results showed that the behavior of the membrane associated enzymes was dictated largely by the physical properties of the membrane phospholipid fatty acids. As membrane viscosity decreased with increased unsaturation, local respiratory enzymes had lower activation energies. The cells respired faster.

With this information, one would expect that a shift in the dietary fat pattern, shifting from primarily saturated to polyunsaturated fats would lead to decreased serum cholesterol. Other lipid levels might decline also. If so, this would have important public health implications.

In the 1970's the typical American diet included 37 - 40% of calories as fat. Obesity had become an obvious public health concern. The incidence of cardiovascular disease (CVD), one of the highest in the world, had been associated with certain risk factors. At the time, elevated serum cholesterol had been identified as an important risk factor (*10*). In our laboratory and elsewhere, evidence had been presented that it could be modulated with dietary intervention to lower an individual's serum cholesterol, thus limiting the risk of cardiovascular disease (*11,12*). Other reports had appeared. Hegsted (*13*) had reported that if saturated fat were replaced with a monounsaturate, such as olive oil, serum cholesterol values decreased by 40 mg/dL. He, and his co-workers, showed further that if corn oil, a polyunsaturate, were fed, serum cholesterol decreased by another 40 mg/dL. Another study, Dayton's now classic study, shored up these observations (*14*).

Dayton and his associates (*15*) had demonstrated that if corn oil, rich in linoleate, was substituted for saturated fat in the daily diet, serum cholesterol levels declined significantly. In this Veterans' Administration study, Dayton and his group showed that the mortality statistics for cardiovascular disease continued to improve after five years. His subjects consumed the corn oil enriched diet for ten years, an oil having two ethylenic sites in most fatty acids. Though the CVD statistics continued to decline, the mortality statistics for malignancy rose significantly. Soon thereafter, controversy ensued when Ederer et al. (*16*) reported that diets providing generous amounts of polyunsaturated fat were not mutagenic. Dayton had shown that in male subjects over 65 years of age (as in his cohort), substitution of PUFA for saturated fat led to lower serum cholesterol levels. The incidence of atherosclerotic heart disease declined. However, Dayton observed in his subjects that after ten years of fat modified diets, the incidence of malignancy rose significantly. In spite of objections from Ederer and others, he stood firm.

In a fat modification study done by Bierenbaum et al. (*11*), corn oil was substituted in part for the usual dietary fat in 'heat and serve' meals ('TV dinners'). In the 200 subjects, serum cholesterol levels decreased. CVD mortality declined concomitantly after ten years. However, after 25 years, the death rates from CVD and malignancy were equal, with the higher malignancy rate in the treatment group. That is, malignancy had displaced CVD in these subjects.

To improve the public health, Senator George McGovern (*17*) convened the Senate Select Committee on Nutrition and Human Needs in 1976 to consider how diet might be altered to improve the quality of life of Americans. Then, as now, the chief public health problem, both in terms of morbidity and mortality, was cardiovascular disease. The Committee collated much evidence about degenerative disease and the purported role of diet in its causation and relief. Rather than delay their report until evidence linking diet and disease began to approach proof, they determined to hear expert testimony and issue recommendations. This they did. In their report (*18*) they recommended that Americans eat less, including less red meat, less fat, less saturated fat, more mono- and polyunsaturated fat, less salt and alcohol in moderation, if it were used. In lieu of the 37% of calories as fat, the Committee recommended that Americans take 30% as fat. Based upon the modulation of serum cholesterol by unsaturated fat, they recommended that Americans divide their fat intake, taking 10% saturates, 10% unsaturates, and 10% polyunsaturates. Over the past twenty years, Americans have increased their use of polyunsaturates. Presently, they consume approximately 6% of calories as polyunsaturates. Polyunsaturates peroxidize readily.

On the basis of chemical principles, we know that the rate of autoxidation of fat increases with the degree of unsaturation (though not linearly). Autoxidation of pure methyl oleate occurs slowly after a long induction period. This can be sped up if the temperature is raised, or a radical source is introduced into the sample, or if the sample is irradiated. In contrast, methyl linoleate autoxidizes at a rate 10-40 times faster, largely because of the methylene group at C11 sitting between two ethylenic bonds. Methyl linolenate reacts 2 - 4 times faster than methyl linoleate (*19*). On this basis, if diets enriched with polyunsaturated fatty acids were fed, one would expect to observe markedly elevated levels of lipid peroxides, whether they arise *in vivo* or *ex vivo* from ingested food. At the least, should these be observed in a tissue, such as the serum, one would not be surprised. As they accumulate in the serum, the risk of cardiovascular disease rises (*20*).

The accumulation of autoxidized lipid in the LDL particle also represents a serious risk factor in the development of cardiovascular disease. Stringer et al. (*21*) have reported that lipid peroxide concentrations were significantly higher in persons with ischemic heart disease and peripheral artery disease. Further, hypertension, obesity, diabetes, smoking, positive family history and use of β-blockers or thiazide diuretics were not associated with significant differences in serum lipid peroxide levels. The sustained presence of lipid hydroperoxides in the serum and their lipoprotein carriers has been associated with cardiovascular risk. Kovacs et al. (*22*) reported that in 70 patients tested three months after coronary artery bypass, plasma hydroperoxides were significantly elevated when compared with levels in controls.

Autoxidation of lipid, a self-perpetuating chain reaction, could be interrupted in the presence of adequate amounts of appropriate antioxidants. Olcott and Van der Veen (*23*) have reported that the lipophilic tocopherols perform effectively as radical scavengers. By self-sacrifice they yield a tocopheryl radical, which is much less reactive than the fatty acid hydroperoxy radical reactant. Hence, the rate of the autoxidation was slowed markedly in the presence of antioxidants, such as the tocopherols in their model studies. Would similar chemistry pertain in the human body, in which tissue is continuously exposed to oxygen?

The purpose of this study was to evaluate dietary trials, both with animals and humans, feeding diets enriched with polyunsaturated fat, noting their influence upon these cardiovascular risk factors: serum cholesterol, LDL and HDL cholesterol, platelet responsiveness, bleeding time, serum fatty acid hydroperoxides and serum α-tocopherol. On the basis of the film data indicating that unsaturation leads to film expansion, one would expect that this would lead to lower activation energies for membrane associated enzymes. In the case of serum lipids, this would be manifested as more facile processing, hence a lower serum cholesterol, its fractions and triacylglycerols. Other systems would be similarly affected. Finally, increased unsaturation would lead to increased oxygen damaged lipids in the circulation, leading to depleted antioxidant reserves. Below a critical level this would lead to platelet aggregation. This also would translate into increased thrombogenic risk.

Materials and Methods

In the animal feeding trials cited, weanling, male pups, mean weight 140 g, were received, weighed and divided into four groups of twelve each, three animals per cage. The animals received a chow based diet, consisting of Purina chow (Ralston-Purina, St. Louis, MO). It contained (weight %): chow, 78; cholesterol, 2; and, fat, 20. The fats were: partially hydrogenated soy oil (Wakefern, Elizabeth, NJ); safflower oil (Lever Brothers, Riveredge, NJ); Canola® oil (Proctor & Gamble, Cincinnati, OH); or, menhaden oil (Zapata-Haynie Corp., Reedsville, VA, stabilized with 50 ppm α-tocopherol). The animals were fed *ad libitum*. The room was maintained at 25±2° C, with a biphasic lighting cycle, with 12 hours of light and 12 hours of dark, the light phase beginning at 7:00 am.

Chow was fed to animals on their arrival in the laboratory over a two-week adaptation period. Afterward, they were shifted to one of the four diets shown in Table I for another eight weeks. Food intake was monitored daily. Body weights were monitored weekly. Animals had access to water *ad libitum*.

Table I. Diet composition for the rat feeding trials.

Component	*PHSO*	*SAFFLOWER*	*CANOLA®*	*MENHADEN*
Chow	78	78	78	78
Corn oil	2.6	-	-	-
Oil	17.4	20	20	20
Cholesterol	2	2	2	2

Data as weight percent. PHSO: partially hydrogenated soya oil.

Blood was taken from rats after light carbon dioxide anesthesia by heart puncture into evacuated silicone separator tubes (Becton Dickinson, Rutherford,

NJ) for serum, citrated tubes for platelet studies, or sodium EDTA tubes for plasma samples. Serum cholesterol, HDL cholesterol and triglyceride was measured with commercial assays (Boehringer Mannheim, Indianapolis). The LDL fraction was computed by difference (24). Serum peroxides were measured by the method of Ohishi et al. (25). Serum vitamin E was measured by the method of Bieri et al. (26).

Statistical differences between groups were evaluated with Student's t test.

In human supplementation trials, the effect of inclusion of Canola® oil or flax seed oil (as milled flax seed) on serum lipids was evaluated. To examine a tissue for functional implications of the polyunsaturated enrichment, platelets were tested for response to the agonists thrombin and arachidonic acid. Either platelet aggregation or bleeding time (the time elapsed until blood flow stopped after a finger was lanced), indicated the tendency of platelets to aggregate when challenged. Further, accumulation of peroxides in serum was measured and the level of α-tocopherol, an indicator of antioxidant reserve.

Thirty-six subjects, 21 M and 15 F, age 39-76 years were recruited. Subjects had moderately elevated serum cholesterol (> 5.6 mmol/L). Fourteen had blood pressure >140/90 mm Hg; four had diabetes; 14 had experienced a cerebrovascular accident; and two had experienced a myocardial infarction. Persons using free radical scavengers, such as dipyridamole were excluded, as were those using β-blockers, anti-lipid drugs, fish oil, aspirin, or vitamins. Subjects ate self-selected diets and consumed, in addition to their usual diets, a daily addendum of 30 g of Canola® oil in one study and three slices of flax fortified bread (3.6 g of α-linolenic acid) and 15 g of milled flax (providing 3.4 g of α-linolenic acid) in the other. Subjects received the supplement for five weeks during each supplementation trial.

Results

The presence of increased levels of unsaturation sites in the dietary fat resulted in decreased serum lipid levels, in general. Cholesterol, LDL and HDL cholesterol decreased, as did triacylglycerol. The relation between the proportion of unsaturation sites in the dietary fat and these serum lipid components appears in Table II.

Cellular membranes enriched with unsaturated fatty acid would be expected to exhibit lower phase transition temperatures and lower activation energies. In

Table II. Influence of degree of unsaturation of dietary fat upon serum lipids.

Lipid	System	Equation*
Cholesterol	Rat	y = -3.40 x + 2.6
HDL cholesterol**	Rat	y = -0.82 x + 0.70
Triacylglycerol	Rat	y = -1.05 x + 0.83
Cholesterol	Human	y = -4.52 x + 6.80
LDL cholesterol***	Human	y = -6.08 x + 4.99

* x: (proportion of dietary fat) [proportion monounsaturated fat x 1) + (proportion of diunsaturated fat x 2) + (proportion of triunsaturated x 3) + ...]. y: the regression line obtained for the indicated blood serum lipid data plotted for each fat fed x of Table I.
** HDL: high density lipoprotein. *** LDL: low density lipoprotein.

the case of cholesterol, processing and moving it from the serum into the tissues would be expected to occur more rapidly. This occurred. The incorporation of unsaturated fatty acids into the diet led to decreased serum cholesterol, whether in a rat model or human subjects, Table II. In the rat model, increasing unsaturation, here tested with partially hydrogenated soya oil, safflower, Canola®, or menhaden oil, led to decreased serum cholesterol and decreased HDL cholesterol. The triacylglycerol level also declined as a function of unsaturation. In the human subject, a similar phenomenon was observed, both serum cholesterol and LDL cholesterol varied inversely with unsaturation of dietary lipid. The human serum cholesterol responded with nearly one-third more sensitivity than the rat to dietary polyunsaturated fat (see slope data, Table II). This may reflect the rat's having eaten vegetable oil over a life time, hence having membrane's largely pre-equilibrated with unsaturated fatty acids.

One might expect that response to various agonists in platelets enriched with polyunsaturates would also be attenuated, especially if n-3 acid were incorporated from the diet. Such was the case in the rat model, Table III. Both ATP charge and release and platelet aggregation decreased with increased dietary unsaturation. The hyper-aggregation tendency of the platelet was nearly four times more sensitive to dietary enrichment with polyunsaturated fatty acid than the ATP accumulation and release upon stimulation with thrombin, as seen in these slope data. As polyunsaturated fat intake rose, the platelet was 'cooled down', less apt to aggregate. This tempered platelet behavior meant less thrombogenic risk. These data also showed the 'cost' of realizing such inhibition of platelet aggregation with diets enriched with polyunsaturates. In the rat model, the loss of serum α-tocopherol (by peroxides) occurred at a rate five times faster than the benefit (the hypocholesterolemia).

Table III. Effect of degree of dietary lipid unsaturation upon platelet function and serum vitamin E in the rat.

Test [agonist]/tissue	Equation*
ATP** release [thrombin]/platelet	$y = -1.01 x + 1.19$
Aggregation [arachidonic acid]/platelet	$y = -4.01 x + 1.80$
Vitamin E/serum	$y = -21.1 x + 12.8$

* See legend of Table II for details of x. y: the regression line of response data for each test lipid x of Table I. ** ATP: adenosine triphosphate.

A pool of fat increasingly enriched in polyunsaturated fatty acids would be prone to increased autoxidation *in vitro*. As Gunstone has pointed out (*19*), the rate would be expected to increase geometrically, not arithmetically, with unsaturation. Thus, these diets enriched in unsaturation sites would be more prone to autoxidation, hence would be expected to lead to depletion of antioxidant reserves *in vivo*; this occurred. In the absence of dietary vitamin E addendum, with increased dietary unsaturation, serum vitamin E fell, Table III.

The functional response of the human platelet obtained from the hypercholesterolemic human subject reflected, in general, the results observed in the rat model. Platelet aggregation was inhibited by polyunsaturated fatty acids, here measured as increased bleeding time to a lancet puncture of the index finger, Table IV. The platelet release of ATP when challenged with thrombin also decreased with enriched unsaturation, here illustrated with flax seed oil enrichment of dietary lipid, compared with mixed dietary lipids.

Human platelet membranes enriched with polyunsaturates also would be expected to show increased susceptibility to autoxidation. Especially in the absence of adequate dietary antioxidant, lipid peroxidation would be expected to rise as the proportion of polyunsaturated fat in total dietary fat rose. This was observed, as illustrated by flaxseed lipid, Table IV. Accumulation of serum hydroperoxides led to depletion of antioxidant reserves, as expected, here measured as α-tocopherol. Without supplementary antioxidant in the diet, an increased proportion of unsaturated lipid intake led to rapid depletion of antioxidant reserves. In fact, the rate of decline of serum α-tocopherol concentration occurred four times faster than the rate of peroxide accumulation, Table IV.

Table IV. Influence of degree of unsaturation of dietary fat upon platelet response, serum fatty hydroperoxides and serum vitamin E in hypercholesterolemic human subjects.

Test response, y [agonist]/tissue	Equation*
Bleeding time, canola®/platelets, sec	y = 1,611 x + 175
ATP release, flax [thrombin]/platelets, μmol/L	y = -5.14 x + 1.33
Fatty hydroperoxides, flax/serum, μmol/L	y = 11.4 x + 1.33
Vitamin E, flax/serum, μmol/L	y = -44.3 x + 21.3

* See legend of Table II for detail about x. y: the regression of the response data of the indicated test plotted against x after feeding each fat of Table I.

Discussion

The functional state of cellular, membrane associated enzymes would be expected to respond to the physical state of membrane lipids. Feeding diets enriched in unsaturates or polyunsaturates would be expected to enrich cellular membranes. Such was observed in yeast cells, and inferred in rat and human cellular membranes. The respiratory enzyme activity in the yeast mutant, dependent upon dietary unsaturated fat for growth, reflected the state of membrane lipids. Similarly, in the rat cholesterol enzyme activity reflected dietary fat. The ability of the rat to process cholesterol into liver and peripheral tissues increased with enrichment of the diet with unsaturated lipids. In like fashion, as expected, the same phenomenon was seen in humans carrying elevated levels of lipid in their blood. Cholesterol and its subfractions were processed more efficiently, if the diet were enriched with unsaturation sites. One would observe similar phenomenon in other tissues using other probes than lipids, it is presumed. Here, the qualitative nature of the relations is shown, since translating the actual intakes of free living human subjects into exact quantities is fraught with some difficulty.

These observations about the clearance of an elevated cholesterol burden from the circulatory system would be of concern in a discussion of cardiovascular risk factors. Saturated or *trans* fat raises serum cholesterol in infants, adults and the very old (*27-31*). Elevated serum cholesterol raises atherogenic risk (*32*). Serum cholesterol has been used as an indicator of risk since the studies of Keys and Parlin (*10*) and others before them. After numerous studies had been published noting the potent hypocholesterolemic effect of dietary polyunsaturates, governmental health agencies issued recommendations that Americans consume a greater proportion of dietary fat as polyunsaturated fat. This the public has done. The consumer has substituted vegetable oil for lard in cooking, as has industry (*e. g.*, a McDonald's-type fast food establishment). This has had a predictable effect on serum cholesterol, as illustrated by these data.

In addition, consumption of diets enriched with unsaturated lipid confers other benefits. Thrombogenic risk factors include hyperaggregation of platelets (*20*). As unsaturation increased in the diet of both the rat, eating a largely vegetarian ration, and the human subject eating an omnivorous diet, both the ATP charge of the platelet, a thrombogenic risk factor, and the aggregation tendency decreased. Thus, unsaturation could be used to curb thrombogenic risk. This has been reported in other tissues, such as the leucocyte (*33*). Yet, eventually the price was increased peroxidation.

As the level of unsaturated sites increases in the dietary, or other lipid pool, especially polyunsaturated sites, the tendency to autoxidation rises. This can be measured, as was shown. Elevated peroxide levels, another well documented cardiovascular risk factor (*20-22*), in particular of thrombogenic risk, rose. Subsequently, antioxidant reserves began to be exhausted. Suboptimal antioxidant reserves has been reported to be perhaps the most powerful predictor of cardiovascular risk (*34*), and a key indicator of thrombogenic risk. In a population such as America, in which antioxidant intakes are modest, at best, being on the order of 10 - 12 I. U. of vitamin E daily, it would behoove us – and the government – to urge increased antioxidant intake. It has been urging increased intake of polyunsaturated fats since 1976 with the issuance of the "Dietary Goals for the United States." If no such antioxidant recommendation be forthcoming, we can expect continued record rates of cardiovascular disease, and also rising rates of malignancy, when antioxidants are limited in the diet.

As expected, whether data were gathered from the rodent model or human subjects, the same pattern was observed. This one would expect, whether the membranes were synthesized by the rat or the human subject. Feeding unsaturated fat led to decreased activation energies for membrane associated events. This was manifested in the case of cardiovascular factors as decreased serum levels of cholesterol, LDL cholesterol, HDL cholesterol and triglyceride.

The platelet was sampled to observe functional effects. The physical influence of increased unsaturation sites resulted in similar chemical effects in both the rat and the human system. Both the bleeding time and the aggregation time (related to the reciprocal of the bleeding time) varied with the degree of unsaturation of the dietary lipid. The ATP release to a thrombin challenge decreased with unsaturation in both models. In these cases, the dietary lipids were enriched in n-3 acids, largely α-linolenic acid. The presence of increased n-3 acid tempered ATP release.

Consumption of increased levels of unsaturated fatty acid in the diet leads to further peroxidation of lipid, and possibly other substrates, when antioxidant reserves are limited. In the hypercholesterolemic human, a dietary addendum of flaxseed oil (as milled flax seed) fed incorporated in the diet in bread, led to increased serum lipid hydroperoxides. Since these react rapidly with antioxidants, such as vitamin E, the serum antioxidant reserve was expected to decline; it did. Increased levels of unsaturation sites in the dietary lipid resulted in a marked decrease in serum α-tocopherol, Table IV.

We posed the question, would we observe the same chemical biomembrane dynamics *in vivo* and *in vitro* (in films)? The example data cited in respiring yeast, the rat and hypercholesterolemic humans showed that increased unsaturation of dietary fatty acids facilitated lipid processing in the tissue (membranes). As unsaturation levels rose, serum lipid values fell. Further, as the degree of polyunsaturation increased, peroxidation rose, increasing the need for antioxidant resources. The implication for human health of elevated peroxides, especially thrombogenic risk, has been noted.

Indeed, membranes behave *in vivo* as films *in vitro*. Dietary unsaturates lower serum lipids. Yet diets rich in polyunsaturates lead to elevated serum peroxides, which exhaust antioxidant reserves. The American diet, rich in unsaturates, typically lacks adequate antioxidants. The consumer deserves a warning. If not warned, these oils, once our boon, will become our bane.

References

1. Atwater, W. O.; Rosa, E. B. *Off. Exp. Sta. Bull. No. 63*. U. S. Dept. Agric., Washington, D.C., 1899.
2. Ghosh, D.; Tinoco, J. *Biochem. Biophys. Acta* **1972**, *266*, 41-49.
3. Ghosh, D.; Williams, M.A.; Tinoco, J. *Biochem. Biophys. Acta* **1973**, *291*, 350-362.
4. Ghosh, D.; Lyman, R. L.; Tinoco, J. *Chem. Phys. Lipids*. **1971**, *7*, 173-184.
5. Smaby, J.; Momsen, M. M.; Brockman, H. L.; Brown, R.E. *Biophys. J.* **1997**, *73*, 1492-1505.

6. Jain, M.; Wagner, R. C. *Introduction to Biological Membranes*. New York, John Wiley & Sons, 1980.
7. Ogston, C. A. Interphases. In: *Biological Membranes*. D. S. Parsons, ed. Oxford, Oxford University Press, 69-80, 1975.
8. Watkins, T. R.; Taylor, D.; Williams, M. A.; Keith, A. *Fed. Proc.* **1973**, *32*, 3938.
9. Eletr, S.; Williams, M. A.; Watkins, T. R.; Keith, A. *Biochem. Biophys. Acta* **1974**, *334*, 190-204.
10. Keys, A.; Parlin, R. W. *Am. J. Clin. Nutr.* **1966**, *19*, 175-181.
11. Bierenbaum, M. L.; Fleischman, A. I.; Raichelson, R.; Hayton, T.; Watson, P.B. *Lancet* **1973**, *1*, 1404-1407.
12. Miettenen, M.; Turpeinen, O.; Karvonen, M. J.; Elosuo, R.; Paavilainen, E. *Lancet* **1972**, *2*, 835-838.
13. Hegsted, D. M.; McGandy, R. B.; Myers, M. L.; Stare, F. J. *Am. J. Clin. Nutr.* **1965**, *17*, 281-295.
14. Dayton, S.; Pearce, M. L.; Hashimoto, S.; Dixon, W. J.; Tomiyasu, U. *Circulation* **1969**, *40*, Suppl. II, 1-63.
15. Pearce, M. L.; Dayton, S. *Lancet* **1971**, *1*, 464-467.
16. Ederer, F.; Leren, P.; Turpeinen, O.; Frantz, I. D., Jr. *Lancet* **1971**, *2*, 203-206.
17. McGovern, G.; Talmadge, H. E.; Percy, Ch. H.; Hart, P.A.; Dole, R.; Mondale, W.F.; Bellmon, H.; Kennedy, E.M.; Schweicker, R.S.; Nelson, G.; Taft, R. Jr.; Cranston, A.; Hatfield, M.O.; Humphrey, H.H. *Diet Related to Killer Diseases*. Senate Select Comm. Nutrition & Human Needs. U. S. Government Printing Office, July 27, 28, 1976, No. 76-554.
18. Select Committee on Nutrition and Human Needs of the United States Senate. 94th Congress. Hearings. 27, 28/7/1976. U. S. G. P. O., Washington, D.C.
19. Gunstone, F. D.; Norris, F. A. *Lipids in Foods*. Oxidation. Pergamon Press, Oxford, 1983.
20. Bierenbaum, M. L.; Reichstein, R. P.; Bhagavan, H. N.; Watkins, T. R. *Biochem. Intl.* **1992**, *10*, 57-66.
21. Stringer, M. D.; Gaorog, P. G.; Freeman, A.; Kakkar, V.V. *Br. Med. J.* **1989**, *298*, 281-284.
22. Kovacs, I. B.; Jahangiri, M.; Rees, G. M.; Gorog, P. *Am. Heart J.* **1997**, *134*, 572-576.
23. Olcott, H. S.; Van der Veen, J. *J. Am. Oil Chem. Soc.* **1958**, *35*, 161-164.
24. DeLong, D. M.; DeLong, R.; Wood P. D.; Lippel, K.; Rifkind, B.M. *JAMA* **1986**, *256*: 2372-2377.
25. Ohishi, N.; Ohkawa, H.; Miike, A.; Tatano, T.; Yagi, K. *Biochem. Intl.* **1985**, *10*, 205-211.

26. Bieri, J. G.; Tolliver, T. J.; Catigniani, C. L. *Am. J. Clin. Nutr.* **1979**, *32*, 2143-2149.
27. Lapinleimu, H.; Viikari, J.; Jokinen, R.; Salo, P.; Routi, T.; Leino, A.; Ronnemaa, T.; Seppanen, R.; Valimaki, I.; Simell, O. *Lancet* **1995**, *345*, 471-476.
28. Christakis, G.; Rinzler, S.H.; Archer, M.; Winslow, G.; Jampel, S.; Stephenson, J.; Friedman, G.; Fein, H.; Kraus, A.; James, G. The Anti-Coronary Club. *Am. J. Public Health* **1966**, *56*, 299-314.
29. Kinsell, L. W.; Partridge, J.; Boling, L.; Margen, S.; Michaels, G.D. *J. Clin. Endocrin.* **1952**, *12*, 909-913.
30. Mensink, R.; Katan, M. *New Engl. J. Med.* **1990**, *323*, 439-445.
31. Weverling-Rijnsburger, A. W. E.; Blauw, G. J.; Lagaay, A. M.; Knook, D.L.; Meinders, A.E. *Lancet* **1997**, *350*, 1119-1123.
32. Katz, L. N.; Stamler, J.; Pick, R. In: *Nutrition and Atherosclerosis.* Katz, L. N., ed., Philadelphia, 1958.
33. Sirtori, C. R.; Gatti, E.; Tremoli, E.; Galli, C.; Gianfranceschi, R.; Franceschini, G.; Colli, S.; Maderna, P.; Marangoni, R.; Perego, P.; Stragliotto, E. *Am. J. Clin. Nutr.* **1992**, *56*, 113-122.
34. Gey, K. F.; Puska, P. *Ann. New York Acad. Sci.* **1989**, *570*, 268-290.

Chapter 7

The Effect of Alpha-Linolenic Acid on Retinal Function in Mammals

Andrew J. Sinclair and Lavinia Abedin

Department of Food Science, Royal Melbourne Institute of Technology University, Melbourne, Victoria 3000, Australia

The retinal function of several different species is compromised by a dietary deficiency of α-linolenic acid. A review of the literature revealed that diets containing less than 0. 1 g/100g diet as α-linolenic acid could not sustain retinal docosahexaenoic acid levels over a prolonged period and that such diets were associated with a reduced response of the retina to light. In these studies the median α-linoleinic acid intake of control animals was 1.25g/100 g diet. A study on the comparative ability of dietary α-linolenic acid and dietary docosahexaenoic acid to provide for retinal docosahexaenoic acid in the guinea pig found that α-linolenic acid was only approximately 10% as effective as docosahexaenoic acid in this regard.

It has been known for many years that the mammalian brain is rich in lipid and that the gray-matter contains a high proportion of polyunsaturated fatty acids (PUFA) (*1*), especially arachidonic acid (20:4n-6), docosatetraenoic acid (22:4n-6) and docosahexaenoic acid (22:6n-3). These PUFA are found in levels of about 12, 6 and 22% of gray-matter phospholipid fatty acids, respectively, for most mammalian species (*2-4*). Subsequent analyses of retinal fatty acids, from a smaller number of mammals, showed that retinal phospholipids were also rich in PUFA, with 22:6n-3 being the predominant fatty acid (*5*). In contrast to the brain and retina, there is a diverse pattern of PUFA in the phospholipids of other tissues, such as liver and muscle, probably reflecting the wide differences in dietary intake of essential fatty acids between species (*4*). In other words, despite a wide variation in the dietary intakes of linoleic acid (18:2n-6) and α-linolenic acid (ALA or 18:3n-3) between different species in their natural or normal environment, there are processing mechanisms (specific carrier proteins, acyl-transferases and 22:6n-3-binding proteins) which regulate the composition of brain PUFA. Subsequent studies in experimental animals showed that alteration of brain and retinal fatty acid patterns can be achieved by manipulation of dietary essential fatty acids intake (6). Diets rich in linoleic acid and poor in ALA from oils such as safflower oil can lead to loss of 22:6n-3 from various tissues including the brain and retina. Other workers have shown that if this manipulation is initiated during pregnancy it is possible to achieve a more substantial deficiency of 22:6n-3, since major accretions of PUFA in the brain occur in utero and in the immediate post-natal period in most species (7). In order to effect maximum change, it has sometimes been necessary to conduct the experiment over several generations (8).

Although the essential fatty acids were discovered in 1930, the essential physiological role of ALA was not convincingly demonstrated until the early 1970s. In terms of retinal function, Benolken et al. (9) showed that the rod responses of the rat become abnormal with electroretinographic (ERG) a- and b-wave amplitudes being dependent on n-3 PUFA supply. Moreover, it was concluded that the specific deficit involved aberrant phototransduction due to changes in photoreceptor membranes of the rod outer segment (9,10). In a subsequent publication, these workers reported a study where rats were fed chow for 14 weeks (ie. adult rats) and then fed test diets for 6 weeks. They found that compared with a fat-free test diet, the diet containing 2% ethyl linoleate gave a 30% increase in ERG amplitude and 2% ethyl linolenate gave an 80% increase in ERG amplitude. These data could be interpreted that both n-6 and n-3 PUFA are biologically active in relation to this physiological parameter (10).

In the rat and the monkey, diets rich in linoleic acid and deprived of n-3 PUFA result in substantial alterations in the retinal and brain PUFA patterns by the first generation, with significant decreases in 22:6n-3 to approximately 40% of normal levels (6, 11). Ward et al. (7) recently achieved greater depletion of 22:6n-3 in rats using a system of artificial rearing. In their first generation animals, brain 22:6n-3 levels were 50% of control values by 8 weeks of age whereas in the second generation, the 22: 6n-3 values were 1 0% of controls by 8 weeks of age. Leat et al. (8) reported that it was possible to almost completely deplete guinea pig retinal lipids of 22:6n-3 to a value of 0.5% of retinal fatty acids by dietary manipulation over three generations.

Concomitant with the 22:6n-3 depletion, there is a compensatory increase in the 22 carbon n-6 PUFA, docosapentaenoic acid (22:5n-6) (6). Interestingly, on such diets the 22:5n-6 accumulates in amounts approximately equal to the losses of 22:6n-3 such that the total 22 carbon PUFA content of the tissues is constant. In the face of the high degree of structural similarity between 22:6n-3 and 22:5n-6, it is surprising to find that substitution of 22:5n-6 for 22:6n-3 is associated with functional changes in the retina and cortex. Replacement of 22:6n-3 by 22:5n-6, as a result of dietary manipulation, is associated with a significant reduction in the activity of Na^+/K^+ dependent AT Phase activity in the nerve terminals of rat brain

(*11*), alterations in ERG responses (*6,9,11*), learning and behavioral abnormalities in rats, mice and monkeys (*12-15*) and reports of ultrastructural changes in the rat hippocampus (*16*).

Since the first reports of Benolken et al. (*9*), other workers have studied the effects of dietary ALA deficiency on retinal function. Neuringer et al. (*17*) investigated in greater detail the relationship between n-3 intake and retinal function in rhesus monkeys. In contrast to the early work of Benolken et al. (*9*) in the rat, they reported that a-wave amplitudes were reduced for both cone and rod responses at 3-4 months of age. However, by 2 years, these differences were no longer present suggesting that the animal may have overcome the effects of the dietary deficiency.

A similar time-dependent recovery has also been reported in rats (*11*), however in the guinea pig the ERG differences due to n-3 deficiency do not recover with time (*18*). While the ERG signal amplitudes normalized in the older but deficient monkeys, both cone and rod ERGs showed delayed b-wave implicit times (b-wave peak time) suggesting an abnormality in the kinetic aspects of visual processing. More importantly, the finding implies abnormality of post-receptoral or higher order neural processing since the b-wave is considered to reflect post-receptoral activity (*19*). The effect that 22:6n-3 deprivation had on recovery from light exposure was also measured using repetitive flashes by Neuringer et al. (*17*). Unlike the effects of n-3 deficiency on ERG amplitudes, which were found to decrease with age, recovery from repetitive flashes was most affected for the b-wave and was found to increase with age. The data leads to only one compelling conclusion that n-3 PUFA, and especially 22:6n-3, is essential for some retinal functions. It also suggests that the locus of the lesion involves more than rod outer segment membranes as first proposed (*9, 10*).

In this chapter we summarize the previous studies on ALA-deficiency and ERG function in animals and consider the effect of two strategies aimed at increasing retinal 22:6n-3 levels in guinea pigs. First we used an increasing dose of ALA relative to a constant linoleic acid intake, and second we used two levels of dietary 22:6n-3 supplements provided in conjunction with dietary arachidonic acid.

Materials and Methods

Animals and Diets

Seventy pigmented female guinea pigs (English Shorthair) were randomly divided into five groups of four-teen animals at 3-weeks of age. The guinea pigs were fed one of five different semi-synthetic diets ad libitum for 12-weeks and were supplemented with fresh carrots and drinking water containing ascorbic acid (400 mg/L) as described by Weisinger et al. (*20*). The diets contained 10% (w/w) lipid, supplied by mixed vegetable oils, and each diet was designed to provide 17 % linoleic acid (as % of total fatty acids). In all diets, the lipids were based on mixed vegetable oils. In diet SAF, the main lipid was provided from safflower oil and the linoleic acid: ALA ratio was 323:1 (ALA 0.05 % total fatty acids). In diet CAN, the main lipid was provided from canola oil with a linoleic acid:ALA ratio of 2.3:1 (ALA 7 % total fatty acids). Diet BASE was based on mixed vegetable oils, with a linoleic acid:ALA ratio of 17.5:1 (ALA I % total fatty acids). Diets BASE+LCPI and BASE+LCP3 were similar to diet BASE but they were designed to contain supplementary levels of arachidonic acid (1%) and 22:6n-3 (0.7 %) (mimicking the levels in human breast-milk) or 3 % arachidonic acid plus 2.1 % 22:6n-3, respectively. Arachidonic acid was obtained from ARASCO oil (Martek Bioscience, Columbia, MD), processed from a common soil fungus widely distributed in nature and 22:6n-3was obtained from DHASCO oil (Martek Bioscience, Columbia, MD), processed from a microalgal organism.

Fatty acid assays of the five diets showed that the linoleic acid levels achieved the desired range of between 16.1 and 17.2 %. Assays of diet BASE+LCP1, gave arachidonic acid and 22:6n-3 values of 0.87 and 0.59 %, respectively, whereas diet BASE+LCP3 gave 2.74 and 1.80 % of arachidonic acid and 22:6n-3, respectively.

At the end of the 12-weeks of feeding, animals were sacrificed by CO_2 asphyxiation. Retinae from each eye were removed and washed in ice-cold phosphate buffered saline and stored in 10 mL of chloroform/methanol (2:1,v/v) containing butylated hydroxytoluene (10 mg/L) as an anti-oxidant.

Lipid Analyses

Following lipid extraction from retina, the total phospholipids were separated from the neutral lipids by thin layer-chromatography (*20*). The methyl esters of the phospholipid fatty acids were separated by capillary gas liquid chromatography using a 50 m x 0.32 mm (I.D.) fused silica bonded phase column (BPX70, SGE, Melbourne, Australia).

Statistical Analyses

Significant differences between dietary groups were tested using a one-way ANOVA for each type of fatty acid. Post-hoc comparisons were made using the Tukey test with a significance level of 0.05.

Results

Previous studies on n-3 deficiency and ERG function in the rat, monkey and guinea pigs are listed in Table I. The experiments used fat-free diets, safflower or sunflower oil to induce a state of n-3 deficiency. The control diets used ALA-rich

Table I. Animal studies demonstrating effects of n-3 PUFA on retinal function

Study reference	Species	Paradigm	Diet lipids	Length of deficiency or age tested	Control ALA %	Deficient ALA %	Retinal DHA, % of control	ERG, % of control
10	Rat, albino, n=12/group	14 weeks chow, 40 days diet	Fat-free versus fat-free+ALA	6 weeks	2	0	ND	63
6	Monkey, rhesus, n=7/group	diet in pregnancy, test offspring	safflower, versus soybean oil	4-12 weeks of age	0.0385	0.015	13	79
11	Rat, wistar, n=12/group	2 generations, test 3rd generation	sunflower versus soybean oil	6 weeks of age	0.13	0.06	39	79
22	Rat, WKY n=12/group	1 generation, test 2nd generation	safflower versus perilla oil	13 & 28 weeks of age	3.2	0.0025	55	89
20	Guinea pig, albino n=12/group	2 generations, test 3rd generation	safflower versus canola oil	6-9 weeks of age	0.82	0.1	12	48
20	as above	as above	safflower + canola+ fish oil versus safflower	6-9 weeks of age	1.25	0.1	8	68
21	Guinea pig, albino n=6/group	16 weeks from weaning	safflower versus canola oil	19 weeks of age	0.82	0.03	42	62

NOTE: Units for ALA are grams per 100g diet. SOURCE: Data from references 6, 10, 11, 20-22.

vegetable oils such as soy, canola or perilla oil and in one case pure ALA was used. The diets were introduced during pregnancy and extended across several generations or commenced at weaning or even were initiated in adult animals. The control ALA intakes ranged from 0.13 to 3.2g/100g diet and the ALA-deficient diets contained from 0 to 0.1 g ALA/100g diet. In the deficient diets, the linoleic acid levels ranged from 0 to 7.2 g/100g diet. Only two studies included more than one level of n-3 PUFA. Both Wheeler et al. (*10*) and Weisinger et al. (*20*) found a dose response relationship existed with increasing ALA level, while the latter study also found that dietary 22:6n-3 did not lead to improved ERG responses over that found with dietary ALA even though the retinal 22:6n-3 level increased by about 30% compared with the ALA-fed group.

These studies showed that reduced retinal 22:6n-3 levels were associated with increased retinal 22:5n-6, almost as a replacement for the lost 22:6n-3. Losses of retinal 22:6n-3 from 40 to 90% of the control value were associated with significantly reduced a-and b-wave ERG amplitudes. The reductions in ERG amplitude varied from 20 to 52%. There was no obvious relationship between the extent of 22:6n-3 loss from the retina and reductions in ERG amplitudes in the various studies, although studies in the same species show losses of up to 30% in the a-wave amplitudes once the retinal 22:6n-3 value decreased below a critical value (*21*).

Strategies to increase retinal 22:6n-3 levels

There was no significant difference in the body weights of animals at the start or finish of the experiment. The mean increase in body weight across all groups was 359 ± 46 g. The proportion of total PUFA in the retinal lipids ranged between 41 and 45% regardless of the diet. However, the fatty acid profiles varied as a function of diet group. The main retinal phospholipid fatty acids were 16:0, 18:0, 18:1n-9, arachidonic acid, 22:5n-6 and 22:6n-3 (Table II). There were significant increases in the proportions of 22:5n-3, 22:6n-3 and 24:6n-3 and significant decreases of linoleic acid, arachidonic acid, 22:4n-6, 24:4n-6, 24:5n-6 and 22:5n-6 as the ALA content of the diet increased (Figure 1). The additions of arachidonic acid and 22:6n-3 supplements (diets BASE+LCPI and BASE+LCP3) were associated with significant decreases in 22:4n-6, 22:5n-6 and 24:4n-6 and significant increases in 22:5n-3 and 22:6n-3. Diet BASE+LCP3 resulted in significantly more 22:6n-3 and less 22:5n-6 than did diet BASE+LCP1 or diet BASE. Supplementing diets with 22:6n-3 and arachidonic acid resulted in significant increases in the proportions of the 24 carbon n-3 PUFA and significant decreases in the 24 carbon n-6 PUFA.

Table II. Polyunsaturated Fatty Acid Composition of Retinal Phospholipids from Guinea Pigs Fed Diets Containing Different Levels of n-3 PUFA

Fatty Acid	Diet Group				
	SAF n=12	CAN n=14	BASE n=14	BASE+LCP1 n=14	BASE+LCP3 n=14
n-6 PUFA					
18:2n-6	1.37±0.15[a]	1.38±0.16[a]	1.50±0.18[a]	1.10±0.12[b]	0.91±0.13[c]
20:3n-6	0.88±0.24[a]	0.74±0.03[ab]	0.73±0.03[b]	0.76±0.03[ab]	0.70±0.01[bc]
20:4n-6	9.09±0.33[a]	8.67±0.22[b]	9.06±0.27[a]	9.21±0.32[a]	8.81±0.39[ab]
22:4n-6	3.50±0.10[a]	2.70±0.10[b]	3.49±0.14[a]	2.85±0.16[b]	2.14±0.12[c]
22:5n-6	17.35±0.67[a]	9.28±0.96[b]	15.16±0.65[c]	9.27±0.73[b]	3.43±0.56[d]
24:4n-6	2.59±0.22[a]	2.02±0.10[b]	2.34±0.16[a]	2.11±0.13[b]	1.31±0.12[c]
24:5n-6	0.31±0.04[a]	0.20±0.01[b]	0.24±0.02[b]	0.23±0.02[b]	0.16±0.02[c]
n-3 PUFA					
20:5n-3	0.09±0.07[a]	0.07±0.0[a]	0.08±0.01[a]	0.07±0.01[a]	0.13±0.01[b]
22:5n-3	0.47±0.05[a]	1.25±0.11[b]	0.65±0.06[cd]	0.59±0.08[c]	0.82±0.08[d]
22:6n-3	8.70±0.76[a]	16.35±1.09[b]	9.64±0.63[c]	17.60±1.05[b]	25.45±0.51[d]
24:5n-3	0.17±0.03[a]	0.65±0.05[b]	0.22±0.02[c]	0.25±0.03[c]	0.41±0.03[d]
24:6n-3	0	0.13±0.01[a]	0.05±0.00[b]	0.14±0.02[a]	0.35±0.05[c]

NOTE: Results expressed as % of total phospholipid fatty acids, mean±SD. Different superscript letters indicate significant differences between the diets at P<0.05. SAF, safflower oil; CAN, canola oil; BASE, mixed vegetable oil; BASE + LCP1, mixed vegetable oil plus 0.9% AA and 0.6% DHA; BASE + LCP3, mixed vegetable oil plus 2.7% AA and 1.8% DHA. The ALA content of the diets (g/100g diet) were SAF 0.01, CAN 0.71, BASE 0.1, BASE+LCP1 0.1 and BASE+LCP3 0.1.
SOURCE: Adapted with permission from reference 30. Copyright 1999 AOCS Press.

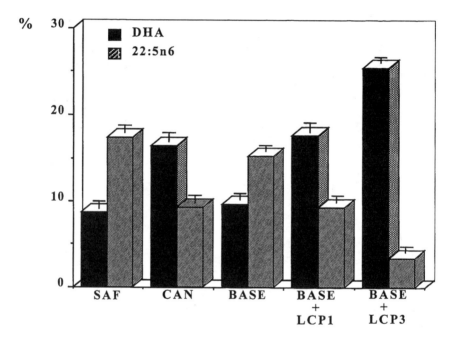

Figure 1. Docosahexaenoic acid and 22:5n-6 proportions in retinal phospholipids in guinea pigs fed diets containing different levels of n-3 fatty acids.

Discussion

Lipid nutrition during pregnancy and early postnatal life is important because rapid brain growth and phospholipid accumulation occurs during this period. Animal studies have shown that extreme modifications of intake of linoleic acid and ALA can lead to alterations in the PUFA composition and function of the developing brain and retina.

Previous studies on ALA deficiency and ERG function

The first aspect under investigation in this chapter involved an examination of the studies in the literature (*6, 10, 11, 20-22*) which have investigated the effect of n-3 deficiency on ERG function in mammals (Table I). The data reveals that:
(i) long and short-term deprivation of ALA in animals in different physiological states (pregnancy, weaning age and adult) induces a loss of 22:6n-3 in retinal lipids.
(ii) the greater the length of time on such diets, the greater the loss of 22:6n-3 in the retina.
(iii) in the retina, 22:6n-3 losses resulting from diets containing little or no ALA resulted in accumulation of another PUFA (22:5n-6) in an apparent attempt to maintain the C22 polyene level.
(iv) diets containing less than 100 mg ALA per 100 gram (approx. 0.2 % energy) were not sufficient to sustain retinal 22:6n-3 levels over a prolonged period, particularly if the diets were rich in linoleic acid. Most studies of n-3 deficiency have been conducted using diets containing high levels of linoleic acid (9-18% energy).
(v) ERG a- and b-wave amplitudes were reduced as a result of animals consuming diets containing these low levels of ALA.
(vi) one study (in rats) showed that 2% by weight of ethyl linoleate improved the ERG amplitudes compared with a fat-free diet, however significantly greater ERG signals were obtained with 1 % ethyl linoleate + 1% ethyl linolenate, and greater again by 2% ethyl linolenate (*10*). That is, linoleic acid is functional, but not as effective as ALA, and there is a dose response to ALA.
(vii) the same study (*10*) was the only one to use adult animals which demonstrates that ERG amplitude changes can result from dietary deficiencies imposed later in life.
(viii) one study (in guinea pigs) examined 4 dietary levels of n-3 PUFA on ERG amplitude responses in a study over 3 generations (*20*). There were three ALA levels (low, medium and high) and another diet containing ALA + EPA + 22:6n-3. The maximum ERG amplitude was found in the group with the highest ALA level (2% energy). The diet containing ALA and long chain n-3 PUFA, and the highest total n-3 PUFA intake (3.1% energy) gave a reduced ERG signal in comparison with the ALA alone diet.

(ix) some studies have shown prolonged n-3 PUFA deficiencies lead to a lessening difference in the ERG signal between ALA-deficient diets and ALA-containing diets, however this appears to be due to reductions in the ERG amplitudes in the control animals rather than an increase in amplitude in the ALA deficient groups *(18)*.

(x) one study showed it was possible to recover ERG amplitude responses and retinal 22:6n-3 levels imposed by an ALA-deficient diet in the first 9 weeks of life *(21)*. Recovery occurred using an ALA-containing diet in the next 10 weeks, however only 5 weeks of the ALA-containing diet was insufficient.

(xi) there was no evidence that dietary 22:6n-3 was required for ERG function in these three different mammalian species. In other words, dietary ALA is able to provide sufficient retinal 22:6n-3 to support ERG function.

(xii) there was insufficient data to establish a dose response relationship for ALA and ERG function in animals. This should be investigated using pure ALA and 22:6n-3 given the finding that the highest retinal 22:6n-3 level was associated with a reduced ERG signal *(20)*.

(xiii) there are studies in human infants which have examined the effect of increasing n-3 PUFA intake and ERG function. These provide support for the role of n-3 PUFA, particularly 22:6n-3, in increasing ERG function, however the studies are with small numbers of infants only *(23)*. Generally, studies in infants have been conducted comparing low ALA-high linoleate formulas made from corn oil with formulas containing soy oil or soy oil plus long chain n-3 PUFA from marine sources. That is, studies are usually not conducted using a high ALA formula with a relatively low linoleate level such as is found in canola oil. The experiment described below compares the effect of dietary ALA versus 22:6n-3 as strategies to increase retinal 22:6n-3 levels.

Strategies to increase retinal 22:6n-3 levels

This study compared dietary ALA with a combined supplement of 22:6n-3 plus arachidonic acid as strategies for raising retinal 22:6n-3. It was found that at the highest level of ALA (7% dietary fatty acids) there were increases in 22:6n-3 in the retina with no decrease in the proportion of arachidonic acid. The other strategy to raise tissue 22:6n-3 levels, which involved using dietary 22:6n-3 plus arachidonic acid, showed there were significant increases in retinal 22:6n-3 levels without alteration in the retinal arachidonic acid level. In the retinal lipids, the proportion of 22:5n-6 varied considerably and was inversely proportional to the proportion of 22:6n-3. This suggests that 22:6n-3 might be an inhibitor of 22:5n-6 synthesis or alternatively a better substrate for reacylation of the phospholipids.

In this study we found that the diet containing 7.1 % ALA gave the same retinal 22:6n-3 values as a diet which contained 0.6 % 22:6n-3 and 1% ALA indicating the potent effect of dietary 22:6n-3 as a source of retinal 22:6n-3. Thus, dietary 22:6n-3 is a more effective precursor of tissue 22:6n-3 than equivalent amounts of ALA, perhaps due to the fact that only a small fraction of ingested dietary ALA is converted to 22:6n-3 (*24*). The high level and the between-species constancy of 22:6n-3 in the brain and retina suggests that this fatty acid plays some essential role in these tissues. It is unclear whether this role involves a single, specific critical function or a range of functions. Research has so far revealed that 22:6n-3 is important in membrane order (fluidity) (*25,26*), membrane excitability (*27,28*) and metabolic processes (*29*). As the retina is such an accessible part of the central nervous system, it provides an excellent tissue for the study of PUFA-related effects on neural function.

Acknowledgments

This project was supported by Meadow Lea Foods Australia and the Grains Research and Development Corporation of Australia. Collaborative support from our colleagues (Drs. Algis Vingrys and Harry Weisinger and Mr. Bang Bui) is gratefully acknowledged since without their support, expertise and friendship none of these studies could have been conducted.

References

1. O'Brien, J.S.; Sampson, E.L. *J. Lipid Res.* **1965**, *6*, 545-551.
2. Crawford, M.A.; Sinclair, A.J. In Elliott, K.; Knight, J. Eds., CIBA Foundation Symposium on Lipids, Malnutrition and the Developing Brain. Associated Scientific Publishers: Amsterdam, 1972; pp 267-287.
3. Sinclair, A.J. *Proc. Nutr. Soc.* **1975**, *34*, 287-291.
4. Crawford, M. A.; Casperd, N.M.; Sinclair, A.J. *Comp. Biochem. Physiol.* **1976**, *54*B, 395-401.
5. Fliesler, S.J.; Anderson, R.E. *Prog. Lipid Res.* **1983**, *22*, 79-131.
6. Neuringer, M.; Connor, W.E.; Lin, D.S., Barstad, L.; Luck, S.J. *Proc. Natl. Acad. Sci. USA.* **1986**, *83*, 4021- 4025.
7. Ward, G.; Woods, J.; Reyzer, M.; Salem, N. *Lipids.* **1996**, *31*, 71-78.
8. Leat, W. M. F.; Curtis, R.; Millichamp, N.J.; Cox, R.W. *Ann. Nutr. Metab.* **1986**, *30*, 166-174.
9. Benolken, R.M.; Anderson, R.E.; Wheeler, T. *Science* **1973**, *182*, 1253-1254.
10. Wheeler, T.; Benolken, R.M.; Anderson, R.E. *Science* **1975**, *188*, 1312-1314.
11. Bourre, J. M.; Francois, M.; Youyou, A.; Dumont, O.; Piciotti, M.; Pascal G.; Durand, G. *J. Nutr.* **1989**, *119*, 1880-1892.
12. Connor, W.E.; Neuringer, M.; Reisbeck, S. *Nutr. Rev.* **1992**, *50*, 21-29.
13. Nakashima, Y.; Yuasa, S.; Hukamizu, Y.; Okuyama, H.; Ohhara, T.; Kameyama, T.; Nabeshima, T. *J. Lipid Res.* **1993**, *34*, 239-247.

14. Wainwright, P. In Essential Fatty Acids and Eicosanoids. Sinclair, A.J.; Gibson, R.A. Eds.; American Oil Chemists' Society, Champaign, IL. 1993; pp 1 69-172.
15. Yoshida, S.; Miyazaki, M.; Takashita, M.; Yuasa, S.; Kobayashi, T.; Watanabe, S.; Okuyama, H. *J Neurochem.* **1997**, *68*, 1269-1277.
16. Yoshida, S.; Yasuda, A.; Kawasato, H.; Sakai, K.; Shimada, T.; Takashita, M.; Yuasa, S.; Kobayashi, T.; Watanabe. S.; Okuyama, H. *J. Neurochem.* **1997**, *68*, 1261-1268.
17. Neuringer, M.; Connor, W.E.; Lin, D.S.; Anderson, G.J. In Essential Fatty Acids and Eicosanoids. Sinclair, A.J.; Gibson, R.A. Eds.;. American Oil Chemists' Society, Champaign, IL. 1993; pp 1 61-164.
18. Vingrys, A.J.; Weisinger, H.S.; Sinclair, A.J. In Lipids and infant nutrition Huang, Y.S.; Sinclair, A.J. Eds.; AOCS Press, Champaign, IL. 1998; pp 85-99.
19. Bush, R.A.; Sieving, P.A. *J. Opt. Soc. Am. A.*, **1996**, *13*, 557-565.
20. Weisinger, H.S.; Vingrys, A.J.; Sinclair, A.J. *Lipids.* **1996**, *31*, 65-70.
21. Weisinger, H.S.; Vingrys, A.J.; Bui, B.V.; Sinclair, A.J. *Invest. Opthalmol. Vis. Sci.* **1999**, *40*, 327-338.
22. Watanabe, I.; Kato, M.; Aonuma, H. *Adv. Biosc.* **1987**, *62*, 563-570.
23. Uauy, R.; Birch, D.; Birch, E.; Tyson, J.; Hoffman, D. *Pediatr. Res.* **1990**, *28*, 485-492.
24. Sinclair, A.J. *Lipids* **1975**, *10*, 175-184.
25. Dratz, E.E.; Holte, L.L. In Essential Fatty Acids and Eicosanoids. Sinclair, A.J.; Gibson, R.A. Eds.; American Oil Chemists' Society, Champaign, IL. 1993; pp 122-127.
26. Litman, B.J.; Mitchell, D.C. *Lipids* **1996**, *31*, S193-S-198.
27. Polin, J.S.; Karanian, J.W.; Salem, N.; Vicini, S. *Mol. Pharmacol.* **1993**, *47*, 381-390.
28. Nishikawa, M.; Kimura, S.; Akaike, N. *J. Physiol.* **1994**, *475*, 83-93.
29. Roberts, L.J; Morrow, J.D. *Prost. Leukotrienes and EFA.* **1997**, *57*, 209.
30. Abedin, L.; Lien, E. L.; Vingrys, A. J.; Sinclair, A. J. *Lipids* **1999**, *34*, 475-482.

Production of Polyunsaturated Fatty Acids and Special Nutraceutical Products

Chapter 8

The Large-Scale Production and Use of a Single-Cell Oil Highly Enriched in Docosahexaenoic Acid

David J. Kyle

Martek Biosciences Corporation, 6480 Dobbin Road, Columbia, MD 21045

Docosahexaenoic acid (DHA), is the primary structural fatty acid in the membranes of neurological tissues and deficiencies of DHA have been correlated with chronic neurological disorders including attention deficit disorder, depression, and senile dementia. Although DHA can be obtained from fish and certain animal organ meats, the primary producers of DHA in the biosphere are the phytoplankton in the sea. We have optimized the growth under heterotrophic conditions of a specific marine algal species to produce a triacylglycerol oil enriched to over 40% with DHA. The scale-up of this process presented many unique problems both in the fermentation of the microalgae, and in the subsequent processing of the oil. Commercial quantities of the DHA-rich oil are now being produced under GMP (Good Manufacturing Practice) conditions at full scale for use in food products ranging from infant formula to nutritional supplements. Double blind clinical intervention trials have demonstrated several unique characteristics of this oil which differentiate the microalgal oil from fish oil. These are consequences of the simplicity of the fatty acid profile, the remarkable stability of the oil, the purity and consistency of this oil, and the exceptionally high level of DHA enrichment.

Docosahexaenoic acid (DHA) is a long chain polyunsaturated fatty acid (LC-PUFA) with twenty-two carbons and six double bonds (C22:6 (Δ4,7,10,13,16,19)) and is the primary fatty acid component of the phospholipids of neuronal cells (Figure 1). It is particularly prevalent in the phospholipids comprising the synaptosomal membranes, making up 20-30 % (wt/wt) of the fatty acyl moieties of phosphatidylethanolamine (PE) and phosphatidylserine (PS) *(1, 2)*. DHA is also found in very high concentrations in other tissues of high electrical activity such as the retina of the eye (3, 4) and cardiac muscle (5, 6).

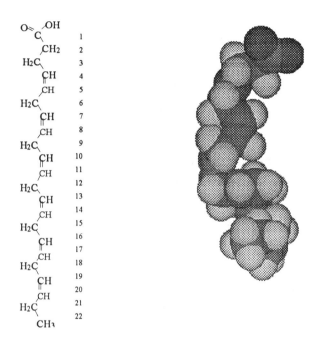

Figure 1. The DHA molecule. (4,7,10,13,16,19)-docosahexaenoic acid

DHA is quite unique among fatty acids in that its biosynthetic pathway involves not only elongation and desaturation of an essential fatty acid precursor, linolenic acid (LNA), but it also requires a peroxisomal beta-oxidation step to produce the final molecule. This unusual biosynthetic pathway was first described by Sprecher and co-workers (7) and is thought to be necessary for the placement of a double bond at the Δ4 position to the carboxyl of the fatty acid. Indeed, it is the Δ4 double bond which distinguishes DHA from all other fatty acids and may be a key to its role in the neuronal membranes.

Recent studies have revealed that DHA has many critical functions in the normal development and metabolism of neuronal cells. These include, but are not limited to, the following: 1) the control of normal migration of neurons from the surface of the ventricles to the cortical plate during brain development (8); 2) the control of the normal resting potential of the neurons and cardiac cells by regulation of sodium and calcium channels (9); 3) the regulation of the density of certain membrane proteins such as rhodopsin in the retina (10); and, 4) the regulation of levels of certain neurotransmitters such as serotonin (11, 12). With such key roles in normal neuronal development and function, it is quite plausible that abnormally low levels of DHA during critical periods of brain development could be one cause of the long term neurological detriments observed in formula-fed infants relative to breast-fed infants (*13-15*).

Importance of DHA in Human Infant Nutrition

Mammals exhibit various degrees of ability to convert LNA to DHA. Nocturnal rodents (*eg.* mice) have a good conversion ability, whereas obligate carnivores, such as cats, are unable to convert dietary LNA to DHA (*16*). Although humans also have the biochemical capacity to convert LNA to DHA, in certain instances such as infancy, the demand for DHA in the developing neural tissues far outpaces the ability of the body to synthesize DHA from dietary LNA (*17, 18*). Carlson and co-workers have referred to this period (the first year of life) as one of "conditional essentiality" for DHA (*19*). This demand for pre-formed DHA in the developing infant is met by the natural source of nutrition for a human infant -- human breast milk. Recent studies have demonstrated that if babies are not provided with preformed DHA during the first months of life they exhibit significantly lower brain DHA levels (*20, 21*) and there is a retarded rate of development of visual and mental acuity (*19, 22-26*).

Over the past twenty years there have been a large number of retrospective studies comparing the neurological outcomes of breast-fed and formula-fed infants. A recent meta-analysis of these studies has indicated that there is a consistent 3-4 IQ point advantage to breast-fed infants even after the contributions of all other confounding factors had been removed (*15*) Figure 2). Breast-fed babies, however, are getting many nutrients from the breast milk in addition to DHA and it can be argued that the contribution of DHA to improved long term IQ is inconclusive. One observation, however, is very clear and consistent. Infants given standard infant formulas have significant deficiencies in red blood cell DHA, plasma phospholipid DHA, and brain DHA compared to breast-fed babies (*21, 22, 24, 27, 28*).

There have been more than 30 well-controlled clinical studies involving over 2,000 infants in the last 15 years that have compared outcomes of standard formula-fed infants with DHA-supplemented formula-fed infants and breast-fed infants. In every study, the DHA status of the infants was returned to normal (as defined by the DHA status of the breast-fed infants) when the formulas were supplemented with DHA. In many of these studies developmental outcomes were also measured and it has been demonstrated that breast-fed babies developed their mental and visual acuities faster than standard formula-fed babies (*19, 25*). Moreover, when formulas were supplemented with DHA at the same levels found in breast milk, this visual and mental developmental delay was prevented (*24, 29*).

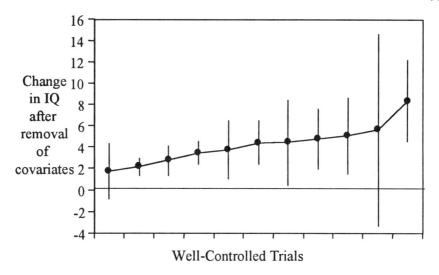

Figure 2. IQ differences between children who were breast-fed vs. formula-fed from eleven well controlled trials reviewed in a meta-analysis by Anderson (15). Values are the mean difference +/- 2SE after removal of the contribution of individual covariates such as social economic status, maternal IQ, etc.

Microalgae - The Primary Producers of DHA

The recognition of the importance of DHA in the development and maintenance of the human nervous system, coupled with a growing knowledge base in human lipid biochemistry, suggests a critical need to maintain adequate amounts of preformed DHA in the diet. Indeed, Crawford and colleagues have suggested that the evolutionary development of the large brain of *Homo sapiens* could not have taken place without a rich source of dietary DHA (*30*). The principal source of DHA is from the marine food web.

From a commercial point of view, fish and fish oils offer a good source of DHA. However, fish reserves are limited and intensive aquaculture practices are producing fish with lower levels of DHA.. The primary sources of DHA in the marine food web are the photosynthetic phytoplankton (Figure 3). Fish cannot produce DHA from precursor fatty acids any better than humans and intensive aquaculture practices involve using grow out feeds that are primarily enriched in products from land-based agriculture. Unlike higher plants, some species of these lower plants or algae have an excellent capacity to elongate and desaturate fatty acids to produce DHA.

There are over 80,000 different species of microalgae in the biosphere and only a fraction of these species have been catalogued. Furthermore, only a small number of the collected microalgae have been studied in terms of fatty acid profiles. Table I provides examples of the range of fatty acids that can be seen in many microalgae. Some of these algae are prolific producers of DHA (*eg. Crypthecodinium*), others produce EPA (*eg. Monodus and Nitzschia*), and still others produce oils which have fatty acid compositions very much like higher plants (*eg. Chlorella*). In general, the marine microalgae predominantly produce fatty acids in the omega-3 family. Most marine fungi and terrestrial fungi, on the other hand, generally produce omega-6 fatty acids. One such fungi (*Mortierella alpina*) is now being grown commercially as a source of arachidonic acid (*31*). Chytrids, which have characteristics of both fungi and algae, make both omega-6 and omega-3 fatty acids. *Thraustochytrium*, for example, makes a significant quantity of the omega-6 DPA along with DHA (*32, 33*). Omega-6 DPA is typically found in higher concentrations in the blood of children with attention deficit hyperactivity disorder (*34*), depression (*35*), and certain other inborn errors of metabolism (*36*) and neurological diseases.

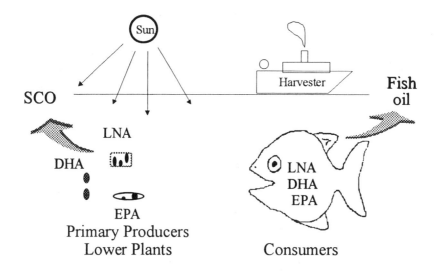

Figure 3. The primary producers of omega-3 fatty acids (DHA, EPA and LNA) in the marine food web are the lower plants or algae. Fish are consumers of these organisms and become a composite mixture of all these essential fatty acids.

One of the challenges in developing a primary source of DHA is that most of the microalgal producers shown in Table I are photosynthetic. Indeed, 99% of all microalgae can only be grown in the presence of light. Large-scale production of photosynthetic microalgae to date has been limited to pond cultivation. This production modality has several inherent problems including the difficulty in maintaining a unialgal culture, and in maintaining good manufacturing controls. This has limited the commercial production of photosynthetic microalgae to only a few species that can been cultured in ponds using certain selective characteristics (eg., high salt concentrations, or high pH). Therefore, we looked at only those organisms that could be grown under classical fermentation conditions such as *C. cohnii* and *T. areum*. Since we were unable to identify a strain of *T. areum* that did not produce the problematic omega-6 DPA, we focused our attention on scaling-up the marine algae, *C. cohnii*.

Table I. Fatty Acid Profiles of Different Microorganisms and Higher Plants.

Org	14:0	16:0	16:1	18:1	18:2	18:3	20:4	20:5	22:5	22:6
Algae										
M. sub.	2	24	24	9	4	-	5	29	-	-
M. lut.	11	15	25	1	1	-	-	16	1	13
N. ang.	5	24	31	3	2	2	-	21	-	-
C. coh.	19	20	1	14	-	-	-	-	-	30
H. vir.	5	18	1	45	7	7	-	-	-	4
C. vul.	-	25	8	3	42	13	-	-	-	-
Fungi										
M. alp.	-	9	10	21	7	3	41	-	-	-
T. aur.	2	27	1	6	2	-	5	6	10	34
Plants										
soy	1	12	-	25	52	6	-	-	-	-
corn	-	12	-	29	56	1	-	-	-	-

Algal fatty acid profiles from *Monodus subterraneus* (Xanthophyceae), *Monochrysis lutheri* (Chrysophyceae), *Nitzschia angularis* (Bacillarophyceae), *Crypthecodinium cohnii* (Dinophyceae), *Halosphaera viridis* (Prasinophyceae), and *Chlorella vulgaris* (Chlorophyceae). Fungal fatty acid profiles from *Mortierella alpina* and *Thraustochytrium aureum*.

Production of a DHA-rich Single Cell Oil (DHASCO)

Crypthecodinium cohnii is a member of the Dinophyta, a phylum of unicellular eucaryotic microalgae comprising an estimated 2,000 species (37). Ecologically, the Dinophyta comprise an extremely important and vastly abundant group of primary producers in both freshwater and marine environments. Among the eukaryotic algae they are second only to the diatoms as primary producers in coastal waters. Most species of the Dinophyta are photosynthetic, but there are also several heterotrophic species including *C. cohnii*. Isolates of *C. cohnii* have been collected from all around the world and are commonly found on the surface of macroalgae or seaweeds. Because this species is easy to cultivate in the laboratory, it has become one of the best studied of all the algae, particularly with regard to ultrastructure, biochemistry and genetics.

Crypthecodinium cohnii, like other algae, reproduces primarily by asexual cell division (mitosis) and is haploid in the vegetative stage (*37*). This organism is capable of sexual reproduction in the laboratory, and genetic studies indicate that *C. cohnii* exhibits genetic stability consistent with other eukaryotic species used in fermentative production of foods and food additives (*37*), that is, there is very little genetic drift in these algae. In the many reports of *C. cohnii* in culture over the last 30 years, there has never been any indication that *C. cohnii* produces any toxin, nor is it related to any toxin-producing species. More broadly speaking, there are no known heterotrophic Dinophyta species, which are either toxigenic or pathogenic.

C. cohnii is a biflagellate and as such was found to be quite shear sensitive. In fact, it was generally thought that this species could not be cultivated in the high shear environment of a conventional fermentor. During our scale-up activities, we did find that the organism was quite shear sensitive and, as a result, sparging and agitation rates needed to be kept at a minimum. Nevertheless, through several years of experimental optimization a fermentation process as shown in Figure 4 was developed. The *C. cohnii* strain used for the production of the DHA-rich oil (DHASCO) is proprietary (US Patents 5,397,591, 5,407,957 and 5,492,938), and the process has been published (*38*). The production strain was selected for rapid growth and high levels of production of the specific oil. Master seed banks of all strains are maintained under liquid nitrogen conditions and working seed stocks, prepared from this master seed bank, are also maintained cryogenically. On initiation of a production run, an individual ampoule from a working seed is used to inoculate a shake flask.

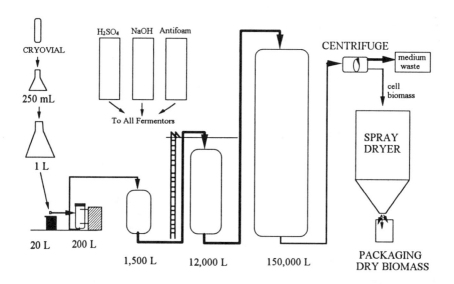

Figure 4. Fermentation process for the production of C. cohnii biomass.

The medium used to grow *C. cohnii* from shake flask to production scale contains dextrose, yeast extract or a hydrolyzed vegetable protein, sodium chloride, magnesium sulfate, and a number of other micronutrients. The cultures are transferred successively to larger vessels based on specific growth parameters. Dextrose concentration, temperature, pH, airflow, pressure, agitation, and dissolved oxygen are constantly monitored and controlled. When the culture reaches a predetermined cell density and DHA content, the culture is harvested by centrifugation and spray dried.

The DHASCO oil is extracted from the spray dried algal biomass and processed using methods and procedures that have been well established in the edible oils industry. The overall oil processing procedure is outlined in Figure 5. In order to protect this long chain polyunsaturated oil, which is much more prone to oxidation than typical vegetable oils, the process has been designed to use the lowest effective temperatures and shortest times for each process step, and the oil is continuously protected from oxygen by nitrogen blanketing or vacuum.

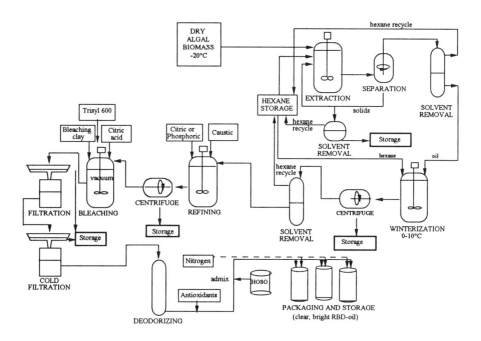

Figure 5. Oil processing flow sheet for the production of DHASCO from C. cohnii.

The oil is first extracted by blending the dried biomass with hexane in a continuous extraction process (Figure 5). The miscella (hexane:oil mixture) is separated from the de-oiled solids, filtered, and desolventized under vacuum to begin removal of the hexane. The oil is then winterized to remove a higher melting oil fraction by placing the miscella in a jacketed vessel, cooled and gently mixed. The chilled miscella is then centrifuged to remove the solids and desolventized again to remove remaining volatiles. This winterized DHASCO is then refined to remove free fatty acids and phospholipids by mixing with citric or phosphoric acid while heating to facilitate removal of phospholipids. The mixture is then neutralized by addition of aqueous sodium hydroxide, heated, and centrifuged to remove the phospholipids and soaps of free fatty acids from the refined oil. The refined DHASCO is transferred to a vacuum bleaching vessel where citric acid, activated silica and bleaching clay are added to adsorb any remaining polar materials and pro-oxidant metals, and to break down lipid oxidation products. The mixture is heated under vacuum and filtered using filter aid. The refined bleached DHASCO is then clarified once again by chilling the oil prior to a filtration step to remove any solids. Finally, the oil is deodorized under vacuum using a thin-film packed-tower continuous deodorizer. The deodorized DHASCO is then diluted to a standard 40% docosahexaenoic acid concentration by the addition of high oleic sunflower oil and mixed with antioxidants (mixed natural tocopherols and ascorbyl palmitate). The oil is finally packaged in nitrogen-purged containers and stored frozen until shipment.

Characterization of DHASCO

DHASCO is triglyceride oil, extracted from the marine microalgae *C. cohnii*, that is enriched to about 40% (wt/wt) in DHA. It is a free flowing, yellow-orange oil, which is greater than 95% triglyceride, with some diglyceride and nonsaponifiable material, as is typical for all food-grade vegetable oils. The fatty acid composition of DHASCO is shown in Table 2. Minor fatty acid components listed as "other" generally constitute about 1-2% of the total fatty acid composition. A recent report (39) indicates the presence of small amounts of C28:8(n-3) in DHASCO oil, as well in other marine and fish oils. This fatty acid is the next expected omega-3 end product of the Sprecher pathway (7) beyond DHA, and is one of the minor components (approximately 1%) of DHASCO. Compositional analyses of other components of the oils compare favorably with typical commercial edible oils. In general, the residual extraction solvent is undetectable (detection limit <0.3 ppm), and there are no detectable cyclic fatty acids or trans fatty acids, pesticide residues, or heavy metals such as arsenic, mercury and lead (all below the 0.1 ppm detection limit).

The nonsaponifiable fraction of DHASCO is generally about 1.5% by weight, and is made up primarily of sterols. The principal sterol has been identified as the 4-methylsterol, dinosterol (40, 41). 4-methyl sterols are found as intermediates in the metabolic pathway of cholesterol biosynthesis in man (42) and have been identified in several other food sources, including fish and shellfish (43). There is no *a priori* reason to believe that these sterols have any significant biological activity, as they would be expected to feed into the normal cholesterol metabolic pathways in humans. A recent study providing large amounts of the isolated nonsaponifiable fraction of crude DHASCO to rats concluded that these sterols had no adverse effects on growth or lipid metabolism (44).

Table II. Fatty acid composition of DHASCO, a DHA-rich Single Cell Oil from C. cohnii.

Fatty Acid	Weight %
C 10:0	1.19
C 12:0	6.02
C 14:0	16.79
C 16:0	13.95
C 16:1	1.46
C 18:0	0.50
C 18:1	14.67
C 18:2	0.49
C 20:5 EPA	<0.10
C 22:6 DHA	43.53
Others	1.44

Triacylglycerols are the predominant component of any natural fat or oil and are present as a family of compounds wherein the various fatty acids may be found attached to any of the three positions on the glycerol backbone. Corn oil, for example, is made up of at least 21 individual triglycerides (45). About 98% of the lipid in human milk is triglyceride, and the vast majority of the DHA delivered to an infant from its mother's milk is in the form of a triglyceride. Certain fatty acids may be found preferentially in either the Sn-1, Sn-2 or Sn-3 positions on the glycerol backbone. In human milk, about 50-60% of the DHA is found on the Sn-2 position (46). Myher et. al. (47) have reported that approximately 45% of the DHA found in DHASCO is also located on the Sn-2 position. Since the triglyceride structure of DHASCO is nearly identical to human milk with respect to the positional specificity of DHA, there would be no reason to believe that the digestion and absorption of DHA from DHASCO should be any different than the DHA from human milk fats. In fact, Carnielli and collegues (48) have shown that the absorption of DHA from DHASCO in infants more closely approximates that of DHA from human milk, than if the DHA is presented to the infant in the form of a phospholipid (48).

Safety Testing of DHASCO

It is important to recognize that DHASCO is a macronutrient, not a micronutrient, vitamin, or drug. Safety testing of macronutrients poses several unique problems (49). For example, it is often difficult to distinguish whether an observed response is related to a toxicological effect of the test material, or to a dietary deficiency of some other component caused by the use of very large amounts of the test macronutrient in the diet. Clearly, one cannot achieve safety margins of greater than 50-fold for a macronutrient that comprises more than 2% of the diet, and one must carefully distinguish a truly toxicological response due to the test material, from a normal physiological response resulting from the high dietary load of that particular macronutrient. A large number of toxicological studies have been undertaken using DHASCO, generating an unprecedented redundancy in general safety assessments. These data, however, provide an extremely valuable assessment of the safety of DHASCO. This is particularly important when we recognize that it is being used in infant formulas around the world.

Over thirty safety studies undertaken in the last ten years have included *in vitro* assays of mutagenicity and genotoxicity, as well as a variety of classical rat studies, including acute, subchronic, developmental, and multigenerational reproductive studies. These studies have been completed on the oil as well as the biomass itself. Since these components represent macronutrients, all studies have included appropriate positive (high fat), as well as negative controls. All studies have also been conducted using Good Laboratory Practice (GLP)-compliant laboratories following guidelines outlined by the FDA in *Toxicological Principles for the Safety Assessment of Direct Food Additives and Color Additives Used in Foods*, commonly referred to as Redbook I, or from their draft Redbook II. These data have also been reported in a number of publications and twenty-four studies were summarized in Kyle and Arterburn (*50*). In addition to the toxicity studies, DHASCO has been used in a large number of experimental diet studies with at least 12 different animal species.

Clinical Studies using a DHASCO

Infant Studies

There have been at least eight well controlled, clinical intervention trials which have been reported wherein DHASCO supplemented infant formulas have been fed to preterm and/or term infants for various periods of time (*24, 51-57*). All of these trials demonstrated that the DHASCO-supplemented formulas elevated the infants' DHA status to that of breast-fed infants. In a few cases, other functional endpoints were also studied. Two studies using DHASCO-supplemented formulas which also included an arachidonic-containing single cell oil (ARASCO), reported significant improvements in the growth rates of the babies fed the supplemented formulas compared to routine formulas (*57, 58*).

Significant improvements in visual and mental acuity have also been reported when infants were fed DHASCO-supplemented formulas. Visual acuity improvements were reported to be equivalent to "one line in an eye chart" at one year of age (*24*). In this same group of infants mental acuity improvements were 7

IQ points better than the standard formula-fed infants by 18 months of age as determined by the Bayley MDI assessment (29). In another study, an extensive analysis of potential adverse events was undertaken and the only statistically significant observations were less anemia and nervousness or irritability in the DHASCO-supplemented, formula-fed infants (59). Although these would generally be categorized as beneficial aspects of the supplemented formula, the incidence levels were low and these were not primary endpoints of the study.

Most importantly, standard formula-fed infants consistently have an altered blood and brain biochemistry, and visual and neurological functional assessments have indicated that there is also a significant deficiency in these infants compared to the breast-fed controls. Although the DHASCO-supplemented, formula-fed babies were significantly better in many of the functional outcomes than the standard formula-fed infants, they were never better than the breast-fed infants. It appears, therefore, that it may be the DHA deficiency in the infants fed standard formulas that is causing the problem, and that resupplying the DHA via the formula (using DHASCO) is sufficient to overcome these problems.

Studies with children and adults

There have been at least nineteen well-controlled, clinical intervention trials reported with adults, adolescents, or children where DHASCO has been used at various dose levels and for various periods of time. Some of these trials have been previously reviewed (50). One of these trials was undertaken under a highly controlled environment of a metabolic ward by the United States Department of Agriculture using a particularly high level (15 g/day) of DHASCO supplementation (60, 61). A vast amount of data were gathered on these subjects and, although the levels of DHA were significantly elevated (*i.e.*, 4-fold higher than placebo control levels), there were no reported adverse responses to the supplements. Other studies represent a cross section of healthy men, women and children of all ages, women who were pregnant or nursing; individuals with certain dietary restrictions (*i.e.*, vegetarians), and individuals with pre-existing metabolic disorders including hyperlipidemia, retinitis pigmentosa, long chain hydroxyacylCoA dehydrogenase deficiency, ADHD, etc.

From all of the well-controlled clinical trials completed with this unique source of DHA, several consistent conclusions can be drawn. DHA can effectively lower triglycerides (about 20% in all studies) and raise HDL cholesterol (about 8-12% in all studies) while not adversely elevating total cholesterol or LDL cholesterol (Figure 6) (50). Although similar responses have been demonstrated with fish oil supplementation (62, 63), the results have not always consistent (64, 65). This suggests that it may be the DHA, not the EPA that is responsible for these clinical effects. Using pure ethyl ester sources, Mori and co-workers (66) has also shown that it was DHA, and not EPA, that was responsible for the hypotensive effects generally seen with fish oils. Other studies have also demonstrated that in many cases when DHA levels are remarkably reduced by diet or pathology, profound neurological or visual implications may ensue, and that when these levels are normalized, some of the symptoms of the pathology may disappear.

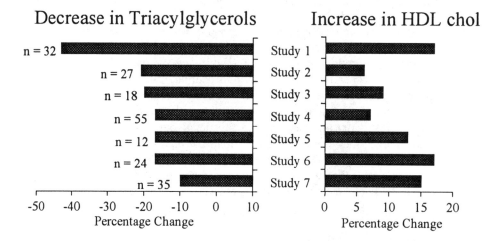

Figure 6. Summary of seven clincial trials with DHASCO supplementation with respect to serum triacylglycerols and HDL cholesterol. Study 1 (67); Study 2 (68); Study 3 (Schaefer, unpublished); Study 4 (62); Study 5 (60); Study 6 (69); Study 7 (Howe, unpublished).

Conclusions

The concept of the large scale production of single cell oils is not new. However, it has only been a theoretical concept until now. With the development of a large-scale production process for a DHA-rich oil using a renewable microbial process, we have provided the first successful embodiment of this concept. The key factors in its success were: 1) the production of an oil with high enough value to offset the high production costs; 2) the use of a microorganism which produces a clean, safe product; 3) the use of an organism with enough plasticity to significantly improve productivities over those in the natural state; and 4) a clearly identified market with a specific requirement for the product produced. The fermentation of *C. cohnii* and the extraction and purification of its triglyceride as well as its use in infant formula fulfilled all of these requirements.

We expect that products with equally high a value will be identified in the future, but it must be recognized that such a value in the pharmaceutical or neutraceutical area is best established by well controlled clinical trials. Clincal trials are costly and scale-up is not very predictable. As a result, it is not a fast process.

The vast genetic diversity of the microalgae, however, still offers a unique oportunity for the exploitaiton of this kingdom for the production of new specialty chemicals, pharmaceuticals and neutraceuticals alike.

References

1. Salem, N.; Kim, H.-Y.; Yergey, J. In *Health Effects of Polyunsaturated Fatty Acids in Seafoods*; Academic Press, Inc., 1986, pp 263-317.
2. Arbuckle, L. D.; Innis, S. M. *Lipids* **1992**, *27*, 89-93.
3. Bazan, N. G.; Rodriguez de Turco, E. B. *J. Ocul. Pharmacol.* **1994**, *10*, 591-604.
4. Louie, K.; Wiegand, R. D.; Anderson, R. E. *Biochemistry* **1988**, *27*, 9014-20.
5. Atkinson, T. G.; Barker, H. J.; Meckling-Gill, K. A. *Lipids* **1997**, *32*, 293-302.
6. Durot, I.; Athias, P.; Oudot, F.; Grynberg, A. *Mol. Cell Biochem.* **1997**, *175*, 253-62.
7. Voss, A.; Reinhart, M.; Sankarappa, S.; Sprecher, H. *J. Biol. Chem.* **1991**, *266*, 19995-20000.
8. Xu, L. Z.; Sanchez, R.; Sali, A.; Heintz, N. *J. Biol. Chem.* **1996**, *271*, 24711-9.
9. Xiao, Y. F.; Gomez, A. M.; Morgan, J. P.; Lederer, W. J.; Leaf, A. *Proc. Natl. Acad. Sci. U S A* **1997**, *94*, 4182-7.
10. Suh, M.; Wierzbicki; Clandinin, M. T. , Seattle, WA 1997.
11. Hibbeln, J. R.; Linnoila, M.; Umhau, J. C.; Rawlings, R.; George, D. T.; Salem, N., Jr. *Biol. Psychiatry* **1998**, *44*, 235-42.
12. de la Presa-Owens, S.; Innis, S. M.; Rioux, F. M. *J. Nutr.* **1998**, *128*, 1376-84.
13. Lucas, A.; Morley, R.; Cole, T. J.; Lister, G.; Leeson-Payne, C. *Lancet* **1992**, *339*, 261-4.
14. Horwood, L. J.; Fergusson, D. M. *Pediatr.* **1998**, *101*, e9.
15. Anderson, J. W.; Johnstone, B. M.; Remley, D. T. *Am. J. Clin. Nutr.* **1999**, *70*, 525-35.
16. Pawlosky, R.; Barnes, A.; Salem, N., Jr. *J. Lipid Res.* **1994**, *35*, 2032-40.
17. Salem, N., Jr.; Wegher, B.; Mena, P.; Uauy, R. *Proc. Natl. Acad. Sci. U S A* **1996**, *93*, 49-54.
18. Greiner, R. C.; Winter, J.; Nathanielsz, P. W.; Brenna, J. T. *Pediatr. Res.* **1997**, *42*, 826-34.
19. Carlson, S. E.; Werkman, S. H.; Peeples, J. M.; Wilson, W. M. *Eur. J. Clin. Nutr.* **1994**, *48*, S27-30.
20. Gibson, R. A.; Neumann, M. A.; Makrides, M. *Lipids* **1996**, *31*, S177-81.
21. Farquharson, J.; Cockburn, F.; Patrick, W. A.; Jamieson, E. C.; Logan, R. W. *Lancet* **1992**, *340*, 810-3.
22. Carlson, S. E.; Ford, A. J.; Werkman, S. H.; Peeples, J. M.; Koo, W. W. *Pediatr. Res.* **1996**, *39*, 882-8.

23. Uauy, R.; Birch, E.; Birch, D.; Peirano, P. *J. Pediatr.* **1992**, *120*, S168-80.
24. Birch, E. E.; Hoffman, D. R.; Uauy, R.; Birch, D. G.; Prestidge, C. *Pediatr. Res.* **1998**, *44*, 201-9.
25. Agostoni, C.; Trojan, S.; Bellu, R.; Riva, E.; Giovannini, M. *Pediatr. Res.* **1995**, *38*, 262-6.
26. Willatts, P.; Forsyth, J. S.; DiModugno, M. K.; Varma, S.; Colvin, M. *Prostaglandins Leukotr. Essent. Fatty Acids* **1997**, *57*, 188-91.
27. Jensen, C. L.; Prager, T. C.; Fraley, J. K.; Chen, H.; Anderson, R. E.; Heird, W. C. *J. Pediatr.* **1997**, *131*, 200-9.
28. Makrides, M.; Neumann, M. A.; Byard, R. W.; Simmer, K.; Gibson, R. A. *Am. J. Clin. Nutr.* **1994**, *60*, 189-94.
29. Birch, E. E.; Garfield, S.; Hoffman, D. R.; Uauy, R.; Birch, D. G. *Dev. Med. Child Neurol.* **2000**, *42*, 174-181.
30. Broadhurst, C. L.; Cunnane, S. C.; Crawford, M. A. *Br. J. Nutr.* **1998**, *79*, 3-21.
31. Kyle, D. J. *Lipid Technology* **1997**, *9*.
32. Kendrick, A.; Ratledge, C. *Lipids* **1992**, *27*, 15-20.
33. Li, Z. Y.; Ward, O. P. *J. Ind. Microbiol.* **1994**, *13*, 238-41.
34. Stevens, L. J.; Zentall, S. S.; Deck, J. L.; Abate, M. L.; Watkins, B. A.; Lipp, S. R.; Burgess, J. R. *Am. J. Clin. Nutr.* **1995**, *62*, 761-8.
35. Hibbeln, J. R.; Salem, N., Jr. *Am. J. Clin. Nutr.* **1995**, *62*, 1-9.
36. Sanjurjo, P.; Perteagudo, L.; Rodriguez Soriano, J.; Vilaseca, A.; Campistol, J. *J. Inherit. Metab. Dis.* **1994**, *17*, 704-9.
37. Spector, D. L. *Dinoflagellates*; Academic Press: Orlando, 1984.
38. Kyle, D. J. *Lipid Technol.* **1996**, *2*, 109-112.
39. VanPelt, C. K.; Huang, M.-C.; Tschanz, C. L.; Brenna, T. *J.Lipid Res.* **1999**, *40*, (in press).
40. Withers, N. W.; Tuttle, R. C.; Holtz, G. G.; Beach, D. H.; Goad, L. J.; Goodwin, T. W. *Phytochemistry* **1978**, *17*, 1987-1989.
41. Piretti, M. V.; Pagliuca, G.; Boni, L.; Pistocchi, R.; Diamante, M.; Gazzotti, T. *J. Phycol.* **1997**, *33*, 61-67.
42. Hashimoto, F.; Hayashi, H. *Biochim. Biophys. Acta* **1994**, *1214*, 11-19.
43. Piretti, M. V.; Viviani, R. *Comp. Biochem. Physiol.* **1989**, *93753-756*.
44. Kritchevsky, D.; Tepper, S. A.; Czarnecki, S. K.; Kyle, D. J. *Nutr. Res.* **1999**, *19*, 1649-1654.
45. Strecker, L. R.; Maza, A.; Winnie, G. F. In: *Corn Oil - Composition, Processing, and Utilization* (Erickson, D. R. ed) American OIl Chemists' Society Press; Champaign, 1990, 309-312.
46. Martin, J. C.; Bougnoux, P.; Antoine, J. M.; Lanson, M.; Couet, C. *Lipids* **1993**, *28*, 637-43.
47. Myher, J. J.; Kuksis, A.; Geher, K.; Park, P. W.; Diersen-Schade, D. A. *Lipids* **1996**, *31*, 207-215.
48. Carnielli, V. P.; Verlato, G.; Pederzini, F.; Luijendijk, I.; Boerlage, A.; Pedrotti, D.; Sauer, P. J. *Am. J. Clin. Nutr.* **1998**, *67*, 97-103.
49. Borzelleca, J. F. *Regul. Toxicol. Pharmacol.* **1992**, *16*, 253-64.

50. Kyle, D. J.; Arterburn, L. M. *World Rev. Nutr. Diet* **1998**, *83*, 116-31.
51. Hoffman, D. R.; Birch, E. E.; Birch, D. G.; Uauy, R. D. *Am. J. Clin. Nutr.* **1993**, *57*, 807S-812S.
52. Foreman-van Drongelen, M. M.; van Houwelingen, A. C.; Kester, A. D.; Blanco, C. E.; Hasaart, T. H.; Hornstra, G. *Br. J. Nutr.* **1996**, *76*, 649-67.
53. Hansen, J.; Schade, D.; Harris, C.; Merkel, K.; Adamkin, D.; Hall, R.; Lim, M.; Moya, F.; Stevens, D.; Twist, P. *Prostaglandins Leukotr. Essent. Fatty Acids* **1997**, *57*, 196.
54. Clandinin, M. T.; Van Aerde, J. E.; Parrott, A.; Field, C. J.; Euler, A. R.; Lien, E. L. *Pediatr. Res.* **1997**, *42*, 819-25.
55. Vanderhoof, J.; Gross, S.; Hegyi, R.; Clandinin, T.; Porcelli, P.; DeCristofaro, P.; Rhodes, T.; Tsang, R.; Shattuck, K.; Cowett, R.; Adamkin, K.; McCarton, C.; Heird, W.; Hook, B.; Pereira, G.; Pramuk, K.; Euler, A. *Pediatr. Res.* **1997**, *41*, 242A.
56. Gibson, R.; Makrides, M.; Neuman, M.; Hawkes, J.; Pramuk, K.; Lein, E.; Euler, A. *Prostaglandins, Leukotr. Essent. Fatty Acids* **1997**, *57*, 198.
57. Carlson, S. E.; Mehra, S.; Kagey, W. J.; Merkel, K. L.; Diersen-Schade, D. A.; Harris, C. L.; Hansen, J. W. *Pediatr. Res.* **1999**, *45*, 278A.
58. Diersen-Schade, D. A.; Hansen, J. W.; Harris, C. L.; Merkel, K. L.; Wisont, K. D.; Boettcher, J. A. In *Essential Fatty Acids and Eicosanoids*; Riemersma, R. A., Armstroma, R., Kelly, R. W., Wilson, R., Eds.; AOCS Press: Champaign, USA, 1999.
59. Vanderhoof, J.; Gross, S.; Hegyi, T.; Clandinin, T.; Porcelli, P.; DeCristolaro, J.; Rhodes, T.; Tsang, R.; Shattuck, K.; Cowett, R.; Adamkin, D.; McCarton, C.; Heird, W.; Hook-Morris, B.; Pereira, G.; Chan, G.; Van Aerde, J.; Boyle, F.; Pramuk, K.; Euler, A.; Lien, E. *J. Pediatr. Gastroenterol.* **1999**, *29*, 318-26.
60. Nelson, G. J.; Schmidt, P. C.; Bartolini, G. L.; Kelley, D. S.; Kyle, D. *Lipids* **1997**, *32*, 1137-46.
61. Nelson, G. J.; Schmidt, P. S.; Bartolini, G. L.; Kelley, D. S.; Kyle, D. *Lipids* **1997**, *32*, 1129-36.
62. Agren, J. J.; Hanninen, O.; Julkunen, A.; Fogelholm, L.; Vidgren, H.; Schwab, U.; Pynnonen, O.; Uusitupa, M. *Eur. J. Clin. Nutr.* **1996**, *50*, 765-71.
63. Zucker, M. L.; Bilyeu, D. S.; Helmkamp, G. M.; Harris, W. S.; Dujovne, C. A. *Atherosclerosis* **1988**, *73*, 13-22.
64. Nestel, P. J. *Am. J. Clin. Nutr.* **2000**, *71*, 228S-31S.
65. Nikkila, M. *Eur. J. Clin. Nutr.* **1991**, *45*, 209-13.
66. Mori, T. A.; Bao, D. Q.; Burke, V.; Puddey, I. B.; Beilin, L. J. *Hypertension* **1999**, *34*, 253-60.
67. Innis, S. M.; Hansen, J. W. *Am. J. Clin. Nutr.* **1996**, *64*, 159-67.
68. Davidson, M. H.; Maki, K. C.; Kalkowski, J.; Schaefer, E. J.; Torri, S. A.; Drennan, K. B. *J. Am. Coll. Nutr.* **1997**, *16*, 236-43.
69. Conquer, J. A.; Holub, B. J. *J. Nutr.* **1996**, *126*, 3032-9.

Chapter 9

Production of Docosahexaenoic Acid from Microalgae

Sam Zeller[1], William Barclay[2], and Ruben Abril[2]

[1]Kelco Biopolymers, Monsanto Company, 8355 Aero Street, San Diego, CA 92123
[2]OmegaTech Inc., 4909 Nautilus Court North, Suite 208, Boulder, CO 80301

A significant body of scientific research has now accumulated pointing to the importance of long-chain omega-3 fatty acids in infant and maternal nutrition and in helping maintain cardiovascular health. Recognizing the fact that microalgae and algae-like microorganisms are the primary producers of eicosapentaenoic acid (EPA) and docosahexaenoic acid (DHA) in the marine environment, we developed a fermentation process for production of a whole-cell microalgae rich in DHA. The microalgae can be used directly as an animal feed ingredient, fed to laying hens to produce DHA-enriched eggs and egg products, and as an aquaculture feed ingredient. Additionally, the oil can be extracted from the whole-cell microalgae and further processed to produce a refined ingredient used in nutritional supplement and food applications. Information related to production, product composition, stabilization, and product delivery systems will be presented.

A significant body of scientific research has now accumulated pointing to the importance of long-chain omega-3 fatty acids in infant and maternal nutrition and in helping maintain cardiovascular health. Fish consumption is associated with reduced cardiovascular disease risk, and fish are major dietary sources of long chain omega-3 polyunsaturated fatty acids, especially docosahexaenoic acid (DHA) and eicosapentaenoic acid (EPA) (*1-4*). In intervention studies, fish oils, which are high in EPA and DHA, are found to reduce various cardiovascular disease risk factors and also disease incidence (*5*). Studies with isolated or highly enriched DHA sources confirm that DHA can specifically reduce important cardiovascular disease markers (*6-9*). DHA is also a normal constituent of cardiac and vascular membranes, and it is an important component of the cell membranes in the nervous system especially in the brain and eyes. Therefore, it is reasonable to assume that DHA is needed to maintain healthy cellular structure and function. Recognizing the fact that microalgae and algae-like microorganisms are the primary producers of EPA and DHA in the marine environment, coupled with the need for new and alternative sources of dietary DHA, we developed a fermentation process for production of a whole-cell microalgae rich in DHA (*10-12*). The microalgae can be used directly as an animal feed ingredient, fed to laying hens to produce DHA-enriched eggs and egg products, fed to broilers, swine and beef cattle to produce meat products that are naturally enriched in DHA and fed to dairy cattle to produce DHA-enriched milk and dairy products. The dried microalgae can also be used as an aquaculture feed ingredient to naturally improve the DHA content of fish and crustaceans. Additionally, the oil can be extracted from whole-cell microalgae and further processed to produce refined ingredients (pure triacylglycerols, ethyl esters, or fatty acid salts of DHA) for use in nutritional supplement, food and feed applications. Egg yolk lipids, dried egg yolk powder and defatted egg yolk powdered products derived from DHA-enriched eggs from laying hens fed whole cell microalgae can also be obtained. These products, rich in DHA, can be used directly or added as an ingredient in a variety of nutritional supplements and food formulations to provide consumers with a wide range of options for incorporating DHA into their diet.

Source Microalgae

The dried DHA-rich microalgae were produced via fermentation using *Schizochytrium* sp., a member of the kingdom Chromista (also called stramenopiles), which includes golden algae, diatoms, yellow-green algae, haptophyte and cryptophyte algae, oomycetes and thraustochytrids. *Schizochytrium* sp. is a thraustochytrid. Thraustochytrids are found throughout the world in estuarine and marine habitats. Their nutritional mode is primarily saprotrophic (obtain food by absorbing dissolved organic matter) and as such are generally found associated with organic detritus (*13,14*), decomposing algal and plant material (*15, 16*) and in sediments (*17*). Microalgae and other microscopic organisms are primarily consumed by filter feeding invertebrates in the marine ecosystem. Filter feeding organisms accumulate the particulate material they filter from water in the foregut before passing the material into their digestive system. By examining foregut contents, scientists at OmegaTech Inc. have demonstrated that thraustochytrids are consumed by a wide variety of filter feeding organisms including mussels that are consumed directly by humans.

Fermentation

Thraustochytrids, especially strains of the genus *Schizochytrium*, are attractive targets for fermentation development for a number of reasons: 1) heterotrophic nature; 2) small size; 3) ability to use a wide range of carbon and nitrogen compounds; 4) ability to grow at low salinities and exhibit enhanced lipid production at low salinities; and 5) the high DHA-content in their lipids even at elevated temperatures (e.g. 30°C) (*10-12*). We have developed a fermentation process for the production of DHA-rich marine microalgae using a strain of *Schizochytrium* sp. (ATCC 20888). *Schizochytrium* sp. microalgae (biomass) was grown via a pure-culture heterotrophic fermentation process. The fermentation process was initiated using a frozen culture to inoculate a shake flask. The shake flask culture was then used to inoculate a seed tank, which in turn, was used to inoculate a final fermentor. The media used to fill the final fermentor consisted of a carbon source, a nitrogen source, bulk nutrients (e.g., sodium, phosphorous, etc.), vitamins and trace minerals. The sugar concentration was monitored during the course of the fermentation and the batch was recovered once the sugar was depleted. An antioxidant was added to the fermentation broth prior to harvesting the cells to ensure oxidative stability of the final dried product. The resulting biomass was harvested by using a centrifuge to separate the cells from the spent broth and the concentrated material was dried to produce the final product. The dried microalgae, in the form of free flowing flakes with a golden coloration, are packaged in bulk bags for use in feed applications (Figure 1).

Composition

Results of proximate analysis conducted on dried microalgae are shown in Table I. Crude fat (66%) and protein (20%) were the predominant components of the dried microalgae with minor amounts of ash (4%), carbohydrate (4%), moisture (3%), and crude fiber (2%) accounting for the balance. DHA content, calculated on a dry weight basis (DWT), in the whole cell microalgae comprised 21% of the cells. Lipid class composition of the dried microalgae was determined by extracting crude lipid from the cells (*18*) and fractionating it into various lipid

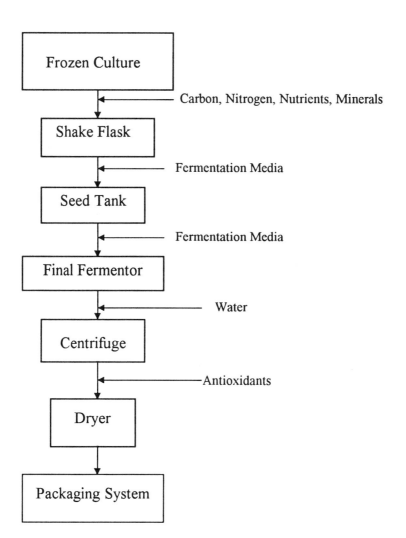

Figure 1. Flow diagram for fermentation and recovery of whole cell microalgae rich in DHA.

class components using a column of silica gel. Triacylglycerols represented the major lipid class fraction in the crude lipids, accounting for 90-92% of the total crude lipid. Minor amounts of fatty acyl sterol esters (0.4%), diacylglycerols (1%), free sterols (1%), and free fatty acids (0.1%) were observed in the crude lipid fraction isolated from algae. A small amount (5%) of polar lipid material was isolated from the crude lipid fraction. Beta-carotene was identified as the primary carotenoid present in crude lipid. Fatty acid profiles of the crude lipids and fractions thereof were determined by gas chromatography following transesterification using 4% HCl in methanol. DHA and DPA (docosapentaenoic acid, the omega-6 isomer) were shown as the major fatty acids in the crude lipid fraction accounting for 41% and 18%, respectively, of the total fatty acid methyl esters. Myristic acid (9%) and palmitic acid (22%) were the other major fatty acids representing >2% of fatty acid methyl esters present in crude lipid isolated from microalgae. Minor fatty acids present in crude lipid as well as fatty acid profiles of the triacylglycerols, steryl ester, diacylglycerols, and polar lipid fractions are shown in Table II. Cholesterol, brassicasterol, and stigmasterol were identified as the major sterol components in the fatty acyl sterol ester and free sterol fraction in the crude lipid isolated from microalgae. Regiospecific fatty acyl position was obtained on the triacylglycerol fraction isolated from the crude oil. ^{13}C NMR analysis was conducted (19) with results demonstrating approximately 23 mol% of combined DHA and DPA esterified to the sn-2 position and 27 mol% esterified to the sn-1 and sn-3 positions of the glycerol molecule.

Table I. Proximate composition of *Schizochytrium* sp. whole cell microalgae produced via a fermentation process.

Proximate Analysis	Content (%)
Moisture	3
Protein	20
Crude, Fiber	2
Crude Fat	66
Carbohydrate	4
Ash	4
DHA (% dry weight)	21

Biovailability

Algae and algae-like organisms have been used as feed ingredients for a variety of animals providing protein, pigments, calcium, trace metals and lipids (*20-22*). One problem limiting the use of algae in feeding applications has been attributed to poor digestibility due to cell walls made of cellulose (*23*). Cell walls of *Schizochytrium* algae are comprised mainly of hemicellulosic material (*24*) and therefore should overcome limitations due to digestibility and nutritional availability of DHA when fed to animals. Bioavailability of DHA in *Schizochytrium* sp. microalgae was previously shown in *Artemia* sp., an animal with a very short gut passage time, by their ability to readily assimilate DHA (*25*). To further demonstrate digestibility and bioavailabilty of DHA in *Schizochytrium* sp. algae, feeding studies were conducted using weanling, Sprague-Dawley rats. Female rats (n=26 per diet) were fed test diets containing three dose levels of DHA (32, 344, and 1520 mg DHA/kg body weight/day) over a 90 day period. A basal diet consisting of Purina Rodent Chow was included as the control diet. At the end of 90 days, the animals were sacrificed and liver and serum samples collected and extracted with chloroform-methanol. Fatty acids in the crude oil were transesterified to methyl ester derivatives which were then separated and quantified by gas-liquid chromatography. Results of fatty acid analysis of the serum (Figure 2) and liver (Figure 3) samples from female rats exhibited an increasing dose response specifically for the long chain polyunsaturated fatty acids, DHA and DPA (n-6), demonstrating the nutritional availability of these fatty acids derived from feeding whole cell microalgae. The increase in DHA and DPA(n-6) in the liver samples was correlated with a corresponding decrease in the arachidonic acid (20:4n-6) and linoleic acid (C18:2n-6). In the serum samples, a similar decrease was only observed in linoleic acid. These data indicate that DHA readily accumulates in cells even in the presence of dietary long chain n-6 fatty acids. Supplying both DHA and DPA(n-6) lead to an increase in DHA and a general decrease in total n-6 polyunsaturated fatty acids. With regard to the saturated and monounsaturated fatty acids, a decrease in the concentration of stearic (18:0), and oleic (18:1 n-9) acids in the liver and serum was associated with an increase in the levels of myristic (14:0), palmitic (16:0) and palmitoleic (16:1) acids.

Egg Enrichment

White leghorn hens were used to evaluate the production of DHA-enriched eggs using a corn-based diet. One hundred hens per treatment were housed in wire mesh commercial cages (6 birds/cage) provided with 16 h of light/day. Feed consumption averaged 110 g/hen/day and water was provided *ad libitum*. DHA

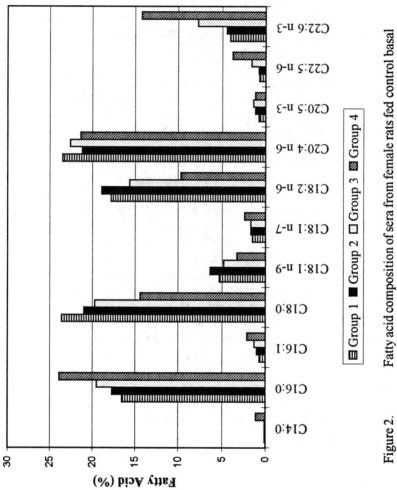

Figure 2. Fatty acid composition of sera from female rats fed control basal diet (Group 1) and admixtures of *Schizochytrium sp.* microalgae in rodent chow at levels 32 (Group 2), 344 (Group 3), and 1520 (Group 4) mg DHA/kg body weight/day after thirteen weeks.

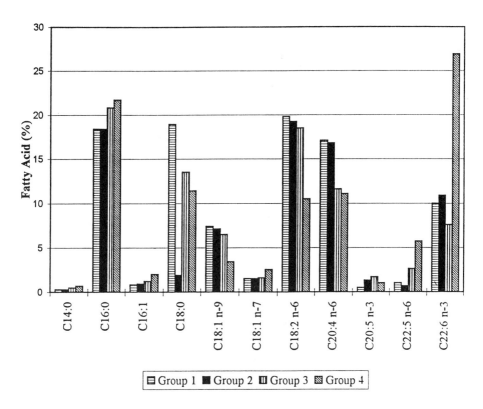

Figure 3. Fatty acid composition of liver tissue from female rats fed control basal diet (Group 1) and admixtures of *Schizochytrium* sp. microalgae in rodent chow at levels 32 (Group 2), 344 (Group 3), and 1520 (Group 4) mg DHA/kg body weight/day after thirteen weeks.

was provided in the form of whole cell algae at an inclusion rate of 165 mg DHA/hen/day with whole flaxseed added as a supplemental source of 18:3n-3. Eggs were collected, cracked, and the yolks separated from the whites using a commercial egg separator. Fatty acid composition was determined in the egg yolk fraction and results compared to eggs from a commercial egg production process. Eggs produced by laying hens fed 165 mg DHA/hen/day from algae incorporated into a corn-based diet contained approximately 153 mg/egg of total long chain omega-3 fatty acids as compared with only 28 mg/egg from control hens (Figure 4). The content of polyunsaturated fatty acids in the treatment eggs was increased by 20% of the total fat as compared to the control. The DHA content of eggs from algae-fed hens was 135 mg/egg as compared to 28 mg/egg in the control eggs, an approximate five-fold increase (26).

Table II. Fatty acid profile of crude oil and lipid fractions isolated from crude oil from *Schizochytrium* sp. microalgae.[a]

Fatty Acid	% Total Fatty Acid				
	Crude Oil	Polar Lipid	TAG	Steryl Ester	DAG
14:0	9.0	7.3	9.2	2.7	6.8
15:0	0.3	0.9	0.3		0.3
16:0	22.2	56.0	22.1	10.9	20.0
16:1	0.2		0.2		
18:0	0.6	2.5	0.5	1.8	1.9
18:1 n-9	0.2	1.9	0.1	1.5	0.3
18:1 n-7	0.2	0.8	0.1		
18:4 n-3	0.4	1.0	0.4	0.4	0.3
20:3 n-6 & 20:4 n-7	1.5		1.5	1.2	1.4
20:4 n-6	0.9	0.9	0.9	1.0	0.9
20:4 n-3	1.0		0.9	1.0	0.9
20:5 n-3	1.9	3.1	1.9	1.2	2.2
22:5 n-6	18.0	7.7	18.2	25.9	18.7
22:6 n-3	41.2	11.6	41.7	48.6	43.0

[a] TAG, triacylglycerols; DAG, diacylglycerols.

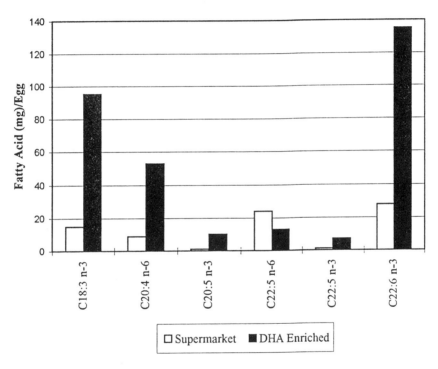

Figure 4. Fatty acid composition of DHA-enriched eggs compared to typical supermarket eggs. The DHA-rich eggs were produced by feeding laying hens 165 mg DHA/hen/day from microalgae incorporated into a corn-based diet.

Enriched Animal Products

Broilers

Broiler chickens (60 birds/treatment and a 50% male:50% female mix) were used to evaluate the production of DHA-enriched broiler meats. Broilers were housed in a commercial setting and fed a pelletized commercial feed containing dried microalgae providing 3.6 g of DHA during the 49-day production cycle. At the end of the trial, the birds were slaughtered, processed, and flash-frozen. Samples of white meat (breast) and dark meat (drum plus thigh) were then analyzed for their DHA content. White meat from control birds contained 14 mg DHA/100 g, while that of birds fed the dried micoalgae diet was enriched to 88 mg DHA/100 g, representing a 6-fold increase in the DHA content. Control dark meat averaged 24 mg DHA/100 g whereas that from birds fed dried microalgae averaged 54 mg DHA/100 g *(27)*.

Swine

A swine feeding trial was conducted whereby 5 swine were fed a standard ration containing approximately 0.6% dried algae in their diet over the course of a production cycle. At the end of the cycle, when the swine had reached approximately 240 lb, they were slaughtered and meat samples collected and lyophilized for subsequent fatty acid analysis. A representative sample of shoulder muscle in swine fed the dried microalge diet was found to contain 51 mg DHA/100g. No DHA was detected in shoulder muscle from control animals.

Beef Cattle

In a pre-study pilot trial, two beef cattle were fed a ration which included 2% dried microalgae during the last 60 days of their production cycle. At the end of the production cycle, the cattle were slaughtered and meat samples were collected and lypohilized for subsequent analysis of their fatty acid constituents. A sirloin cut from a sample (100 g) of back muscle contained 18 mg of DHA and 15 mg of EPA in cattle fed dried microalgae. Similar samples from the control animals contained no detectable amounts of either DHA or EPA.

Nutritional Supplements and Food Ingredients

Nutritional supplement and food ingredient products with high levels of DHA can be produced from either dried microalgae or from DHA-enriched eggs.

Refined Oil

Oil can be extracted from dried microalgae and purified using conventional and well established methods applied by the edible oil industry. A process scheme to produce a refined oil product is outlined in Figure 5. This oil is stabilized with antioxidants and then encapsulated in soft gelatin capsules for use as a dietary supplement. Each 1 g capsule delivers approximately 385 mg DHA with a demonstrated shelf life of over two years. The fatty acid profile for this product in shown in Table III.

Egg Yolk and Defatted Egg Yolk Powders

The availability of eggs containing high levels of DHA presents opportunities for production of new enriched food products that use eggs, yolks, egg lipids, egg yolk powders, and defatted egg yolk powders. Ingredient options are obtained by processing DHA-enriched eggs to prepare egg yolk lipids, egg yolk powders and defatted egg yolk powders. To this end we have demonstrated processing capability to produce these ingredients options (Figure 6). Egg yolks are isolated from the whole egg, pasteurized and dried to produce egg yolk powders rich in DHA. Egg yolk powders demonstrate many of the same functionality's of whole eggs and can be used in a variety of food applications. Egg yolk powders can also be extracted with supercritical carbon dioxide to remove triacylglycerols and cholesterol, thus producing a defatted egg yolk powder containing high levels of DHA and low levels of cholesterol. The fatty acid profile of egg yolk and defatted egg yolk powders are outlined in Table III.

Conclusions

The epidemiological and clinical data relating consumption of fish and fish oil and several indices of cardiovascular disease are consistent with a beneficial role of long chain omega-3 fatty acids, including DHA. Research on DHA supports specific benefits of this fatty acid in maintaining normal, healthy cardiovascular function. The increased awareness of the health benefits of long chain omega-3 fatty acids coupled with a consumer need for alternative dietary sources has led to the development of a technology for naturally enriching a wide range of food

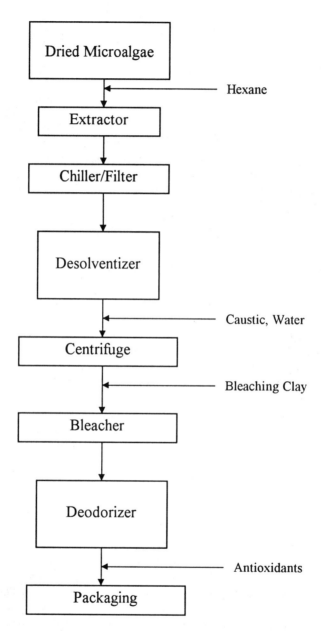

Figure 5. Oil extraction and purification process used to produce DHA-rich oil.

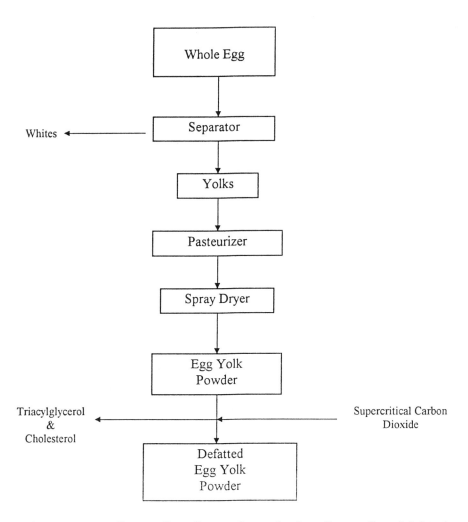

Figure 6. Process flow diagram for production of egg yolk and defatted egg yolk powder.

Table III. Fatty acid profile of refined oil derived from dried microalgae and egg yolk powder and defatted egg yolk powder derived from DHA-enriched eggs.

Fatty Acid	Fatty Acid (mg/g)		
	Refined Triacylglycerol Oil	Egg Yolk Powder	Defatted Egg Yolk Powder
14:0	83	3	<1
16:0	211	137	73
16:1	7	12	1
18:0	<4	44	34
18:1	<4	181	52
18:2 n-2	<4	83	30
18:3 n-3	<4	11	<1
20:4 n-6	9	6	12
20:5 n-3	20	<1	<1
22:5 n-6	167	1	2
22:6 n-3	387	16	24

products with DHA. By producing microalgae containing high levels of DHA through controlled fermentation, and feeding the dried, stabilized material to laying hens, broiler chickens, swine and cattle as part of their normal diets, DHA-enriched eggs, meat and milk products can be produced. The eggs and milk can also be used directly as ingredients in foods or further processed to extract the DHA-rich lipids which are then used as food ingredients. Alternatively, refined oil can be obtained by direct extraction of oil from the microalgae with subsequent purification of the isolated oil using standard commercial oil refining operations. Dietary supplements and food ingredients can then be produced in the form of refined triacylglycerols, concentrated ethyl esters, microencapsulated oils, salts of fatty acids, and phospholipids. These various food and ingredient options arising from this microalgal–based technology are summarized in Table IV. These ingredient options provide a wide range of dietary alternatives to fish and fish oil for those consumers interested in maintenance of cardiovascular health by including long chain omega-3 fatty acids in their diets.

References

1. Bang, H.O.; Dyerberg, J.; Sinclair, H.M. *Am. J. Clin. Nutr.* **1980**, *33*, 2657-2661.
2. Kagawa, Y.; Nishizawa, M.; Suzuki, M.; Miyatake, T.; Hamamoto, T.; Goto, K.; Motonaga, E.; Izumikawa, H.; Hirata, H.; Ebihara, A. *J. Nutr. Sci. Vitaminol.* **1982**, *28*, 441-453.

Table IV. DHA ingredient options derived from microalgae and DHA-enriched eggs.

DHA-enriched Ingredient	DHA Content (mg/g)	Potential Application
Microalgal Oil		
Refined Oil	385	nutritional supplements, broad-based food & dairy
Ethyl Esters	600	nutritional supplements, pharmaceuticals
Microencapsulated Oil	varied	nutritional supplements, broad-based food & dairy
Dried Microalgae	220	nutritional supplements, foods, animal feed supplement
Egg-based Products		
Egg Yolk Powder	14	pasta, baked goods, ready mixes, batters
Defatted Egg Yolk Powder	25	sport/nutritional powders, baked goods, ready to eat foods, pasta, salad dressings
Egg Phospholipids	100	nutritional supplements, broad-based food & dairy, cosmetics

3. Leaf, A.; Weber, P.C. *New Engl. J. Med.* **1988**, *318*, 549-557.
4. Schmidt, E.B.; Dyerberg, J. *Drugs* **1994**, *47*, 405-424.
5. Burr, M.L.; Fehily, A.M.; Gilbert, J.F.; Rogers, S.; Holliday, R.M.; Sweetnam, P.M.; Elwood, P.C.; Deadman, N.M. *Lancet* **1989**, *2 (8666)*, 757-761.
6. Agren, J.J.; Hanninen, O.; Julkunen, A.; Fogelholm, L.; Vidgren, H.; Schwab, U.; Pynnonen, O.; Uusitupa, M. *Eur. J. Clin. Nutr.* **1996**, *50*, 765-771.
7. Conquer, J.A.; Holub, B.J. *J. Nutr.* **1996**, *126*, 3032-3039.
8. Davidson, M.H.; Maki, K.C.; Kalkowski, J.; Schaefer, E.J.; Torri, S.A.; Drennan, K.B. *J. Am. Coll. Nutr.* **1997**, *16*, 236-243.

9. Grimsgaard, S.; Bonaa, K.H.; Hansen, J.B.; Nordoy, A. *Am. J. Clin. Nutr.* **1997**, *66*, 649-659.
10. Barclay, W.R. 1992, US Patent 5, 130, 242.
11. Barclay, W.R. 1994, US Patent 5,340,742.
12. Barclay, W.R.; Abril, J.R.; Meager, K.M *J. Appl. Phycol.* **1994**, *6*, 123-129.
13. Findlay, R.H.; Fell, J.W.; Coleman, N.K. In *Biology of Marine Fungi.* Moss, S.T. ed. Cambridge University Press: London, 1980, pp. 91-103.
14. Raghukumar, S.; Basasubramanian, R. *Indian J. Mar. Sci.* **1991**, *20*, 176-181.
15. Bremer, G.B. *Hydrobiologia,* **1995**, *295*, 89-95.
16. Sathe-Pathak, V.; Raghukumar, S.; Raghukumar, C.; Sharma, S. *Indian J. Mar. Sci.* **1993**, *22*, 159-167.
17. Bahnweg,G.; Sparrow, F.K. *Am. J. Bot.* **1974**, *61*, 754-766.
18. Folch, J.; Lees, M.; Stanley, G.H.S. *J. Biol. Chem.* **1957**, *226*, 497-509.
19. Aursand, M.; Rainuzzo, J.R.; Grasdalen, H. *J. Am. Oil Chem. Soc.* **1993**. *70*, 971-981.
20. Gouveia, L.; Veloso, V.; Reis, A.; Fernandes, H.; Novals, J.; Empis, J. *J. Sci. Food Agric.* **1996**, *70*, 167-172.
21. Lipstein, B.; Hurwitz, S. In *Algae Biomass*, Shelef, G.; Soeder, J., Eds. Elsevier, New York. 1980, pp. 667-685.
22. Soeder, C.J. In *Algae Biomass*, Shelef, G.; Soeder, C.J., Eds. Elsevier, New York. 1980, pp. 9-20.
23. Becker, E.W. *Microalgae: Biotechnology and Microbiology.* Cambridge University Press, London (1994).
24. Darley, W.M.; Porter, D.; Fuller, M.S. *Arch. Mikrobiol.* **1973**, *90*, 89-106.
25. Barclay, W.; Zeller, S. *J. World Aqua. Soc.* **1996**, *27*, 314-322.
26. Abril, J.R.; Barclay, W.R.; Abril, P.G. In *Egg Nutrition and Biotechnology.* Sim, J.S.; Nakai, S.; Guenter, W., Eds. Cab International: New York, 2000, pp. 197-202.
27. Abril, R.; Barclay, W. In *The Return of omega-3 Fatty Acids into the Food Supply. I. Land-Based Animal Food Products and Their Health Effects.* Simopoulos, A.P., Ed. *World Rev. Nutr. Diet.* Basel, Karger, 1998, Vol 83, pp. 77-88.

Chapter 10

Blue-Green Alga *Aphanizomenon flos-aquae* as a Source of Dietary Polyunsaturated Fatty Acids and a Hypocholesterolemic Agent in Rats

Christian Drapeau[1,3], Rafail I. Kushak[2], Elizabeth M. Van Cott[2], and Harland H. Winter[2]

[1]Cell Tech, 1300 Main Street, Klamath Falls, OR 97601
[2]Pediatric GI/Nutrition, Massachusetts General Hospital, Harvard Medical School, 55 Fruit Street, VBK 107, Boston, MA 02114–2698
[3]Current address: Research and Development, Desert Lake Technologies, 12750 Keno-Worden Road, Keno, OR 97627

The blue-green alga *Aphanizomenon flos-aquae* is shown to be a good source of omega-3 polyunsaturated fatty acids in rats. It was able to partially reverse the effects on digestive functions induced by a deficiency in polyunsaturated fatty acids. *Aphanizomenon flos-aquae* is also reported to significantly decrease total cholesterol and triacylglycerol levels.

Algae has been used worldwide as a food source or as a remedy for various physical ailments for thousands of years (*1,2*). In coastal regions of the Far East, notably in Japan, there is evidence that algae were used as a food source 6000 BC, and there are records of many species of seaweed used as food around 900 AD (*3*). Many reports at the time of the Spanish conquest reveal that the natives of Lake Tezcoco, near the city of Tenochtitlan (today's Mexico city), collected blue-green algae from the waters of the lake to make sun-dried cakes called Tecuitlatl, or "excrement of stone" as they believed stones were the source of algae (*4,5*).

Still today in Africa, local tribes harvest blue-green algae growing in Lake Chad which is used to make hard cakes called Dihé (*4*). Similar to the inhabitants of Tenochtitlan, patches of floating microalgae are collected and sun-dried in shallow holes dug in the sand along the shore of the lake.

© 2001 American Chemical Society

For the past seventeen years *Aphanizomenon flos-aquae* (*Aph. flos-aquae*), a naturally-occurring blue-green alga growing in Upper Klamath Lake, Oregon, has also been harvested and sold as a high protein nutritional source. Nearly one million consumers have been eating *Aph. flos-aquae* over the past few years, and many anecdotal reports suggest that *Aph. flos-aquae* might have health-promoting properties. Recently, it was shown that consumption of *Aph. flos-aquae* triggers within two hours the migration of natural killer cells from the blood to the tissues, as well as a significant increase in the expression of adhesion molecules on the surface of natural killer cells (6). Such an effect was later confirmed and shown to be much more pronounced in regular consumers than in people exposed to *Aph. flos-aquae* for the first time (7). *Aph. flos-aquae* was also shown to be useful as an adjunct to the treatment of mild traumatic brain injury, likely by facilitating neuroplasticity (8). Finally, a retrospective epidemiological study suggested that *Aph. flos-aquae* may be beneficial in the treatment of various conditions such as attention deficit disorders, fibromyalgia, hypertension and chronic fatigue (9).

Currently, the market for *Aph. flos-aquae* as a health food supplement is close to $100 million in sales with an annual production exceeding 100 metric tons of dried algae. The present contribution would provide a brief review of the techniques of harvesting and processing of naturally-growing *Aph. flos-aquae* and reports the results pertaining to the effect of *Aph. flos-aquae* on blood lipids.

Klamath Lake Ecology and *Aph. flos-aquae* Bloom

Upper Klamath Lake is the largest freshwater lake (324 km^2) in Oregon and drains a watershed of 9800 km^2. This shallow lake (ave. depth = 2.4 m) is flanked by the Cascade mountains to the west and the Great Basin to the east. The lake has two major tributaries, the Williamson and Wood Rivers, as well as many smaller springs and stream inflows.

Historically, the lake has always been extremely productive and has supported not only a tremendous biomass of algae, but also fish, waterfowl, and predatory bird species. When ice was first collected from the lake (1906) it was reported to be green with algae (10). Lake suckers were so common that people used pitchforks to harvest them. Ospreys were reported in densities of up to 10 nests per square mile. Today the Klamath Basin is still home to the largest wintering congregation of bald eagles in the lower 48 states, and is the largest stopover for waterfowl in the Pacific flyway.

Records of the dynamics of algal blooms in Klamath Lake have been maintained over the past 8 years and have established *Aph. flos-aquae* as the predominant algal species growing in the lake (11,12). Typically, *Aph. flos-aquae* starts blooming in early June with an initial biomass of 3-15 mg/L and remains the dominant species until November, with a biomass sometimes reaching more than 100 mg/L.

Harvesting and Processing of *Aph. flos-aquae*

One of the main harvesting facilities is located along an aqueduct system, approximately 14 kilometers from the southern end of Klamath Lake, at the site of a previous low head hydroelectric turbine facility with a 3m drop. This feature is utilized to feed water onto screens by gravity. The flow at this location averages 1.8 million liters per minute and is entirely processed for algae removal by dispersing the water over a total of 48 screens. This is the site of initial dewatering. A water spray assembly is located atop and across each screen and is manually operated to move the collected algae to a trough where it is pumped to a secondary vibrating filter screen. This screen is used to remove any unwanted macroscopic debris. The algae is then pumped to a series of slow speed horizontal centrifuges for the final removal of small extraneous material. This processing stage utilizes the fact that *A. flos-aquae* has a specific gravity almost that of water, allowing for the separation of sand, silt, and any other light debris.

The liquid algae concentrate is then pumped to a vertical centrifuge that applies high G-force to separate algal cells and colonies from water, removing virtually 90% of the remaining water. Once concentrated, the algae is immediately chilled to 5°C and stored in a chilled and agitated tank before being pumped to flake freezers to be instantaneously frozen. When needed, the frozen algae is shipped to an external freeze drying facility to be freeze dried into an algal powder containing 3-5% water content. This final product is processed into consumable capsules or tablets as needed.

The other main harvesting process takes place directly in the algal blooms in the deep waters of Upper Klamath Lake, toward the Northern end of the Lake. Harvesting directly in the lake allows for a harvesting season that endures until late November. In brief, small harvesting platforms make passages in newly formed algal blooms that gather in large, thick patches at the surface. These harvesters are equipped with rotating screens that lift the algae from the water surface and transfer it on a screened conveyor system to perform the initial dewatering step. The algae is then transported to a 800 m² processing barge located in close proximity to the blooms. On this platform the algae is slowly pumped through a tube and shell heat exchanger that brings the concentrated algae to a temperature of 5°C. The algae is then stored in refrigerated tanks. At this point, the algae is filtered through a centrifugal sieve to remove debris and certain species of algae such as *Microcystis*. The algae is then passed through a concentrating system to further concentrate the liquid algae. The concentrated algae is then transported to shore in a refrigerated tank and then trucked to a drying facility to be dried fresh. The powder is finally processed into consumables.

Ability of *Aph. flos-aquae* to Provide Essential Lipids

Over the past 15 years, many empirical reports have been made on the beneficial effect of *Aph. flos-aquae* on conditions such as depression, attention deficit disorders, epilepsy, skin diseases, rheumatism, Crohn's disease and other inflammatory disorders. The occurrence of many of these conditions have been associated with a deficiency in polyunsaturated fatty acids (PUFA), mostly omega-3 fatty acids.

In brief, omega-3 fatty acids have been shown to improve arthritic and other inflammatory conditions (*13*). Epidemiological studies suggest that decreased omega-3 fatty acid consumption correlates with increasing rates of depression (*14*). Attention deficit disorder has been associated with a decrease in blood PUFA (*15*). Skin conditions such as eczema have been associated with altered blood levels of PUFA and oral PUFA has been shown to improve such conditions (*16*). Omega-3 fatty acids prevent platelet aggregation (*17,18*), arrhythmia and ventricular fibrillation (*19,20*), and protect myocardial cells against hypoxia-reoxygenation-induced injury following ischemic heart disease (*21*). Eicosapentaenoic acid (EPA) was also shown to prevent epileptic seizures in a rat model of epileptogenesis (*22,23*). Finally, omega-3 PUFA have been shown to lower cholesterol, possibly by stimulating its excretion into bile (*24,25*).

PUFA constitute 30% to 50% of the lipid content of dried *Aph. flos-aquae* (5% to 9% of total dry weight), with the main contributor being alpha- linolenic acid. It was hypothesized that the PUFA content of *Aph. flos-aquae* might be mediating some of the reported health benefits. Experiments were designed to assess the bioavailability of the PUFA contained in *Aph. flos-aquae*. Here we report the results of experiments using rats, in which *Aph. flos-aquae* proved to be not only a good source of PUFA, but also lowered blood cholesterol and triglyceride levels.

Methods

Animals. Thirty-two adult male Sprague-Dawley rats were randomly distributed into 4 groups. Animals were placed into individual wire cages, and maintained at 22°C with a 12-hour light-dark cycle. Food and water were supplied *ad libitum*. Animals were fed for 32 days with the following semipurified test diets (Table I) based on the American Institute of Nutrition (AIN-76) standards: 1) Standard diet

Table I. Fatty acid profiles of selected oils.

Fatty Acid (%)	Soybean	Coconut	Alga
Caprylic acid	0	9.7	
Capric acid	0	7.5	
Lauric acid	0	42.1	
Myristic acid	0	22.4	9.1
Palmitic acid	14.69	18.2	36.6
Palmitoleic acid	0		11.9
Heptadecanoic acid	0		0.89
Stearic acid	5.4		2.7
Oleic acid	26.8		6.7
Linoleic acid (LA)	44.4		7.4
Linolenic acid (LNA)	8		22.3
Arachidic acid	0.35	0.14	
Arachidonic acid (AA)	0		0.65
Eicosapentaenoic acid (EPA)	0		0.08
Behenic acid	0.33		
Polyunsaturated (PUFA)	52.4	0	30.43
Monounsaturated (MUFA)	26.8		18.6
Saturated (SFA)	20.77	100.04	49.29

SOURCE: From ref. 28.

containing 5% soybean oil (SBO); 2) PUFA deficient diet containing 5% coconut oil (PUFA-D); 3) PUFA-D diet containing 10% algae (0.50% algal lipids) (Alg10%); and 4) PUFA-D diet containing 15% algae (0.75% algal lipids) (Alg15%). The algal material used in this study was supplied by Cell Tech (Klamath Falls, OR) and contained 6.29% lipids. Feed was provided by Purina Test Diets (Richmond, IN).

The weight of the animals was monitored prior to and during the 32 day feeding trial. After the feeding trial, the animals were fasted overnight and euthanased by carbon dioxide inhalation. Viscera (liver, kidney, small and large intestine, pancreas, cecum and spleen) were collected and frozen in liquid nitrogen and kept at -80° C for further biochemical testing. Plasma was collected by heart puncture in a tube containing 100 μl 0.5 M ethylenediaminetetraacetic acid (EDTA; pH 8.0), spun at 3,000 g for 15 minutes, and stored at -80°C.

Fatty Acid Analysis. Blood plasma fatty acid analysis was performed using direct transesterification method (*26,27*). Fatty acid identification was performed using external and internal standards on a Hewlett-Packard 5890 series II model gas chromatograph-mass spectrometer GC-MS with a Hewlett-Packard 5971 mass spectrometer (Hewlett-Packard, Wilmington, DE). Soybean and coconut oils were methylated before analysis. The algal material was soaked in methanol, extracted and then methylated before analysis. Plasma triglycerides and cholesterol were measured using Boehringer Mannheim kits on the automated clinical chemistry analyzer Roche BHO/H917.

Statistical Analysis. For the various parameters, statistical difference between groups was determined using unpaired Student's T test. Differences in fatty acid profiles were determined using univariate and multivariate repeated measures analysis. For both analyses, differences of $p<0.05$ were considered statistically significant.

Results

Dietary Fatty Acids. Table I shows the lipid content for the various sources of lipids. Soybean oil is rich in linoleic acid (LA, 18:2ω6; 44.4% of total lipids) and contains a substantial amount of α-linolenic acid (LNA, 18:3ω3; 8%). *Aph. flos-aquae* is richer in LNA (22.3%) and contains less LA (7.4%) than soybean oil. *Aph. flos-aquae* has also small amount (0.65%) of arachidonic acid (AA, 20:4ω6) and traces (0.08%) of eicosapentaenoic acid (EPA, 20:5ω3). Coconut oil is free of both ω3 and ω6 fatty acids.

Fatty acid composition of the various diets is represented in Figure 1. All diets had a similar amount of lipids provided by soybean oil (5%), coconut oil (5%) or *Aph. flos-aquae*. The lipid content of the algae used in this study was 6.29% d.w., therefore Alg10% and Alg15% diets contained 5.13% and 5.20% of lipids, respectively. Ratios of PUFA to saturated fatty acids (SFA) and ω6 to ω3 varied considerably between the diets.

Calculations showed that *Aph. flos-aquae* contains 1.40% LNA and 0.46% LA of total algal dry weight. Diets containing 10% and 15% of algae (corresponding to 0.63% and 0.94% lipids) provided a total dietary intake of 0.14% LNA and 0.047% LA for the Alg10% diet, and 0.21% LNA and 0.07% LA for the Alg15% diet. Thus, the SBO diet contained 2.9 times more LNA and 47 times more LA than the Alg10% diet, and 1.9 times more LNA and 32 times more LA than the Alg15% diet (Figure 1). Therefore, amounts of w3 and w6 PUFA in algae-supplemented diets were significantly lower than in the positive SBO control diet. Furthermore, the ω6/ω3 ratio varied significantly between the algal (0.36) and the SBO (5.5) diets.

Because of their content in coconut oil, algae supplemented diets contained more SFA then the SBO diet (Figure 1). Respectively, Alg10% and Alg15% diets contained 2.1 and 2.0 times more SFA than the SBO diet. Therefore, the PUFA/SFA ratio was significantly lower in the Alg10% (0.047) and Alg15% (0.07) diets compared to the SBO (2.6) diet. The main SFA in the Alg10% and Alg15% diets were lauric acid (12:0, »1.8% of total diet), myristic acid (14:0, »1.0%) and palmitic acid (16:0, »1.1%). Algae supplemented diets contained virtually no stearic acid (18:0, »0.02%). Conversely, the SBO diet contained no lauric and myristic acids but significant amounts of palmitic (0.73%) and stearic (0.27%) acids.

Effect of Inducing PUFA Deficiency. Among the various parameters measured, the PUFA-deficient diet significantly increased the protein content of the pancreas and decreased the activity pancreatic amylase ($p < 0.03$). The protein content of the intestinal mucosa showed a tendency to increase whereas the aminopeptidase activity of the mucosa showed a tendency to decrease. Protein content of the intestinal mucosa, mucosal aminopeptidase activity and pancreatic amylase activity were restored to control value when *Aph. flos-aquae* was added to the PUFA-D diet (Figure 2).

Plasma Lipids. Figure 3 shows a full plasma lipid profile. Plasma palmitate levels increased with palmitate dietary intake ($r=0.60$), reaching the highest level in the Alg15% group ($p<0.01$). Paradoxically, plasma oleic acid (18:1) decreased with increasing oleic acid dietary intake ($r=-0.97$), being lowest in the SBO group ($p<0.001$).

Plasma LA increased with dietary LA intake ($r=0.67$), being highest in the SBO group ($p<0.001$), which correlates with the high amount of LA in this diet. In rats fed coconut oil (deficient in LA), the plasma LA level was 36% ($p<0.0005$) of the SBO control level. When the PUFA-deficient diet (PUFA-D) was supplemented with *Aph. flos-aquae*, plasma LA level was restored to 71% (Alg10%) and 67% (Alg15%) of the SBO control level, in spite of the fact that algae supplemented diets contained less than 3% the amount of LA present in the SBO diet.

Plasma arachidonic acid (AA, 20:4ω6) also decreased with increasing AA dietary intake ($r=-0.88$) and was highest in the plasma of the SBO group ($p<0.001$). However, the plasma AA level correlated positively with the dietary level of the AA precursor LA ($r=0.64$), which was most abundant in the SBO diet.

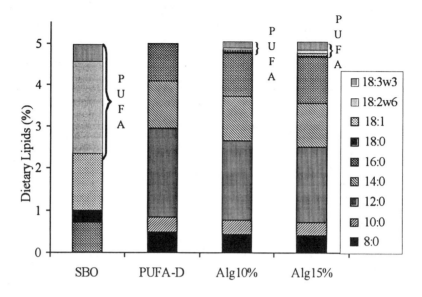

Figure 1. Lipid content for the various diets. Brackets indicate the proportion of PUFA in each diet. (Adapted from reference 28.)

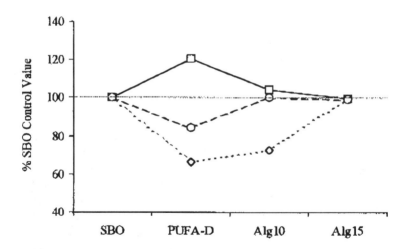

Figure 2. Effect of PUFA-deficiency on mucosal protein content (--□--), mucosal aminopeptidase (—O—) and pancreatic amylase (—◊—) activity, and ability of Aphanizomenon flos-aquae to restore control value. Values are expressed as a percent of the control values (n=8). Grey line indicates 100% of control value. Bars indicate SEM.

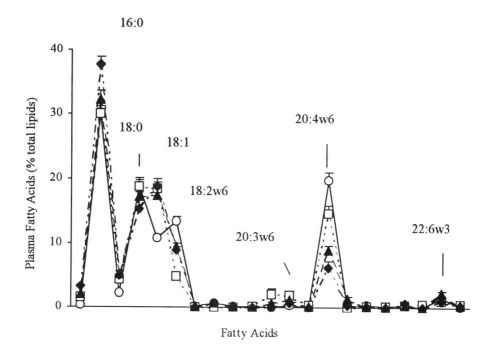

Figure 3. Complete plasma lipid profile for animals fed diets containing soybean oil (—O—), coconut oil (PUFA-deficient; —□—), algae 10% (—▲—), and algae 15% (— -♦—). Values are expressed as percent of total plasma fatty acids. Each data point is the mean of the value for 8 animals. Bars indicate SEM. (Adapted from reference 28. Copyright 2000 JANA.)

Rats fed the PUFA-D diet, which contains no LA, had a plasma AA level significantly lower than controls fed the SBO diet ($p<0.01$). However, supplementing PUFA-D diet with *Aph. flos-aquae* further decreased plasma AA levels in a dose-dependent manner ($r=-0.99$), in spite of the small LA and AA content of Alg10% and Alg15% diets.

In order to appreciate better the variation in plasma PUFA levels. Figure 4 shows the plasma lipid profile for PUFAs on a different scale than shown in Figure 3. Rats fed the PUFA-D diet had no plasma LNA, which is consistent with the absence of LNA in coconut oil. However, algae supplementation of the PUFA-D diet restored plasma LNA to the SBO (control) level, in spite of the fact that the algal diets contained only 35% (Alg10%) and 52% (Alg15%) of the LNA present in the SBO diet.

EPA ($20:5\omega3$) was absent in the plasma of rats fed the PUFA-D diet. However, when this diet was supplemented with 10% and 15% algae, plasma EPA increased 6 times ($p<0.005$) and 1.7 times ($p>0.1$) above the SBO control level, respectively. The docosahexaenoic acid (DHA, $22:6\omega3$) concentration in the plasma of rats fed the PUFA-D diet was 35% lower than in controls, although this difference did not reach statistical significance. Supplementation with 10% algae increased the plasma DHA level by a factor of 2 ($p<0.05$). Supplementation with 15% algae did not affect the plasma DHA level. The effect of algae on EPA and DHA levels in rat blood plasma may not be dose-dependent.

Cholesterol and Triacylglycerol. Algae affected not only free fatty acids but also other lipids in the blood. The PUFA-D diet did not significantly decrease the triacylglycerol level relative to the SBO controls (Figure 5). However, supplementation of the PUFA-D diet with 15% algae decreased plasma triacylglycerols to 24% of the SBO control level ($p<0.005$). The PUFA-D diet supplemented with 10% algae did not significantly affect the plasma triacylglycerol level. Plasma levels of triglycerides were positively correlated with PUFA/SFA ratio ($r^2=0.87$).

Cholesterol concentration in plasma was very sensitive to diet supplementation with algae (Figure 5). Rats fed the PUFA-D diet had a lower cholesterol level ($p<0.05$) than the SBO controls. The PUFA-D diet supplemented with algae caused a further dose-dependent decrease in the plasma cholesterol level. PUFA-D diet supplementation with 10% and 15% algae decreased the plasma cholesterol level to 54% and 25% of the SBO control level ($p<0.0005$), respectively. Cholesterol and triglyceride levels were positively correlated ($r^2=0.91$).

The cholesterol levels positively correlated to the plasma PUFA/SFA ratio ($r^2=0.84$) and to plasma stearic acid ($r^2=0.86$). On the other hand, blood cholesterol was strongly negatively correlated with plasma myristic acid ($r^2=-0.99$). From

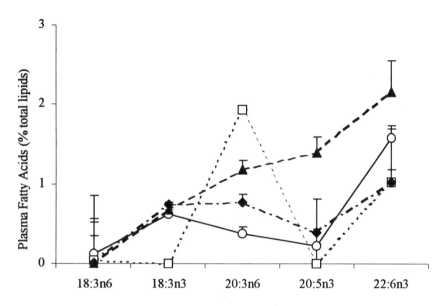

Figure 4. Plasma lipid profiles for animals fed the same diets on a different scale (n=8). Soybean oil (—O—), coconut oil (PUFA-deficient; --☐--), algae 10% (—▲—), and algae 15% (— - -♦—). Values are expressed as percent of total plasma fatty acids. Bars indicate SEM. (Adapted from reference 28. Copyright 2000 JANA.)

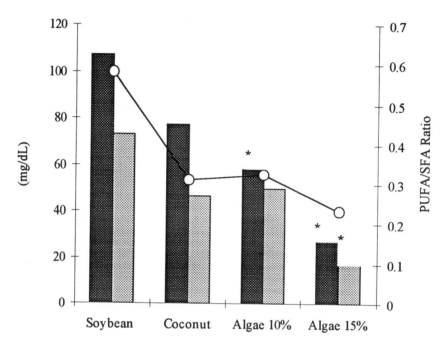

Figure 5. Levels of cholesterol (■; left axis) and triacylglycerol (▩; left axis), as well as PUFA/SFA ratio (gray line; right axis) associated with the various diets (n=8). Bars indicate Mean ± SEM. Asterisks indicate $p < 0.01$. (Reproduced from reference 28).

a dietary standpoint, blood cholesterol was only related to dietary palmitic acid ($r^2 = -0.95$).

Discussion

The results reported here demonstrate that, in rats, *Aph. flos-aquae* appears to be a good source of PUFA and to have significant hypocholesterolemic properties (28). In addition, these results indicate that a deficiency in PUFA may affect digestive functions.

Inducing a deficiency in PUFA led to a statistically significant decrease in pancreatic amylase activity, and a tendency toward increased mucosal aminopeptidase activity and mucosal protein content. This study does not allow determining the significance of this effect on digestive functions. However, it is possible that a deficiency in PUFA may lead to a decreased ability to digest complex sugar with a consequent decrease in blood glucose. The ability of *Aph. flos-aquae* to partially compensate negative effects of PUFA deficiency in rat diets has been reported elsewhere (29).

There was a good correlation between dietary and serum levels of saturated fatty acids (SFA), except for stearic acid whose serum level was similar in all groups despite low levels in the algae diets. There was also a good correlation between dietary and serum levels of linoleic acid (LA). However, the correlation between dietary and serum LNA was poor. Rats fed a PUFA-deficient diet supplemented with *Aph. flos-aquae* had blood levels of LNA comparable to levels found in rats fed soybean oil diet containing nearly three times the amount of LNA. This suggests possible higher bioavailability of LNA in *Aph. flos-aquae* compared to soybean oil.

In the animals fed the PUFA-D diet, serum levels of LA and LNA were 4.83% and 0.0% of total plasma lipids, respectively. This study was performed on adult rats; therefore animals were introduced to the various diets after a period of a few months of being fed a standard diet containing both LA and LNA. The fact that the PUFA-D diet led to a complete disappearance of LNA and EPA, while keeping LA and AA level at 36% and 75% of the SBO control value, may indicate that w3 fatty acids get depleted much faster than ω6 fatty acids during deficiency. The fact that the only measurable ω3 fatty acid in the PUFA-deficient group was DHA may further indicate that during deficiency, ω3 fatty acids are fully metabolized down their metabolic pathway.

In the rats fed the *Aph. flos-aquae* diet, blood level of EPA (20:5ω3) was significantly higher than in rats fed the SBO diet, in spite of the fact that the blood level of LNA was similar in both groups. It has been previously suggested that increased dietary SFA increased the rate of conversion of LNA to EPA, whereas increased dietary ω6 PUFA decreased this conversion by 40-50% (*30*). This dual effect could explain the fact that rats fed algae supplemented diets, which contained significantly more SFA, had higher blood levels of EPA than rats fed the SBO diet, which contained significantly more LA.

When the two main plasma ω6 PUFA (LA and AA) were analyzed, there was a very good positive correlation between LA dietary intake and the total level of ω6 PUFA. Using multivariate repeated measures analysis, the w6 profiles were significantly different between the various groups. Supplementing diets with algae led to a dose-dependent decrease in plasma AA and a concomitant accumulation of LA. This could be due to *Aph. flos-aquae*'s content in phycocyanin. Phycocyanin is the blue pigment in blue-green algae and was recently shown to have significant anti-inflammatory properties (*31,32*), which seemed to be mediated by an inhibition of arachidonic acid (AA) metabolism (*33,34*). The presence of phycocyanin in the algae supplemented diets may have inhibited AA synthesis and consequently promoted the accumulation of LA.

This study suggests that *Aph. flos-aquae* has significant hypocholesterolemic properties when compared to soybean oil. Many studies have demonstrated the hypocholesterolemic properties of PUFA (*24,25,35,36*) and the negative correlation between PUFA/SFA ratio and blood cholesterol levels (*37,38*). In this study, cholesterol levels were positively correlated with the PUFA/SFA ratio. Furthermore, the main SFA present in the diet of the algae-treated groups were lauric, myristic and palmitic acids, which were all demonstrated to promote hypercholesterolemia to some degree (*36,39-44*). This suggests that the hypocholesterolemic effect of *Aph. flos-aquae* is likely to be mediated by factors other than its fatty acid content. *Aph. flos-aquae* contains a significant amount of chlorophyll (1-2% dw) which was shown to stimulate liver function, and increase bile secretion (*45*). A synthetic derivative of chlorophyll was shown to reduce blood cholesterol (*46*). It is possible that the chlorophyll content of *Aph. flos-aquae* is responsible for the increased liver function and secretion of cholesterol into bile. Two species of green algae, *Chlorella vulgaris* and *Scenedesmus acutus*, were shown to decrease blood cholesterol levels by stimulating excretion into bile (*47,48*). *Nostoc commune*, a blue-green algae, was also shown to decrease cholesterol (*49*). *Spirulina spp.*, another blue-green alga, was also shown to affect cholesterol metabolism by increasing HDL (*50*).

In conclusion, this study demonstrated that *Aph. flos-aquae* is a good source of PUFA and has hypocholesterolemic properties. *Aph. flos-aquae*'s ability to promote high serum level of LNA and EPA, and lower level of AA makes it a good candidate for future research in therapeutic nutritional support.

Acknowledgment

We are indebted to Dr. David J. Schaeffer, Professor at University of Illinois, for assistance in statistical analysis. Authors are also grateful for the grant provided by Cell Tech and the Clinical Nutrition Research Center of the Massachusetts General Hospital (P30 DK40561).

References

1. Hoppe, H. A.; Levring, T.; Tanaka, Y. Marine Algae in Pharmaceutical Science; Walter de Gruyter Eds.: New York, NY, **1979**, 807 p.
2. Richmond, A. Handbook of Microalgal Mass Culture; CRC Press: Boca Raton, FL, **1990**, 528 p.
3. Cannell, R.J.P. *Appl. Biochem. Biotechnol.* **1990**, *26*, 85-105.
4. Ciferri, O. *Microbiol. Rev.* **1983**, *47*, 551-578.
5. Farrar, W.V. *Nature* **1966**, *5047*, 341-342.
6. Manoukian, R.; Citton, M.; Huerta, P.; Rhode, B.; Drapeau, C.; Jensen, G.S. *Phytoceuticals: Examining the health benefits and pharmaceutical properties of natural antioxidants and phytochemicals*; IBC Library Series 1911, **1998**, 233-241.
7. Jensen, G.S.; Ginsberg, D.I.; Huerta, P.; Citton, M.; Drapeau, C. *JANA* **2000**, *2 (3)*, 50-58.
8. Valencia, A.D.; Walker, J.E. Proceedings of the 3rd World Congress on Brain Injury, Quebec City, June, **1999**.
9. Krylov, V.S.; Drapeau, C.; Hassell, C.; Millman, J.D.; Strickland, J.W.; Schaeffer, D.J. Retrospective epidemiological study using medical records to determine which diseases are improved by *Aphanizomenon flos-aquae*, submitted.
10. Bortleson, G.C.; Fretwell, M.O. A review of possible causes of nutrient enrichment and decline of endangered sucker populations in the Upper Klamath Lake, Oregon. U.S.G.S. Water-Resources Investigations Report **1993**, *93*, 4087
11. Kann, J. Ecology and water quality dynamics of shallow hypereutrophic lake dominated by cyanobacteria (*Aphanizomenon flos-aquae*), PhD Dissertation, Univ. North Carolina, Chapel Hill, NC, **1997**.
12. Carmichael, W.W.; Drapeau, C.; Anderson, D.M. *J. Appl. Phycol.* **2000**, in press.
13. Kremer, J.M. *Lipids* **1996**, *31 Suppl*. S243-7.
14. Hibbeln, J.R.; Salem, N.J. *Am. J. Clin. Nutr.* **1995**, *62*, 1-9.

15. Stevens, L.J.; Zentall, S.S.; Deck, J.L.; Abate, M.L.; Watkins, B.A.; Lipp, S.R.; Burgess, J.R. *Am. J. Clin. Nutr.* **1995**, *62*, 761-768.
16. Wright, S.; Burton, J.L. *Lancet* **1982**, *2 (8308)*, 1120-1122.
17. Siess, W.; Scherer, B.; Bohlig, B.; Roth, P.; Kurzmann, I.; Weber, P.C. *Lancet* **1980**, *1 (8166)*, 441-444.
18. Smith, D.L.; Wills, A.L.; Nguyen, N.; Conner, D.; Zahedi, S.; Fulks, J. *Lipids* **1989**, *24*, 70-75.
19. Kinoshita, I.; Itoh, K.; Nishida-Nakai, M.; Hirota, H.; Otsuji, S.; Shibata, N. *Jpn. Circ. J.* **1994**, *58*, 903-912.
20. Leaf, A. *Prostaglandins Leukot Essent Fatty Acids* **1995**, *52(2-3)*, 197-198.
21. Hayashi, M.; Nasa, Y.; Tanonaka, K.; Sasaki, H.; Miyake, R.; Hayashi, J.; Takeo, S. *J. Mol. Cell. Cardiol.* **1995**, *27*, 2031-2041.
22. Voskuyl, R.A.; Vreugdenhil, M.; Kang, J.X.; Leaf, A. *Eur. J. Pharmacol.* **1998**, *341*, 145-152.
23. Vreugdenhil, M.; Bruehl, C.; Voskuyl, R.A.; Kang, J.X.; Leaf, A.; Wadman, W.J. *Proc. Natl. Acad. Sci. USA*, **1996**, *93*, 12559-12563.
24. Ramesha, C.S.; Paul, R.; Ganguly, J. *J. Nutr.* **1980**, *110*, 2149-2158.
25. Balasubramaniam, S.; Simons, A.; Chang, S.; Hickie, J.B. *J. Lipid Res.* **1985**, *26*, 684-689.
26. Lepage, G.; Roy, C.C. *J. Lipid Res.* **1984**, *25*, 1391-1396.
27. Moser, H.W.; Moser, A.B. Measurements of very long fatty acids in plasma. Techniques in diagnostic human biochemical genetics: A laboratory manual; Willey-Liss, Inc., **1991**; pp 177-191.
28. Kushak, R.I.; Drapeau, C.; Van Cott, E.M.; Winter, H.H. *JANA* **2000**, *2 (3)*, 59-65.
29. Kushak, R.I.; Drapeau, C.; Winter, H.H. Abstract of the *Fourth International Conference on Nutrition and Fitness: Plan of Action for the 21st Century*. Greece, May 25-29, **1999**.
30. Gerster, H. *Int. J. Vitam. Nutr. Res.* **1998**, *68*, 159-173.
31. Romay, C.; Armesto, J.; Remirez, D.; Gonzalez, R.; Ledon, N.; Garcia, I. *Inflamm. Res.* **1998**, *47(1)*, 36-41.
32. Gonzalez, R.; Rodriguez, S.; Romay, C.; Gonzalez, A.; Armesto, J.; Remirez, D.; Merino, N. *Pharmacol. Res.* **1999**, *39*, 1055-1059.
33. Romay, C.; Ledon, N.; Gonzalez, R. *Inflamm. Res.* **1998**, *47*, 334-338.
34. Romay, C.; Ledon, N.; Gonzalez, R. *J. Pharm. Pharmacol.* **1999**, *51*, 641-642.
35. Ikeda, I.; Wakamatsu, K.; Inayoshi, A.; Imaizumi, K.; Sugano, M.; Yazawa, K. *J. Nutr.* **1994**, *124*, 1898-1906.
36. Kris-Etherton, P.M.; Yu, S. *Am. J. Clin. Nutr.* **1997**, *65*, 1628S-1644S.
37. Lee, J.H.; Sugano, M.; Ide, T. *J. Nutr. Sci. Vitaminol.* (Tokyo) **1988**, *34*, 117-129.

38. Hirai, K.; Nakano, T.; Katayama, Y.; Amagase, S. *J. Nutr. Sci. Vitaminol.* (Tokyo) **1985**, *31*, 279-289.
39. Yu, S.; Derr, J.; Etherton, T.D.; Kris-Etherton, P.M. Plasma cholesterol-predictive equations demonstrate that stearic acid is neutral and monounsaturated fatty acids are hypocholesterolemic. *Am. J. Clin. Nutr.* **1995**, *61*, 1129-39.
40. Nicolosi, R.J.; Wilson, T.A.; Rogers, E.J.; Kritchevsky, D. *J. Lipid Res.* **1998**, *39*, 1972-1980.
41. Watts, G.F.; Jackson, P.; Burke, V.; Lewis, B. *Am. J. Clin. Nutr.* **1996**, *64*, 202-209.
42. Nagaya, T.; Nakaya, K.; Takahashi, A.; Yoshida, I.; Okamoto, Y. *Ann. Clin. Biochem.* **1994**, *31*, 240-244.
43. Salter, A.M.; Mangiapane, E.H.; Bennett, A.J.; Bruce, J.S.; Billett, M.A.; Anderton, K.L.; Marenah, C.B.; Lawson, N.; White, D.A. *Br. J. Nutr.* **1998**, *79*, 195-202.
44. Hassel, C.A.; Mensing, E.A.; Gallaher, D.D. *J. Nutr.* **1997**, *127*, 1148-1155.
45. Dashwood, R.; Guo, D. *Princess Takamatsu Symp.* **1995**, *23*,181-189.
46. Vlad, M.; Bordas, E.; Caseanu, E.; Uza, G.; Creteanu, E.; Polinicenco, C. *Biol. Trace Elem. Res.* **1995**, *48(1)*, 99-109.
47. Sano, T.; Kumamoto, Y.; Kamiya, N.; Okuda, M.; Tanaka, Y. *Artery* **1988**, *15*, 217-224.
48. Rolle, I.; Pabst, W. *Nutr. Metabol.* **1980**, *24(5)*, 291-301.
49. Hori, K.; Ishibashi, G.; Okita, T. *Plant Foods Hum. Nutr.* **1994**, *45(1)*, 63-70.
50. de Caire, G.Z.; de Cano, M.S.; de Mule, C.Z.; Steyerthal, N.; Piantanida, M. *Int. J. Exp. Bot.* **1995**, *57(1)*, 93-96.

Chapter 11

Seal Blubber Oil and Its Nutraceutical Products

Fereidoon Shahidi and Udaya N. Wanasundara[1]

Department of Biochemistry, Memorial University of Newfoundland, St. John's, Newfoundland A1B 3X9, Canada
[1]Current address: POS Pilot Plant Corporation, Saskatoon, Saskatchewan S7N 2R4, Canada

Seal blubber oil is an important source of omega-3 fatty acids, namely eicosapentaenoic acid (EPA), docosapentaenoic acid (DPA) and docosahexaenoic acid (DHA). The total content of omega-3 fatty acids in seal blubber oil is about 20% and the oil is characteristically different from fish oil in its high content of DPA (about 5%) and the dominance of the occurrence of long-chain polyunsaturated fatty acids (LC PUFA) in sn-1 and sn-3 positions as opposed to the sn-2 position in fish oil. The oil is now available in the capsule form and concentrates of it in the free fatty acid, alkyl ester or acylglycerol forms may be produced commercially. An overview of the topic is provided.

Harp seal (*Phoca groenlandica*) is a marine mammal found abundantly in the ice-cold waters off Newfoundland and Labrador. Because of their natural habitat, harp seals have unique biological characteristics which make them interesting as a potential source of food and nutraceutical products for humans (*1*).

Based on the current harvest of 275,000 animals per year, some 11 million kilograms of blubber is annually available for further processing (*1*). The blubber of seals is a rich source of long-chain omega-3 polyunsaturated fatty acids (PUFA) which have attracted much attention in recent years due to their beneficial health effects (2). The interest in marine oils stemmed from the observation of the diet of Greenland Eskimos in which fish as well as seal meat and blubber was important. The incidence of cardiovascular disease (CVD) in Eskimos was considerably lower than that of the Danish population, despite their high fat consumption (3-8).

The beneficial health effects of omega-3 PUFA have been attributed to their ability to lower serum triacylglycerol and cholesterol (2). In addition, omega-3 fatty acids are essential for normal growth and development of the brain and retina (9-12). They also play a role in the prevention and treatment of hypertension, arthritis, inflammatory and autoimmune disorders.

Unlike saturated and monounsaturated fatty acids which can be synthesized by all mammals, including humans, the omega-3 PUFA cannot be easily synthesized in the body and must be acquired through the diet. These omega-3 fatty acids are abundant in the oil obtained from the body of fatty fish species such as mackerel and herring, oil from the liver of lean fish such as cod and halibut, and the oil from the blubber of marine mammals such as seals, whales and walruses (1). The long-chain omega-3 fatty acids, namely eicosapentaenoic acid (EPA), docosahexaenoic acid (DHA) and, to a lesser extent, docosapentaenoic acid (DPA) in marine oils originate from unicellular sea algae and phytoplanktons and eventually pass through the food web and become incorporated into the body of fish and higher marine species (13).

Seal blubber oil has been used traditionally as an industrial oil or in the hydrogenated form for incorporation into margarine. However, recent research findings have shown its potential for application in foods and nutraceuticals; a summary of which is provided in this review.

Compositional Characteristics of Seal Blubber Oil (SBO)

Seal oil may be obtained from the blubber of the harvested animals via rendering using steam injection or a low temperature process. The resultant oil consists of over 98% neutral components and small amounts of polar lipids (1). The oil is rich in monounsaturated fatty acids and contains approximately 20% long-chain omega-3 PUFA consisting mainly of DHA, EPA and DPA in a decreasing order (2). The fatty acid profile of SBO is provided in Table I.

The distribution of fatty acids in triacyglycerols of SBO is different from that of fish oils (14). The PUFA in SBO occur mainly in the sn-1 and sn-3 positions while they are present mainly in the sn-2 and sn-3 position in fish oils (see Table II). This difference in positional distribution of fatty acids in SBO is responsible for existing differences in their absorption and assimilation into the body.

Table I. Major Fatty Acids of seal blubber oil.

Fatty Acid	Weight %
14:0	3.7
14:1	1.1
16:0	6.0
16:1	18.0
18:1ω9	20.8
18:1ω11	5.2
18:2ω6	1.5
20:1ω9	12.2
20:5ω3	6.4
22:1ω11	2.0
22:5ω3	4.7
22:6ω3	7.6

Table II. Fatty acid distribution in different position of triacylglycerols of seal blubber oil.[1]

Fatty acid	sn-1	sn-2	sn-3
Saturates	6.34	25.56	4.32
Monounsaturates	62.91	65.98	51.09
Polyunsaturates	27.60	7.27	43.23
EPA	8.36	1.60	11.21
DPA	3.99	0.79	8.21
DHA	10.52	2.27	17.91
Total Omega-3	25.65	5.56	38.87

[1]EPA, eicosapentaenoic acid; DPA, docosapentaenoic acid; and DHA, docosahexaenoic acid.

In addition to its major components, seal blubber oil contains small amounts of squalene as well as α-tocopherol (*14*). The content of squalene is raw seal blubber oil was 0.59% and decreased to about 0.28% in refined-bleached and deodorized (RBD) oil. Meanwhile α-tocopherol was present in approximately 24-32 ppm in the rendered and processed oils from seal blubber.

Application of Seal Blubber Oil in Food and Nutraceuticals

Seal blubber oil has traditionally been used in the treatment of leather following the tanning process or has been hydrogenated and used in production of margarines. However, appreciation of the nutritional value and health benefits of marine oils has resulted in the innovative use of SBO in a number of foods. Thus, SBO may be used in fabricated seafoods, bread, crackers and pasta as well as dairy products, soups, salad dressings, mayonnaise and snack bars. In addition, SBO may be used in infant formulas and attempts have been made to employ it in dermaceuticals to alleviate skin disorders. The oil may be used in the encapsulated or microencapsulated form (*15*) or consumed as a nutraceutical. Thus, SBO capsules may be used by pregnant and lactating women as well as patients with arthritis, CVD and related ailments (*1*). However, individuals with bleeding problems and those on blood thinners must consult their physicians prior to consuming marine oil supplements.

For pharmaceutical applications, however, seal blubber oil needs to be processed in order to obtain omega-3 concentrates. Omega-3 concentrates are used in many countries around the world as non-prescription over-the-counter (OTC) and sometimes as prescription drugs for applications mainly related to CVD (*13*). Methodologies for production of omega-3 concentrates are diverse and afford products with different levels of omega-3 fatty acids and in different forms.

Omega-3 Concentrates from Seal Blubber Oil as Pharmaceuticals

Omega-3 concentrates from seal blubber oil may be obtained using the same methodologies that are applicable to fish oils. Thus, the resultant concentrates may be in the form of free fatty acids, their methyl or ethyl esters, or in the form of acylglycerols (including triacylglycerols). Methodologies that may be used include:

- Urea complexation
- Low temperature crystallization
- Enzymatic hydrolysis
- Enzymatic esterification with glycerol
- Distillation
- Chromatography
- Supercritical fluid extraction

Each of the methodologies listed above has its own advantages and disadvantages as well as operational costs and effectiveness (*13*). Thus, the method of choice would depend on a variety of factors, but generally dictated by the final use of product and its market value.

Urea complexation

This is a well established methodology for elimination of saturated and monounsaturated fatty acids from hydrolyzed oils (*16*). Hydrolysis of the oil is achieved using alcoholic sodium or potassium hydroxide. The free fatty acids so formed are then subjected to complexation in an ethanolic solution of urea. Urea usually crystallizes in a tetragonal pattern with a cavity of ca 5.67Å, but in the presence of long-chain alkyl chains, it crystallizes in a hexagonal pattern with a cavity large enough (about 8-12Å) to accommodate the hydrocarbon chain of saturated and monounsaturated fatty acids (*16*). This fraction is referred to as urea complexing fraction (UCF). The polyunsaturated fatty acids remain in the liquid and are referred to as non-urea complexing fraction (NUCF). However, short-chain saturated fatty acids may not complex with urea (*13*). Table III shows the fatty acid composition of NUCF of seal blubber oil. The condition for complexation with urea as well as the number of unit operations may be manipulated to obtain a high concentration of a desired long-chain omega-3 fatty acid.

Table III. Content of Omega-3 fatty acids in urea complexing fraction (UCF) and non-urea complexing fraction (NUCF) of seal blubber oil.[1]

Fatty acids	*UCF*	*NUCF*
EPA	6.77	10.9
DPA	3.92	2.38
DHA	5.27	67.6
Total omega-3	17.0	88.2

[1]EPA, eicosapentaenoic acid; DPA, docosapentaenoic acid; and DHA, docosahexaenoic acid.

Low temperature crystallization

The melting point of fatty acids depends mainly on their chain length and degree of unsaturation. In general, as the chain length increases, there is an increase in the melting point of fatty acids (*17*). However, for longer chain fatty acids, unsaturation results in a decrease in the melting point of oils. Therefore, fractional crystallization may provide a means by which highly unsaturated fatty acids (HUFA) could be separated from the rest of the fatty acids. Although neat fatty acids might be used, generally organic solvents are employed in order to facilitate separation and concentration of HUFA from marine oils (*18*). Among solvents that are generally used for this purpose, hexane, acetone and possibly acetonitrile may be considered as being most suitable.

Enzymatic Hydrolysis

It is well known that PUFA in triacylglycerol molecules are generally resistant to in-vitro lipolysis by pancreatic enzymes (*19*). This might be caused by the inhibitory effect for the approach of enzyme to the ester group because of proximity of the double bonds. Microbial lipases have also been found to discriminate against PUFA in both hydrolysis and esterification reactions (*20*). Therefore, it is possible to concentrate omega-3 fatty acids of seal blubber oil (*2*). Amongst enzymes tested, Candida cylindracea was most effective for production of omega-3 concentrates. Impregnated silica gel column may be used to further separate EPA and DHA from each other.

Enzymatic Esterification with Glycerol

If desired, omega-3 fatty acids obtained via urea complexation and/or other concentration techniques may be reacted with glycerol under enzymic or chemical reaction conditions in order to produce concentrates in the acylglycerol form (*21*). The acylglycerol omega-3 concentrates are preferred over free fatty acids and their methyl/ethyl esters (*13,16*). While, triacylglycerols could be prepared, presence of partial acylglycerols cannot be easily avoided. A number of enzymes are known to catalyze the formation of acylglycerols (*21*), however, methoxide-assisted formation of acylglycerols from the reaction of methyl and ethyl esters with glycerol may also be practiced. However, in the latter methodology one must be aware of the possibility of formation of oxidation products, trans isomerization and formation of conjugated products.

Distillation

This method may be used to preferentially remove certain fatty acids and/or their alkyl esters from mixtures (*13*). The fatty acids and/or their alkyl esters generally have different boiling points and if subjected to heat under reduced pressure would distill off according to factors such as chain length and degree of unsaturation (*13*). Short-path and molecular distillation may be used for this purpose in order to achieve reasonable separations at lower temperatures. This methodology has gained popularity in recent years by the industry due to its simplicity and its minimum environmental effect as there were no chemicals required for it. In such cases, use of a nitrogen blanket would be essential in order to deter hydrolysis, thermal oxidation, isomerization and polymerization of long-chain PUFA.

Chromatography

Similar to other methods described above, fatty acids may be separated chromatographically according to their chain length and degree of unsaturation (*22*). High performance liquid chromatography and silver resin chromatography have been described as possible techniques for preparation of omega-3 concentrates. In addition, silver nitrate impregnated silica gel column has been used to separate EPA and DHA from squid-liver oil fatty acid methyl esters (*23*). Isocractic elution from silver resin column also led to the enrichment of the omega-3 content of concentrates from 76.5 to 99.8% (*24*). However, practical problems associated with this approach as related to scale up, extensive use of organic solvents and possible loss of resolution of the column upon repeated use have not allowed its commercial application.

Supercritical fluid extraction (SFE)

This is a relatively new separation process that may be used for concentration of polyunsaturated fatty acids. The separation of PUFA by SFE is dependent on the molecular size of the components involved rather than their degree of unsaturation (*13*). Therefore, a prior concentration step may be required if attaining a high concentration of PUFA is essential (*25*). The use of SFE for extraction of oil and concentration of ω3-PUFA from fish oil and seaweed has been reported (*25-27*). SFE has also been used to remove cholesterol and polychlorinated biphenyls from fish oils (*28*).

Protection of Seal Blubber Oil and its Products from Oxidation

Marine oils, including seal blubber oil, contain highly unsaturated fatty acids with five and six double bonds and thus are prone to rapid oxidation. Protection of SBO from oxidation may be achieved using synthetic as well as natural antioxidants (*29*). In addition, encapsulation/ microencapsulation of the oil might be used to prolong the shelflife of products (*30*). In general, oils intended for use in foods might be protected from oxidation only for a limited period of time as most products of oxidation of PUFA have low threshold values and could be easily perceived at ppm or even ppb levels. Among synthetic antioxidants, tert-

Table IV. Inhibition of oxidation of seal blubber oil by antioxidants under accelerated storage condition.

Antioxidant (ppm)	Seal blubber oil
α-Tocopherol (200)	14.2
BHA (200)	23.0
BHT (200)	35.5
TBHQ (200)	56.3
DGTE (200)	32.9
DGTE (1000)	47.1
EC (200)	39.5
EGC (200)	40.5
ECG (200)	58.6
EGCG (200)	50.0

Abbreviations are: BHA, butylated hydroxyanisole; BHT, butylated hydroxytoluene; TBHQ, tert-butyl ydroquinone; DGTE, dechlorophyllized green tea extract; EC, epicatechin; EGC, epigallocatechin; ECG, epicatechin gallate; and EGCG, epigallocatechin gallate.

butylhydroquinone was found to be most effective and its effect was close to that of some of the catechins found in green tea (29). However, results presented in Table IV are only for bulk oil. In emulsion system, the stability of the oil depends on many factors. Effectiveness of antioxidants in emulsion systems is also dictated by the hydrophobic/hydrophilic balance of the compounds involved.

In ω3 concentrates, stabilization of products is even more crucial because of their high content of highly unsaturated fatty acids. In addition, the chemical nature of the concentrate, that is free fatty acid, simple alkyl ester or triacylglycerols form, might influence the stability of the products.

Conclusions

Seal blubber oil might be used as a functional food ingredient or as a nutraceutical product. These products should be consumed within a reasonable period of time. Stabilization of the oil is also essential in order to minimize the formation of flavor active lipid oxidation products. Concentrated ω3 PUFA might also be prepared using different techniques. These concentrates are often used as prescription or over-the-counter (OTC) drugs.

References

1. Shahidi, F. In *Seal Fishery and Product Development*. F. Shahidi, Ed. ScienceTech. Publishing Co., St. John's, NF, pp. 99-146.
2. Wanasundara, U.N.; Shahidi, F. *J. Am. Oil Chem. Soc.* **1998**, *75*, 945-951.
3. Beare-Rogers, *J. Am. Oil Chem. Soc.* **1988**, *65*, 91-94.
4. Bang, H.O.; Bang, J. *Lancet* **1956**, *1*, 381-383.
5. Dyerberg, J.; Bang, H.O.; Hjorne, N. *Am. J. Clin. Nutr.* **1975**, *28*, 958-966.
6. Bang, H.O.; Dyerberg, J.; Hjorne, N. *Acta Med. Scand.* **1976**, *200*, 69-73.
7. Dyerberg, J.; Bang, H.O.; Stofferson, E. *Lancet* **1978**, *2*, 117-119.
8. Dyerberg, J.; Bang, H.O. *Lancet* **1979**, *2*, 433-435.
9. Birch, D.G.; Birch, E.E.; Hoffman, D.R.; Uauy, R. *Invest. Opthamol. Vis. Sci.* **1995**, *32*, 3224-3232.
10. Uauy, R.; Birch, D.G.; Birch, E.E.; Tyson, J.E.; Hoffman, D.R. *Pediatr. Res.* **1990**, *28*, 484-492.
11. Lucas, A.; Morley, R. *Lancet* **1992**, *339*, 261-264.
12. Shahidi, F.; Wanasundara, U.N. *Trends Food Sci. Technol.* **1998**, *9*, 230-240.
13. Shahidi, F.; Wanasundary, U.N. *J. Food Lipids* **1999**, *6*, 159-172.
14. Wanasundara, U.N.; Shahidi, F. *J. Food Lipids* **1997**, *4*, 51-64.
15. Wanasundara, U.N.; Shahidi, F. *Food Chem.* **1999**, *65*, 683-688.
16. Senanayake, S.P.J.N.; Shahidi, F. *J. Food Lipids* **2000**, *7*, 51-61.
17. Wanasundara, U.N. In *Marine Oils: Stabilization, Structural Characterization and Omega-3 Fatty Acid Concentration*. Ph.D. Thesis, Memorial University of Newfoundland, St. John's, NF, Canada.
18. Yokochi, T.; Usita, M.T.; Kamisaka, Y.; Nakahara, T.; Suzuki, O. *J. Am. Oil Chem. Soc.* **1990**, *67*, 846-851.
19. Bottino, N.R.; Vandenburg, G.A.; Reiser, R. *Lipids* **1967**, *2*, 489-493.
20. Mukherjee, K.D.; Kiewitt, I.; Hills, M.J. *Appl. Microbiol. Biotechnol.* **1993**, *40*, 489-493.
21. Mishra, V.K.; Temelli, F.; Ooraikul, B. *Food Res. Int.* **1993**, *26*, 217-226.
22. Yamagouchi, K.; Murakami, W.; Nakano, H., Konosu, S.; Kokura, T.; Yamamoto, H.; Kosaka, M.; Hata, K. *J. Agric. Food Chem.* **1986**, *34*, 904-907.
23. Choi, K.J.; Nakhost, Z.; Krukonis, V.J.; Karel, M. *Food Biotechnol.* **1987**, *1*, 263-271.
24. Rizvi, S.S.H.; Daniels, L.A.; Bernado, A.L.; Zollweg, L.A. *Food Technol.* **1986**, *40*, 57-64.
25. He, Y.; Shahidi, F. *J. Am. Oil Chem. Soc.* **1997**, *74*, 1133-1136.
26. Brown, L.B.; Kolb, D.X. *Prog. Chem. Fats Lipids* **1955**, *3*, 57-94.
27. Teshima, S.; Kanazawa, A.; Tokiwa, S. *Bull. J. Soc. Sci. Fish.* **1978**, *44*, 927.
28. Adlof, R.O.; Emiken, E.A. *J. Am. Oil Chem. Soc.* **1985**, *62*, 1592-1595.
29. Wanasundara, U.N.; Shahidi, F. *J. Am. Oil Chem. Soc.* **1995**, *73*, 1183-1190.
30. Wanasundara, U.N.; Shahidi, F. *J. Food Lipids* **1995**, *2*, 73-86.

Chapter 12

Structured Lipids Containing Omega-3 Highly Unsaturated Fatty Acids

Casimir C. Akoh

Department of Food Science and Technology, The University of Georgia, Athens, GA 30602–7610

Structured lipids (SL) containing n-3 highly unsaturated fatty acids (n-3 HUFA) were produced with immobilized sn-1,3 specific and nonspecific lipases as biocatalysts. HUFA such as eicosapentaenoic (EPA, 20:5n-3), docosahexaenoic (DHA, 22:6n-3), linolenic (18:3n-3) and gamma linolenic (GLA, 18:3n-6) acids are important in foods, nutrition, and pharmaceutical applications. SL containing these fatty acids and medium chain fatty acids (MCFA) may be desirable as "nutraceuticals" for supplementation in infant formula or as food supplement for adults to enhance overall health. For the most part, the position of the HUFA in the glycerol moiety is key to their functionality in foods and absorption when consumed. Perhaps, these designer lipids may replace conventional fats and oils in certain specialty applications because of their structure-health (nutraceutical or medical lipids) and structure-function (functional lipids) attributes. In most cases insertion of the desired HUFA at the sn-2 position will provide maximum nutritional benefits.

Structured lipids (SL) are triacylglycerols (TAG) that has been modified or synthesized to contain short or medium chain and long chain fatty acids (LCFA) attached preferably to the same glycerol backbone for maximum benefits and functionality. The short (SCFA) and medium chain fatty acids (MCFA) are metabolized differently than LCFA and have been used as a source or rapid energy for preterm infants and patients with lipid malabsorption-related diseases (*1, 2*). SCFA (C2:0–C6:0) and MCFA (C6:0–C12:0) are transported through the portal system while LCFA (C14:0–C24:1) are transported through the lymphatic system. They are synthesized enzymatically or chemically by direct esterification, acidolysis, or interesterification as reviewed recently (*1, 3-5*), with the later two processes being the preferred route. To achieve a desired structural configuration, the enzymatic process with stereoselective lipases is often employed. The products of enzymatic process are likely to be more desirable than the chemically produced SL. Depending on the desired molecules and use, the potent fatty acids can be selected from the available fats and oils, and used in the preparation of the SL. A proper balance of the n-6/n-3 fatty acids is important for maintaining certain physiological functions and good health. In designing SL for certain applications, efforts should be made to maintain a good balance of the n-6/n-3 ratio.

It is well known that approximately 30% of brain glycerophospholipids are composed of docosahexaenoic acid (DHA), preterm infants require diets rich in DHA or n-3 HUFA, and sometimes supplementation is needed especially for infants fed formula diets instead of mother's breast milk. SL containing DHA and/or eicosapentaenoic acid (EPA) may be the best way to deliver these fatty acids. Their potency will be enhanced if DHA and /or EPA are located at the sn-2 position of the SL where they are easily metabolized. DHA is also known to improve the visual acuity of preterm infants. Supplementation of DHA and arachidonic acid (20:4n-6) also improves the visual outcome in term infants (6, 7). Other reasons to supplement SL of HUFA in infant formula will be to replenish the level of DHA in mother's breast milk, which depletes with increase in the number of pregnancy events. Direct consumption of preformed DHA is 45 times better than dietary 18:3n-3 found in the breast milk or other vegetable oils. DHA and EPA-rich oils and SL can decrease the level of prostaglandin-2 (PGE2), thromboxane–2 (TBX2), tumor necrosis factor (TNF), and interleukin-2 (IL-2) in the serum and may boost the immune system (4, 8).

Supplementation of SL rich in HUFA in foods will require stabilization with antioxidants and proper packaging to prevent oxidation, off flavor, and deterioration. Tertiarybutylhydroxyquinone (TBHQ), ascorbyl palmitate, lecithin, alpha-tocopherol are excellent antioxidants for stabilization of fish oils and SL containing DHA, EPA, and other highly unsaturated fatty acids.

Various food products were enriched with n-3 HUFA to raise the content of n-3 fatty acids in the diet without affecting the taste of the food significantly (9). Fish oil containing 30% combined EPA and DHA and microencapsulated powdered form (10% EPA and DHA) was used to design functional foods or "nutraceuticals" for the prevention of many diseases. Bread, pasta, milk, and salad oils can be enriched with SL containing HUFA. Fish oil consumption can alleviate cardiovascular diseases and certain kinds of cancer. It is commonly recommended that we increase our fish consumption to 200-300 g/week or even increase the EPA and DHA consumption to 500-700 mg/day (9). For adults, incorporation of EPA, DHA, or GLA into the sn-2 position of the SL will provide health benefits. Infant formula should target incorporation of linoleic, arachidonic, and palmitic acids, as well as DHA at the sn-2 position of SL for maximum absorption and nutraceutical benefits. Having SCFA and MCFA at the sn-1,3 positions of the SL will accelerate the absorption of the HUFA at the sn-2 position.

Types of Reactions Catalyzed by Lipases

Lipases catalyze ester hydrolysis, direct esterification and transesterification reactions. Acidolysis, alcoholysis, interesterification, and aminolysis are types of transesterification reactions. Acidolysis is the exchange of acyl groups between an acid and an ester; alcoholysis is acyl group exchange between an alcohol and an ester; interesterification is the exchange of two acyl groups; and aminolysis is the exchange of an acyl group between an acid and an amine. Aminolysis is not used in SL synthesis.

Acidolysis in the Production of Structured Lipids. Acidolysis can be used to incorporate different fatty acids into triacylglycerols producing structured lipids. The free acids of the n-3 HUFA, EPA and DHA being incorporated into vegetable oils and fish oils to increase their nutritional value are examples (*3, 5, 10, 11*).

For general small scale synthesis of modified fish oil, 100 mg of fish oil was mixed with 36 mg of capric acid (C10:0). The molar ratio was 1:2 fish oil TAG to C10:0 free fatty acid (FFA) in 3 ml of hexane for reactions conducted in organic solvent. The reaction was also performed in the complete absence of organic solvent. Immobilized IM60 lipase (10% by weight reactants) were added to the mixture and incubated in an orbital shaking water bath at 55 °C for 24 h at 200 rpm.

Structured Lipids from Fish Oil

EPA from fish oil can reduce the level of very-low-density lipoprotein (VLDL) and low-density lipoprotein (LDL) in humans (*12, 13*). EPA is used for the synthesis of eicosanoids such as series-3 prostaglandins. They compete with arachidonic acid (AA) as substrates for cyclooxygenase and lipoxygenase to produce different series of eicosanoids. EPA can increase the levels of leukotriene (LTB5), IL-2, and high-density lipoprotein (HDL) while decreasing the levels of LTB4, IL-1, and AA (*4*). We modified fish oil with C10:0 to produce SL in an acidolysis reaction catalyzed by an immobilized sn-1,3 specific *Rhizomucor miehei* lipase, IM60 (Table I). The SL products contained 25.0 and 22.6 mol% EPA and DHA, respectively, and 40.8 mol% capric acid. This type of SL can be used to deliver EPA and DHA in the diet. Nutrition experts consider a level of 2-5% of n-3 HUFA in SL as optimum for enhanced immune function (*2*). Table II shows the use of a different substrate, tricaprylin for the production of SL. In this case, n-3 HUFA (free fatty acids) was the acyl donor for the acidolysis reaction.

Tricaprylin was transesterified with EPA-rich fatty acids (EPAX 6000) at 1:2 molar ratio in haxane catalyzed by a nonspecific immobilized SP435 lipase from *Candida antarctica*. A diet containing this SL was compared with a diet containing soybean oil (fat at 17% by weight of diet) in female mice for 21 days. The SL diet resulted in the reduction of total cholesterol by 49%, LDL cholesterol by 35.4%, and triacylglycerol by 53.2% (*8*).

Table I. Enzymatically Modified Fish Oil with Capric Acid to Produce Structured Lipids

Reaction	Major Fatty Acids	Before (mol%)	SL (mol%)
IM60 lipase, 1,3-specific	10:0	-	40.8
	20:5n-3	40.9	25.0
	22:6n-3	33.0	22.6

Table II. Modification of Tricaprylin with EPAX 6000 Free Fatty Acids to Produce Structured Lipids

	Mol%		
Major Fatty Acids	EPAX 6000	SL	Sn-2 Position
8:0	-	46.9	64.3
20:5n-3	33.8	23.2	17.8
22:6n-3	26.0	21.7	15.0

NOTE: Molar ratio of tricaprylin to EPAX 6000 = 1:2. EPAX 6000 contained 33.8% EPA and 26.0% DHA.

Structured Lipids Containing Gamma Linolenic Acid and n-3 HUFA

The nutritional value of fats and oils and their physicochemical properties are determined by the type of fatty acids esterified to the glycerol backbone and the positional distribution of the acyl groups within the glycerol. Gamma linolenic acid (GLA, 18:3n-6) is found in many plant seed oils such as evening primrose (7-10 g/100 g GLA), borage (18-26 g/100 g GLA), and blackcurrant oils (15-20 g/100 g GLA), fungal oil (23-26 g/100 g GLA), and human milk (16, 19). Dietary γ-linolenic acid may have "nutraceutical" property in addition to its use in curing skin-related and other diseases (14, 15). Dietary GLA can increase the content of its elongation product, dihomo-γ-linolenic acid (DGLA) in cell membranes and often without changing the content of AA (16). We recently ezymatically synthesized SL containing capric acid, GLA and EPA in the same glycerol backbone using borage oil as the main substrate (17). Both IM60 and SP435 lipases were used as biocatalysts with free capric and eicosapentaenoic acids as acyl donors. High incorporation of EPA (10.2%) and capric acid (26.3%) was obtained with IM60 lipase, compared to 8.8 and 15.5%, respectively, with SP435 lipase (*17*). However, only SP435 was able to incorporate both capric and EPA at the sn-2 position (Table III). Approximately 14.5% of GLA were retained at the 2-position. The modified borage oil may be useful in the treatment of certain clinical disorders, which at present involves use of individual sources of GLA, EPA, and medium chain fatty acids (MCFA), or physical mixtures of their triacylglycerols. These structured lipids may help ameliorate inflammatory response and modulate eicosanoid biosynthesis. The MCFA will be a good source of easily metabolizable fat for quick energy. The presence of EPA and/or GLA at the sn-2 position of the SL will ensure that absorption is enhanced as 2-monoacylglycerols (2-MAG) rather than used as energy substrate. We previously modified borage oil to incorporate n-3 HUFA (*18*).

Table III. Enzymatic Modification of Borage Oil with Capric Acid to Produce Structured Lipids

Reaction	Major Fatty Acids	Before (mol%)	SL (mol%)
IM60 Lipase	10:0	-	31.6
Ratio of borage oil:	18:2n-6	38.6	32.3
10:0:20:5, 1:3:3	18:3n-6	20.1	14.0
	20:5n-3	-	11.2
Sn-2 Position			
SP435 Lipase reaction	10:0	-	7.5
Molar ratio of reactants,	18:3n-6	20.2	14.5
1:2:2	20:5n-3	-	5.2

Other Types of Structurally Modified Lipids Containing n-3 HUFA

Structurally modified lipids may not always contain short and/or medium chain fatty acids. Indeed, the native state of fat or oil can be changed by removing or introducing new fatty acids for improved nutrition and physicochemical properties. Various vegetable oils have been modified to incorporate EPA and DHA. We used *Mucor miehei* (Tables IV) and *Candida antarctica* (Tables V) lipases to add EPA and DHA to soybean and canola oils containing 18:3n-3 (*10*). The ethyl esters of EPA and DHA or their free acids were used as acyl donors for the interesterification and acidolysis reactions, respectively, at a molar ratio of soybean oil to acyl donor of 1:2. The nonspecific Candida lipase gave better incorporation of EPA (34.7%) and DHA (32.9%) when the ethyl esters were used than the Mucor lipase which gave 29.2 and 14.6%, respectively. We were able to incorporate these n-3 HUFA into other vegetable oils such as canola, peanut, melon seed, and hydrogenated soybean (*10, 20*).

Table IV. Fatty Acid Composition (mol%) of Linolenic Acid Containing Vegetable Oils Modified by *Rhizomucor miehei* Lipase to Contain n-3 HUFA

Fatty Acids	Canola Oil			Soybean Oil		
	Before	After		Before	After	
		EEPA	EDHA		EEPA	EDHA
14:0	0.1	0.3	0.4	0.2	0.6	0.7
14:1	-	0.3	0.5	0.1	1.1	1.3
16:0	5.2	3.3	4.0	11.6	9.6	12.1
16:1n-7	0.3	0.3	0.3	0.1	0.4	0.2
18:0	1.7	1.1	1.5	3.5	2.1	2.9
18:1n-9	55.2	37.1	45.7	21.9	15.5	19.0
18:2n-6	23.6	15.5	18.5	53.9	36.8	43.5
18:3n-6	1.1	0.6	0.8	0.3	-	-
18:3n-3	11.4	6.6	7.6	8.4	5.1	-
20:5n-3	-	32.9	-	-	29.2	-
22:6n-3	-	-	18.1	-	-	14.6

NOTE: Molar ratio of EPA or DHA ethyl ester: vegetable oil = 2. (Adapted from reference 10. Copyright 1994 American Oil Chemists' Society).

In another study (*21*), we were successful in incorporating EPA, DHA, or both, into pure trilinolein using the above lipases (Table VI). A 1:2 molar ratio of trilinolein to EPA and DHA ethyl ester was interesterified by *C. antarctica* or *M. miehei* lipase in hexane. The n-3/n-6 ratio was increased to 0.6 after reacting a 1:1:1 molar ratio of EPA:DHA:trilinolein. Very high incorporations of EPA and DHA were obtained in these experiments. This type of SL can be designed to alter the n-3/n-6 ratio of the diet, and thus control the synthesis of various eicosanoids. It is known that excess or imbalance of n-6 HUFA can lead to arachidonic acid production which favors thrombotic events mediated through thromboxane A2 (TXA2) such as vasoconstriction and platelet aggregation. However, consumption of n-3 HUFA or SL containing n-3 HUFA may tip the physiological balance toward production of eicosanoids that will promote vasodilation and antiaggregation of platelets. In general, n-3 HUFA are competitors of arachidonic acid cascade and exert their action by modulating eicosanoid production.

Stability

The stability of n-3 HUFA is a big issue as far as adding SL containing them into food products. We found from our work that the SL containing n-3 HUFA are very susceptible to oxidation probably because they lost some of the protective natural antioxidants present in the fish oil used for the synthesis (*22*).

Table V. Fatty Acid Composition (mol%) of Linolenic Acid Containing Vegetable Oils Modified by *Candida antarctica* Lipase to Contain n-3 HUFA

Fatty Acids	Canola Oil			Soybean Oil		
	Before	After		Before	After	
		EEPA	EDHA		EEPA	EDHA
14:0	0.1	-	0.2	0.2	0.2	0.1
14:1	-	-	0.3	0.1	0.1	0.1
16:0	5.2	3.1	3.3	11.6	6.9	7.3
16:1n-7	0.3	0.3	0.2	0.1	0.2	0.1
18:0	1.7	1.2	1.2	3.5	2.5	2.6
18:1n-9	55.2	43.0	39.3	21.9	15.7	16.2
18:2n-6	23.6	15.6	15.8	53.9	35.1	36.0
18:3n-6	1.1	0.6	0.7	0.3	-	-
18:3n-3	11.4	6.6	6.8	8.4	5.0	5.0
20:5n-3	-	29.5	-	-	34.7	-
22:6n-3	-	-	29.0	-	-	32.9

NOTE: Molar ratio of EPA or DHA ethyl ester: vegetable oil = 2. (Adapted from reference 10. Copyright 1994 American Oil Chemists' Society).

However, addition of appropriate antioxidants will restore some oxidative stability to the SL. These antioxidants work better when two or more of them are added at 100-200 ppm levels. We studied the effects of food grade α-tocopherol (TOC, 95% pure) and tert-butylhydroxyquinone (TBHQ) alone and in combination on the oxidative stability of SL from fish and canola oils.

SL (2 kg each) from canola and fish oils were synthesized in a packed bed bioreactor in our laboratory. SL of fish and canola oils containing caprylic acid (C8:0) were produced by acidolysis using immobilized Lipozyme IM from *Rhizomucor miehei* (Novo Nordisk A/S, Denmark). The SL products were purified using a KDL-4 short-path distillation system (UIC Inc., Joliet, IL) with 0.04 m^2 heated evaporation surface, 500 ml 2-neck receivers with valve adapters for semi-continuous operation as we previously reported (*23*). Briefly, SL was passed through short-path distillation twice to obtain FFA (as oleic acid) concentration less than 1%. They were further purified at 185 °C for canola oil and 170 °C for fish oil. FFA was removed from SL under a vacuum of 0.2 mTorr. The flow rate was 500 ml/h. SL samples were cooled at room temperature and placed in amber bottles (500 ml), then flushed with nitrogen and stored at –96 °C for further analysis.

Table VI. Fatty Acid Composition (mol%) of Enzymatically Modified Trilinolein to Contain n-3 HUFA

Acyl Donor	Fatty Acid	IM60 Lipase	SP435 Lipase
EPA Ethyl ester	18:2n-6	65.1	65.5
	20:5n-3	34.9	34.5
	n-3/n-6 ratio	0.5	0.5
DHA Ethyl ester	18:2n-6	81.0	73.2
	22:6n-3	19.0	26.8
	n-3/n-6 ratio	0.2	0.4
EPA and DHA Ethyl esters	18:2n-6	-	64.5
	20:5n-3	-	20.2
	22:6n-3	-	15.3
	n-3/n-6 ratio	-	0.6

NOTE: Molar ratio of EPA or DHA ethyl ester to trilinolein = 2:1; and molar ratio of EPA ethyl ester: DHA ethyl ester:trilinolein = 1:1:1. (Adapted from reference 21. Copyright 1995 American Oil Chemists' Society).

The Oxidative Stability Index (OSI) of structured lipids treated with antioxidants were determined with an Oxidative Stability Instrument manufactured by Omnion (Rockland, MA) using the AOCS method Cd12b-92 (24) at 110 °C for canola oil and at 80 °C for fish oil SL. Additional treatments included controls (SL) of canola and fish oils containing no antioxidants. Samples of unmodified fish and canola oils were prepared under the same conditions for comparison purposes. All tests were performed in duplicate and average values reported.

Although the unmodified or fresh fish and canola oils used in this study were of good initial quality, important physicochemical changes had occurred after acidolysis and downstream processing of the SL. There was a significant difference between unmodified oils and SL. The OSI values of SL are reported in Table VII. Unmodified samples were more stable than controls (SL). For example, the unmodified canola oil had an OSI value of 9.65 h against 3.50 h for SL. In addition to the fatty acid composition of triacylglycerols, other factors such as the loss of non-triacylglycerol components (tocopherols and phospholipids), during short-path distillation, may have contributed to the low stability of SL (25). Table VII also shows the effect of TOC and TBHQ, and their combination (TBHQ/TOC) on the OSI time of SL. TBHQ alone at 100 and 200 ppm, increased the induction period of SL from canola oil from 3.50 h (SL) to 4.25 h (SL1) and 5.52 h (SL2), respectively, whereas OSI values of TOC were 3.72 h (SL3) and 3.90 h (SL4). However, for fish oil, neither TBHQ nor TOC at 100 ppm concentration improved the oxidative stability. This could be due to the fact that they were too low to compensate for the loss caused by the distillation. Mixtures of TBHQ and TOC (SL5, SL6), at concentrations of 100 and 200 ppm proved to be effective against autoxidation of SL. In both cases, the addition of antioxidants significantly improved the OSI values with TBHQ at 200 ppm being the most effective.

Table VII. Effects of Food Grade Antioxidants on the Oxidative Stability Index (OSI) Values (h) of Enzymatically Synthesized Structured Lipids Containing HUFA

Sample	Canola Oil (OSI at 110 °C)	Fish Oil (OSI at 80 °C)
Unmodified	9.65	14.33
Control SL	3.50	1.99
SL1 (100 ppm TBHQ)	4.25	1.99
SL2 (200 ppm TBHQ)	5.52	8.36
SL3 (100 ppm TOC)	3.72	1.99
SL4 (200 ppm TOC)	3.90	3.98
SL5 (50 TBHQ ppm +50 ppm TOC)	4.10	2.31
SL6 (100 ppm TBHQ +100 ppm TOC)	4.57	3.58

NOTE: TBHQ = tert-butylhydroxyquinone and TOC = α-tocopherol. (Unpublished results of Akoh and Muossata, 1999).

Conclusions

Use of SL containing HUFA in food formulations as nutraceuticals or medical lipids must consider the following issues: consumer acceptance (Europe and Japan are adding n-3 HUFA to bread and margarine) (26), target market, regulatory concerns (U.S. has no separate regulations while Japan has well established guidelines and regulations for nutraceuticals), the purported health benefits must be proven by diligent research and clinical trials, scale-up of production with enzymes and cost, must address the oxidation problem associated with highly unsaturated fatty acids by adding the appropriate antioxidants and using the best packaging to retard light penetration to the product, and sensory issues. Because of the relatively high HUFA content of the SL produced and increased risk of stability, the need to stabilize them before nutritional applications cannot be overemphasized. Therefore, inhibition of oxidation would be a major criterion when SL is to be incorporated into food products.

Enrichment or supplementation of existing food products with acceptable amounts of SL will ensure that the general public consumes enough n-3 HUFA for it to be beneficial and effective. However, in parental, enteral nutrition and some other specialty needs, the dose has to be determined. For structure–function applications in food formulations, these SL can be used to replace conventional fats and oils and zero calorie fat substitutes, such as olestra, in various food products, including salad dressing, frozen desserts, margarine, confectionery fats, and baked goods. However, in the short run, cost will continue to be a limiting factor. For clinical applications as nutraceuticals, or as pharmaceutical lipids, especially for enteral, parental nutrition, and patients with pancreatic insufficiency, the cost of SL will be competitive or cheaper than the cost of drugs. Use of genetic engineering to produce desirable SL for various niche markets will bring the cost close to conventional vegetable oils. However, the controversy over genetically modified organisms (GMO) is yet to be resolved to every country's satisfaction. The question is, will the algal- or plant-derived SL have the desired n-3 or n-6 HUFA at the sn-2 position for the intended use? The prospect for designing SL to contain SCFA/MCFA, GLA, monounsaturated, conjugated linoleic acid (CLA), arachidonic acid, EPA, DHA, linolenic acid, and saturated fatty acids, or any combinations thereof, at the appropriate position, preferably at sn-2 position for most, continue to be promising.

References

1. Willis, W.M.; Lencki, R.W.; Marangoni, A.G. *Crit. Rev. Food Sci. Nutr.* **1998**, *38*, 639-674.
2. Kennedy, J.P. *Food Technol.* **1991**, *45*, 76-83.
3. Akoh, C.C. In *Food Lipids: Chemistry, Nutrition, and Biotechnology;* Akoh, C.C.; Min, D.B., Eds.; Marcel Dekker, Inc.: New York, NY, 1998; pp 699-727.
4. Akoh, C.C., Lee, K.-T.; Fomuso, L.B. In *Structural Modified Food Fats: Synthesis, Biochemistry, and Use;* Christophe, A.B., Ed.; American Oil Chemists' Society Press: Champaign, IL, 1998; pp 46-72.
5. Willis, W.M.; Marangoni, A.G. In *Food Lipids: Chemistry, Nutrition, and Biotechnology*; Akoh, C.C.; Min, D.B., Eds.; Marcel Dekker, Inc.: New York, NY, 1998; pp 665-698.
6. Carlson, S.E.; Ford, A.J.; Werkman, S.H.; Peeples, J.M.; Koo, W.W.K. *Pediatr. Res.* **1996**, *39*, 882-888.
7. Crawford, M.A.; Costeloe, K.; Ghebremeskel, K.; Phylactos, A.; Skirvin, L.; Stacey, F. *Am. J. Clin. Nutr.* **1997**, *66*, 1032S-1041S.
8. Lee, K.-T.; Akoh, C.C.; Dawe, D.L. *J. Food Biochem.* **1999**, *23*, 197-208.
9. Kolanowski, W.; Swiderski, F.; Berger, S. *Intl. J. Food Sci. Nutr.* **1999**, *50*, 39-49.
10. Huang, K.-H.; Akoh, C.C. *J. Am. Oil Chem. Soc.* **1994**, *71*, 1277-1280.

11. Lee, K.-T.; Akoh, C.C. *J. Am. Oil Chem. Soc.* **1996**, *73*, 611-615.
12. Nestel, P.J. *Ann. Rev. Nutr.* **1990**, *10*, 149-167.
13. Akahane, N.; Ohba, S.; Suzuki, J.; Wakabayashi, T. *Thrombosis Res.* **1995**, *5*, 441-450.
14. Horrobin, D.F. *Prog. Lipid Res.* **1992**, *31*, 163-194.
15. Horrobin, D.F. In *Antioxidants, Free Radicals, Free Radicals and Polyunsaturated Fatty Acids in Biology and Medicine;* Diplock, A.T.; Gutterridge, J.M.C.; Shukla, V.K.S., Eds.; International Food Science Center A/S, Lystrup, 1993: pp 181-198.
16. Fan, Y.-Y.; Chapkin, R.S. *J. Nutr.* **1998**, *128*, 1411-1414.
17. Akoh, C.C.; Moussata, C.O. *J. Am. Oil Chem. Soc.* **1998**, *75*, 697-701.
18. Akoh. C.C.; Sista, R.V. *J. Food Lipids* **1995**, *2*, 231-238.
19. Horrobin, D.F. *Rev. Contemp. Physiol.* **1990**, *1*, 1-41.
20. Huang, K.-H.; Akoh, C.C. Erickson, M.C.; *J. Agric. Food Chem.* **1994**, *42*, 2646-2648.
21. Akoh, C.C.; Jennings, B.H.; Lillard, D.A. *J. Am. Oil Chem. Soc.* **1995**, *72*, 1317-1321.
22. Lee, K.-T.; Akoh, C.C. *J. Am. Oil Chem. Soc.* **1998**, *75*, 495-499.
23. Moussata, C.O.; Akoh, C.C. *J. Am. Oil Chem. Soc.* **1998**, *75*, 1155-1158.
24. AOCS; *Official Methods and Recommended Practices of the American Oil Chemists' Society;* Firestone, D., Ed.; American Oil Chemists' Society, Champaign, IL, 1992, Method Cd 12-B-92.
25. Akoh, C.C. *J. Am. Oil Chem. Soc.* **1994**, *71*, 211-215.
26. Garcia, D.J. *Food Technol.* **1998**, *52*, 44-49.

Chapter 13

Modified Oils Containing Highly Unsaturated Fatty Acids and Their Stability

S. P. J. Namal Senanayake and Fereidoon Shahidi

Department of Biochemistry, Memorial University of Newfoundland, St. John's, Newfoundland A1B 3X9, Canada

Modification of oils with specific fatty acids, especially short-, medium- and long-chain fatty acids, using various enzymes in organic media, may be achieved. Various reactions such as hydrolysis, random reesterification or interesterification are involved in oil modification. Modified lipids are known to possess potential health benefits; however, products containing highly unsaturated fatty acids are highly susceptible to oxidation. Hence, their protection against autoxidation is essential in order to counterbalance any harmful effects arising from their oxidation and to take the maximum advantage of their health benefits.

Oils containing highly unsaturated fatty acids (HUFA), especially those of the ω3 family, are very susceptible to oxidative deterioration, mainly *via* autoxidation. Autoxidation is a natural chemical reaction that takes place between molecular oxygen and unsaturated fatty acids (*1*). This process proceeds *via* a free-radical chain mechanism (*2*) involving initiation, propagation and termination steps, forming primary and secondary oxidation products (*3*). Primary products of lipid autoxidation are hydroperoxides which decompose readily to a variety of secondary products such as aldehydes, ketones, hydrocarbons and alcohols, among others, which may give undesirable flavor to oils (*2*).

The rate of autoxidation of fatty acids depends on their degree of unsaturation and physical states. The relative rate of autoxidation of oleate, linoleate and linolenate is reported to be in the order of 1:40-50:100 (*4*). Arachidonic acid was reported to oxidize 2.9 times faster than linoleic acid (*5*). Ethyl esters of eicosapentaenoic acid (EPA) and docosahexaenoic acid (DHA) showed induction periods of 3 to 4 days compared to 20 to 60 days for lineate and linolenate, respectively (*6*). Oxygen uptake of EPA and DHA after induction period was 5.2 and 8.5 times faster than that of ethyl linolenate (*6*). Therefore, oils that contain relatively large amounts of polyunsaturated fatty acids (PUFA) experience stability problems.

In order to determine the oxidative state and quality of edible oils, a number of stability tests are generally employed. Methods reported in the literature include chemical, instrumental and sensory techniques (*1,3,7*). No standard method to evaluate the oxidative stability of oils containing HUFA has been recommended (*3*). However, it is advisable to use a variety of tests to assess both primary and secondary oxidative changes in oils. In this contribution, production of modified oils containing HUFA and their oxidative stability are discussed.

Modified Oils

Much attention has been paid to the use of microbial lipases to produce modified lipids by interesterification reaction of plant and marine oils. Modified lipids are triacylglycerols (TAG) with mixtures of short-, or medium and/or long-chain fatty acids esterified to the same glycerol molecule (*8*). There are several possible methods to synthesize them; hydrolysis and esterification, interesterification, traditional chemical methods and genetic manipulation (*9*). However, only enzymatic interesterification and genetic manipulation can create targeted modified lipids, resulting in specific placement of the fatty acids on the glycerol backbone. The other methods produce randomized modified lipids. Enzyme-assisted interesterification has become the preferred method for small scale modified lipid synthesis because reaction conditions tend to be milder and less side reactions may occur. Furthermore, some lipases are highly specific, allowing targeting of the fatty acids to a particular position on the glycerol molecule (*10*). Modified lipids provide an effective means for delivering fatty acids with particular properties desirable either in overall nutrition or in specific diseases of humans (*8*).

Benefits of Modified Oils

The benefits of modified lipids are derived from the short-, medium- and long-chain fatty acids and the uniqueness of the modified TAG molecules themselves. Long-chain fatty acids, especially fatty acids of the ω3 and ω6 family, may be included in modified lipids to promote health and nutrition (11).

Modified lipids containing medium chain and long-chain essential fatty acids meet the nutritional needs of hospital patients and those with special dietary requirements. When medium chain fatty acids such as capric and caproic acids are consumed, they are not incorporated into chylomicrons and are therefore not likely to be stored, but will be used for energy. They are readily oxidized in the liver and constitute a highly concentrated source of energy for premature babies and patients with fat malabsorption (12).

A modified lipid made from fish oil and medium-chain triacylglycerol was found to decrease tumor growth in mice (13). In another study, the tumor growth rate was slowed in the rats fed with modified lipids containing medium-chain fatty acids and fish oil (14). Jandacek et al. (15) showed that a modified lipid containing caprylic acid (8:0) at the sn-1 and sn-3 positions and a long-chain fatty acid at the sn-2 position is more rapidly hydrolyzed and effectively absorbed than typical long-chain TAG molecules.

Impact (produced by Novartis Nutrition, Minneapolis, MN) is a medical product containing randomized modified lipids created by interesterification of high-lauric acid oil and high-linoleic acid oil. It is used for patients who have suffered from trauma or surgery, sepsis, or cancer (9).

Synthesis of Modified Oils Containing Highly Unsaturated Fatty Acids

Several research groups have successfully modified plant oils by incorporating long-chain ω3 PUFA (16-22). Huang and Akoh (17) studied the ability of immobilized lipases IM60 from *Mucor miehei* and SP435 from *Candida antarctica* to modify the fatty acid composition of soybean oil by incorporation of ω3 PUFA. The interesterification was carried out using free fatty acids and ethyl esters of EPA and DHA as acyl donors. With free EPA as acyl donor, IM60 gave higher incorporation of EPA than SP435. However, when ethyl esters of EPA and DHA were the acyl donors, SP435 gave higher incorporation of EPA and DHA than IM60.

Huang et al. (18) were also able to modify crude melon seed oil by incorporating EPA using two immobilized lipases, IM60 from *Mucor miehei* and SP435 from *Candida antarctica* as biocatalysts. Higher EPA incorporation was obtained using EPA ethyl ester than its free acid counterpart for both enzyme-catalyzed reactions. Increasing the mole ratio of acid or ester to TAG, significantly increased EPA incorporation, especially when EPA ethyl ester was used.

Recently, EPA and capric acid (10:0) have been incorporated into borage oil with two immobilized lipases, SP435 from *Candida antarctica* and IM60 from *Rhizomucor miehei* as biocatalysts (*21*). Higher incorporation of EPA (10.2%) and 10:0 (26.3%) was obtained with IM60 lipase, compared to 8.8 and 15.5%, respectively, reported with SP435 lipase.

In our studies, we have shown that lipase from *Pseudomonas sp.* effectively incorporated EPA and DHA into borage and evening primrose oils. Table 1 shows the fatty acid composition of borage and evening primrose oils, before and after interesterification with EPA and DHA by the nonspecific *Pseudomonas sp.* enzyme. In this study, the substrate mole ratio of 1:0.5:0.5 (oil/EPA/DHA) was kept constant because incorporation of EPA and DHA was satisfactory at this mole ratio. Predominant fatty acids found in borage oil before enzymatic modification were 18:2ω6 (38.4%) and 18:3ω6 (24.4%). The concentration of these acids was comparable to those reported by Akoh and Sista (*19*). After modification, 18:2ω6 and 18:3ω6 decreased by 15.9 and 7%, respectively. The amounts of EPA and DHA incorporated into borage oil were 25.9 and 9.6%, respectively. The ratio of ω3 PUFA to ω6 PUFA increased from 0.002 to 0.89. On the other hand, the main fatty acid found in evening primrose oil before enzymatic modification was 18:2ω6 (73.6%). The content of GLA found in this oil was 9.1%. After interesterification reaction, the content of 18:2ω6 was decreased drastically by 28%. However, the content of γ-linolenic acid (GLA) was reduced by only 2%. The amount of EPA and DHA incorporated into evening primrose oil was 25.7 and 7.9%, respectively. The corresponding ω3/ω6 ratio increased from 0 to 0.64. Ju *et al.* (*22*) selectively hydrolysed borage oil using immobilized *Candida rugosa* and then used the hydrolysed product and ω3 PUFA as substrates, in the presence of immobilized Lipozyme IM60 from *Mucor miehei*, to produce modified oils. The total content of ω3 and ω6 PUFA found in acylglycerols was 72.8% following the interesterification reaction. The contents of GLA, EPA and DHA were 26.5, 19.8 and 18.1%, respectively, with a corresponding ω3/ω6 ratio changing from 0 to 1.09 after modification.

Previously, Akoh and Sista (*19*) modified the fatty acid composition of borage oil using the ethyl ester of EPA and in the presence of an immobilized nonspecific SP435 lipase from *Candida antarctica* as a biocatalyst. The highest incorporation (31%) of EPA was obtained with 20% (w/w of substrates) SP435 lipase. At a substrate mole ratio of 1:3, the corresponding ratio of ω3 to ω6 PUFA was 0.64. Under similar conditions, Akoh *et al.* (*20*) were able to increase the ω3 PUFA (upto 43%) of evening primrose oil with a corresponding increase in the ω3/ω6 ratio from 0.01 to 0.60. Sridhar and Lakshminarayana (*16*) were able to effectively modify groundnut oil by incorporating EPA and DHA using a sn-1,3 specific lipase from *Mucor miehei* as the biocatalyst. The resultant contents of EPA and DHA of the modified oil were 9.5 and 8.0%, respectively.

Oxidative Stability of Modified Oils

The fatty acid composition of borage and evening primrose oils and their modified counterparts containing EPA and DHA are shown in Table I. These oils were assessed for their oxidative stability under Schaal oven conditions at 60°C over a 96h period. The content of conjugated dienes, as reflected in the absorption readings at 234 nm, during storage of both modified and unmodified borage and evening primrose oils is given in Table II. Higher conjugated diene values for modified oils may arise from their higher content of omega-3 PUFA as compared to that of their unmodified counterparts. Lipid radicals formed during the initial step may undergo rearrangement, thus the methylene-interrupted feature of PUFA of oils are lost in favor of formation of conjugated dienes, including those of hydroperoxides. Conjugated diene values may be used to determine the initial rate of lipid oxidation (23). Formation of hydroperoxides normally coincides with that of the conjugated dienes upon autoxidation of lipids. Therefore, conjugated diene content reflects the formation of primary oxidation products of these oils.

Production of thiobarbituric acid reactive substances (TBARS) of both modified and unmodified oils is shown in Table 2. TBARS values of these oils increased progressively over the entire storage period. As expected, modified oils had significantly higher TBARS values as compared to unmodified oils. The main compounds in the oils reacting with the 2-thiobarbituric acid (TBA) reagent are alkenals and alkadienals as well as malonaldehyde.

Oxidative stability of oils may also be evaluated by measuring carbonyl compounds formed during storage using gas chromatographic techniques. Carbonyl compounds have been identified as significant contributors to the flavor of a variety of foods. Hexanal is by far the most prominent volatile aldehyde in food lipids containing linoleic acid, but this is perhaps not unexpected. Hexanal is the only aldehyde that is formed from both 9- and 13-hydroperoxides of linoleic acid and from other unsaturated aldehydes produced during oxidation of this fatty acid (24,25). Autoxidation of dominant fatty acids present in modified borage and evening primrose oils has been reported to produce many aldehydes, some of which are listed in Table III.

Similar to other sources of ω6 fatty acids, borage and evening primrose oils degrade and produce hexanal as their dominant volatile. Upon enrichment of these oils with EPA and DHA *via* enzyme-catalyzed acidolysis, they produce hexanal and propanal as their main volatiles during degradation; propanal being a major breakdown product of ω3 fatty acids such as linolenic acid as well as EPA and DHA (26,27). The other volatile compounds identified in modified oils were acetaldehyde, butanal, pentanal, heptanal, octanal and nonanal (unpublished

Table I. Fatty acid profile of borage (BO) and evening primrose oils (EPO) before and after enzymatic modification by *Pseudomonas sp.* lipase.

Fatty acid[a]	Unmodified BO	Modified BO[b]	Unmodified EPO	Modified EPO[c]
10:0	0.08 ± 0.01	0.03 ± 0.01	ND[d]	0.05 ± 0.04
12:0	0.07 ± 0.02	0.04 ± 0.02	ND[d]	0.04 ± 0.01
14:0	0.06 ± 0.01	0.04 ± 0.01	0.05 ± 0.01	0.03 ± 0.02
16:0	9.81 ± 0.12	5.06 ± 0.37	6.16 ± 0.09	3.35 ± 0.06
18:0	3.12 ± 0.26	2.05 ± 0.01	1.72 ± 0.12	0.99 ± 0.02
18:1	15.2 ± 0.74	10.7 ± 0.09	8.65 ± 0.56	6.08 ± 0.05
18:2ω6	38.4 ± 0.89	22.5 ± 0.33	73.6 ± 0.91	45.9 ± 0.11
18:3ω6	24.4 ± 0.90	17.4 ± 0.19	9.12 ± 0.38	7.03 ± 0.01
18:3ω3	0.17 ± 0.02	0.07 ± 0.05	ND[d]	0.12 ± 0.02
20:0	0.20 ± 0.07	0.10 ± 0.01	0.25 ± 0.06	0.20 ± 0.03
20:1	4.09 ± 0.25	2.60 ± 0.01	ND[d]	0.18 ± 0.02
20:5ω3	ND[d]	25.9 ± 0.54	ND[d]	25.7 ± 0.56
22:1	2.49 ± 0.10	1.69 ± 0.30	ND[d]	ND[d]
24:1	1.52 ± 0.20	0.83 ± 0.01	ND[d]	ND[d]
22:6ω3	ND[d]	9.60 ± 0.82	ND[d]	7.9 ± 0.08

[a] As area percentage
[b] Modified borage oil was prepared under optimum reaction conditions (278 enzyme units, 42°C, 26h).
[c] Modified evening primrose oil was prepared under optimum reaction conditions (299 enzyme units, 43°C, 24h).
[d] Not detected

Table II. Conjugated dienes and TBARS values of borage (BO) and evening primrose oils (EPO) during accelerated storage at 60°C.

Stability test	Storage time (h)	Unmodified BO	Modified BO	Unmodified EPO	Modified EPO
Conjugated dienes	0	1.50 ± 0.7	7.08 ± 0.1	9.00 ± 0.2	9.71 ± 0.3
	24	3.43 ± 0.2	10.2 ± 0.1	12.9 ± 0.5	14.3 ± 0.1
	48	4.78 ± 0.2	10.8 ± 0.1	15.1 ± 0.2	18.9 ± 0.1
	96	12.9 ± 0.6	21.0 ± 0.7	23.0 ± 0.6	31.0 ± 1.0
TBARS (μmol/g)	0	2.0 ± 0.1	6.00 ± 0.1	0.4 ± 0.2	6.00 ± 0.3
	24	2.5 ± 0.3	7.50 ± 0.5	0.9 ± 0.5	7.00 ± 0.1
	48	3.0 ± 0.1	8.20 ± 0.3	2.0 ± 0.2	8.10 ± 0.2
	96	4.3 ± 0.3	14.7 ± 0.1	3.4 ± 0.2	13.2 ± 0.4

Table III. Some volatile aldehyde products from autoxidation of selected unsaturated fatty acids, their corresponding hydroperoxide and reactive methylene group sites.

Fatty acid	Methylene group	Hydroperoxide	Aldehyde
18:2ω6	11	9	2,4-Decadienal
		13	Hexanal
18:3ω6	8,11	6	2,4,7-Tridecatrienal
		9	2,4-Decadienal
		10	3-Nonenal
		13	Hexanal
20:5ω3	7,10,13,16	5	3,6,9,12-Pentadecatetraenal
		8	2,4,7,10-Tridecatetraenal
		9	3,6,9-Dodecatrienal
		11	2,4,7-Decatrienal
		12	3,6-Nonadienal
		14	2,4-Heptadienal
		15	3-Hexenal
		18	Propanal
22:6ω3	6,9,12,15,18	4	2,4,7,10,13,16-Nonadecahexaenal
		7	2,4,7,10,13-Hexadecapentaenal
		8	3,6,9,12-Pentadecatetraenal
		10	2,4,7,10-Tridecatetraenal
		11	3,6,9-Dodecatrienal
		13	2,4,7-Decatrienal
		14	3,6-Nonadienal
		16	2,4-Heptadienal
		17	3-Hexenal
		20	Propanal

results). The content of hexanal produced in these modified oils was significantly higher than that produced in their unmodified counterparts (Table IV). The amount of propanal produced in modified oils also increased with time (Table IV). Under similar conditions, unmodified oils showed no significant formation of propanal perhaps due to their low content of ω3 PUFA as compared to the amounts present in the modified oils. The initial products of autoxidized linolenic acid are dominated by 9-, 12-, 13- and 16-hydroperoxides because the diallylic radicals formed favor the attack of oxygen on these specified positions (*28*). The hydroperoxides so formed degrade to a variety of products, including propanal, ethane and 2,4,7-decatrienal (*29*). Propanal is a predominant product of linolenate oxidation and lipids containing a large proportion of this fatty acid or those containing long-chain PUFA such as EPA and DHA (*30,31*).

Moussata and Akoh (*32*) measured the oxidative stability of melon seed oil, interesterified with high oleic sunflower oil, and found that oleic acid could enhance the stability of melon seed oil. This observation was in accordance with previous studies of sardine oil, interesterified with oleic acid (*33*). Miyashita *et al.* (*34,35*) investigated the oxidative stability of free HUFA and HUFA-containing lipids in an aqueous system and reported that EPA and DHA, which were highly oxidizable in the air, were more stable than linoleic and linolenic acids. However, Endo *et al.* (*36*) showed that HUFA were very unstable in both nonaqueous and aqueous systems when they were highly concentrated in a single triacylglycerol molecule.

PUFA are among the most easily oxidizable components of foods and many of the oxidized products including peroxides, free radicals and aldehydes are potentially toxic and mutagenic (37). The ease of autoxidation of fatty acids is proportional to the number of methylene groups between double bonds, thus modified borage and evening primrose oils with a higher content of EPA and DHA are more prone to oxidation than their unmodified counterparts. Thus, modified oils rich in PUFA compared to unmodified oils, must be protected against oxidation in order to counterbalance any harmful effects from production of oxidation products and to take full advantage of their nutritional and health-related benefits.

Conclusions

Oils may be successfully modified by incorporating highly unsaturated fatty acids, especially PUFA of the ω3 family, using enzyme-catalyzed reactions. The resultant oils are known to possess nutritional and health benefits. Modified oils may exhibit lower oxidative stability than their unmodified counterparts, mainly due to the fatty acid constituents of the molecules involved.

Table IV. Main volatile aldehydes of borage (BO) and evening primrose oils (EPO) stored at 60°C.

Stability test	Storage time (h)	Unmodified BO	Modified BO	Unmodified EPO	Modified EPO
Hexanal content (mg/kg)	0	7.91 ± 0.7	7.28 ± 1.0	- 3.00 ± 1.0	-
	24	11.0 ± 2.5	18.0 ± 2.6	9.84 ± 1.0	5.62 ± 2.3
	48	12.4 ± 4.5	31.1 ± 6.9	13.8 ± 2.2	41.0 ± 6.6
	96	55.4 ± 6.5	189.2 ± 8.5	60.0 ± 5.9	316.4 ± 9.9
Propanal content (mg/kg)	0	-	2.0 ± 0.5	-	2.80 ± 0.1
	24	-	3.9 ± 2.4	-	36.0 ± 8.0
	48	-	26.4 ± 2.3	-	52.1 ± 8.6
	96	2.2 ± 1.5	337.4 ± 6.8	2.2 ± 1.0	602.9 ± 15.9

References

1. Shahidi, F.; Wanasundara, U.N. In *Food Lipids: Chemistry, Nutrition and Biotechnology.* Akoh, C.C.; Min, D.B., Eds.; Marcel Dekker, Inc.: New York, 1998, pp. 377-396.
2. Labuza, T.P. *CRC Crit. Rev. Food Technol.* **1971**, *2,* 355-405.
3. Shahidi, F.; Wanasundara, U.N. *Food Sci. Technol. Int.* **1996**, *2,* 73-81.
4. Hsieh, R.J.; Kinsella, J.E. *Adv. Food Nutr. Res.* **1989**, *33,* 233-241.
5. Porter, N.A.; Lehman, L.S.; Weber, B.A.; Smith, K.J. *J. Am. Oil Chem. Soc.* **1981**, *103,* 6447-52.
6. Cho, S.-Y.; Miyashita, K.; Miyazawa, T.; Fujimoto, K.; Keneda, T. *J. Am. Oil Chem. Soc.,* **1987**, *64,* 876-80.
7. Rossell, B. *Lipid Technol.* **1991**, October-December, 122-126.
8. Akoh, C.C. *INFORM* **1995**, *6,* 1055-1061.
9. Haumann, B.F. *INFORM,* **1997**, *8,* 1004-1011.
10. Marangoni, A.G.; Rousseau, D. *Trends Food Sci. Technol.* **1995**, *6,* 329-335.
11. Kennedy, J.P. *Food Technol.* **1991**, November, 76-83.
12. Willis, W.M.; Marangoni, A.G. In *Food Lipids.* Akoh, C.C.; Min, D.B., Eds.; Marcel Dekker, Inc.: New York, 1998, pp. 665-698.
13. Ling, P.R.; Istfan, N.W.; Lopes, S.M.; Babayan, V.K. *Am. J. Clin. Nutr.* **1991**, *53,* 1177-1184.
14. Mendez, B.; Ling, P.R., Istfan, N.; Babayan, V.K.; Bistrain, B.R. *J. Parent. Enter. Nutr.* **1992**, *16,* 545-551.
15. Jandacek, R.J.; Whiteside, J.A.; Holcombe, B.N. *Am. J. Clin. Nutr.* **1987**, *45,* 940-945.
16. Sridhar, R.; Lakshminarayana, G. *J. Am. Oil Chem. Soc.* **1992**, *69,* 1041-1042.
17. Huang, K.; Akoh, C.C. *J. Am. Oil Chem. Soc.* **1994**, *71,* 1277-1280.
18. Huang, K.; Akoh, C.C.; Erickson, M.C. *J. Agric. Food Chem.* **1994**, *42,* 2646-2648.
19. Akoh, C.C.; Sista, R.V. *J. Food Lipids* **1995**, *2,* 231-238.
20. Akoh, C.C.; Jennings, B.H.; Lillard, D.A. *J. Am. Oil Chem. Soc.* **1996**, *73,* 1059-1062.
21. Akoh, C.C.; Moussata, C.O. *J. Am. Oil Chem. Soc.* **1998**, *75,* 697-701.
22. Ju, Y.; Huang, F.; Fang, C. *J. Am. Oil Chem. Soc.* **1998**, *75,* 961-965.
23. Gray, J.I. *J. Am. Oil Chem. Soc.* **1978**, *55,* 539-546.
24. Shahidi, F.; Pegg, R.B. *J. Food Lipids* **1994**, *1,* 177-186.
25. Frankel, E.N.; Hu, M.-L.; Tappel, A.L. *Lipids* **1989**, *24,* 976-981.
26. Frankel, E.N.; Selke, E.; Neff, W.E.; Miyashita, K. *Lipids* **1992**, *27,* 442-446.
27. Frankel, E.N.; Huang, S.-W. *J. Am. Oil Chem. Soc.* **1994**, *71,* 225-229.
28. Frankel, E.N. *J. Am. Oil Chem. Soc.* **1984**, *61,* 1908-1917.

29. Ho, C.-T.; Chen, Q., Zhou, R. In *Bailey's Industrial Oil and Fat Products, Edible Oil and Fat Products: General Applications*. 5th Edition, Volume 1, Wiley Interscience: New York, 1996, pp. 83-104.
30. Shahidi, F.; Spurvey, S.A. *J. Food Lipids* **1996**, *3*, 13-25.
31. He, Y.; Shahidi, F. *J. Agric. Food Chem.* **1997**, *74*, 1133-1136.
32. Moussata, C.O.; Akoh, C.C. *J. Am. Oil Chem Soc.* **1998**, *75*, 1155-1159.
33. Endo, Y.; Kimoto, H.; Fujimoto, K. *Biosci. Biotech. Biochem.* **1993**, *57*, 2202-2204.
34. Miyashita, K.; Nara, E.; Ota, T. *Fisheries Sci.* **1994**, *60*, 315-318.
35. Miyashita, K.; Hirano, M.; Nara, E.; Ota, T. *Fisheries Sci.* **1995**, *61*, 273-275.
36. Endo, Y.; Hoshizaki, S.; Fujimoto, K. *J. Am. Oil Chem Soc.* **1997**, *74*, 1041-1045.
37. Pearson, A.M.; Gray, J.I.; Wolzak, A.M.; Horenstein, N.A. *Food Technol.* **1983**, *37*, 121-129.

Aroma Effects, Stabilization, and Analytical Procedures

Chapter 14

Effect of Lipase Hydrolysis on Lipid Peroxidation in Fish Oil Emulsion

Junji Terao[1] and Takashi Nagai[2]

[1]Department of Nutrition, The University of Tokushima School of Medicine, Tokushima 770-8503, Japan
[2]Tsukuba Research Laboratories, Kyowa Hakko Kogyo Company, Ltd., Tsukuba 300-4247, Japan

Lipase-hydrolysis treatment was applied to fish oil emulsion in order to improve its oxidative stability. Sardine oil was emulsified with Triton X-100 in Tris-HCl buffer (pH 7.4) and partially hydrolyzed by porcine pancreatic lipase. The emulsion was subjected to lipid peroxidation induced by myoglobin, Fe^{2+} and ascorbic acid, free radical generating azo compounds or autoxidation. In all cases, treated fish oil emulsion was extremely resistant to lipid peroxidation as compared with its nontreated counterpart. Free eicosapentaenoic acid (20:5, EPA), but not its methyl ester, was found to possess significantly lower peroxidizability than fish oil in emulsion system. Furthermore, free EPA suppressed lipid peroxidation of fish oil emulsion in a concentration-dependent manner. The suppressive effect of free fatty acid against antioxidation decreased in the order of 20:5 > 20:4 >18:2 > 10:0. These results strongly suggest that stabilization of fish oil emulsion by lipase-hydrolysis is derived from, at least partly due to the appearance of peroxidation-resistant and peroxidation-inhibiting nonesterified polyunsaturated fatty acids.

It is well known that lipid peroxidation in edible oils is a critical factor leading to the loss of their nutritional value and the generation of off-flavors. Furthermore, peroxidation products may be harmful to the human body because they are toxic and potentially carcinogenic (*1-3*). In particular, fish oils rich in polyunsaturated fatty acids (PUFA) are undoubtedly susceptible to lipid peroxidation and thus systems to improve their oxidative stability need to be developed to facilitate the practical delivery of the health benefits of the oils.

In general, non-enzymatic lipid peroxidation proceeds *via* a radical chain reaction and the peroxidizability of PUFA increases with the number of double bonds in their structures (*4,5*). However, Bruna et al. (*6*) and Miyashita et al. (*7*) claimed that eicosapentaenoic acid (EPA) and docosahexaenoic acid (DHA), common fatty acids present in fish oils, are unexpectedly stable against lipid peroxidation when present in aqueous micelles. Yazu et al. (*8*) suggested that low peroxidizability of EPA in aqueous micelles is attributed to the enhancement of chain-termination reaction due to the diffusion of peroxidized EPA from core to the micelle surface. Our preliminary study indicated that oxidative stability of fish oil emulsion was increased by the treatment with pancreatic lipase. This prompted us to investigate the effect of lipase hydrolysis on the oxidative stability of fish oil emulsion, because nonesterified EPA and DHA are released from fish oil triacylglycerols by the action of this enzyme. Here we prepared fish oil emulsions using Triton X as an emulsifier and estimated the effect of porcine pancreatic lipase on the oxidative stability of the emulsion. The effect of nonesterified EPA on the stability of fish oil emulsion was also studied.

Materials and Methods

Materials

Fish oil (antioxidant-free sardine oil, peroxide value (PV): 51.0 meq/kg) was kindly supplied by Tsukishima Food Industrial Co. Ltd. (Tokyo, Japan). Fish oil was purified with silica gel column chromatography according to the method described previously and then PV of purified oil was lower than 1.0 meq/kg. It was then kept in n-hexane solution at -20 °C until use. Lipase (Type II) from porcine pancreas and metmyoglobin were purchased from Sigma Chemicals Co. (St. Louis, MO). Triton X-100, 2,2'-azobis(2-amidinopropane) hydrochloride (AAPH) and 2,2'-azobis(2,4-dimethylvaleronitrile)(AMVN) were purchased from Wako Pure Chemicals (Osaka, Japan). Eicosapentaenoic acid (20:5, EPA), methyl eicosapentaenoate (MeEPA), arachidonic acid (20:4), linoleic acid (18:2) and capric acid (10:0), each 99% grade, were purchased from Nakaral Tesque (Kyoto, Japan). Other reagents were of analytical grade and used without any further purification.

Hydrolysis of fish oil emulsion

Fish oil emulsion was prepared by the same procedure as described previously (*9*). Briefly, fish oil (46 mmol) was dispersed in 4.0 ml Tris-HCl buffer (10 mM, pH 7.4) containing 10 mM Triton X-100 and mixed with a vortex mixer for 30 sec followed by ultrasonic irradiation with an Astrason sonicator XL2020 (20 kHz, Heat system-Ultrasonics, Inc., NY) for 6 min in an ice bath. Lipase solution (0.8 ml; 1,900 units in 10 mM Tris-HCl buffer, pH 7.4 containing 50 mM $CaCl_2$) was mixed with fish oil emulsion (3.2 ml) and incubated at 37 °C for 1 h in the dark. After the reaction was completed, total lipids were extracted with chloroform and methanol according to the method of Bligh and Dyer (16). Lipase hydrolytic activity was monitored by a method described previously (11). In some experiments, lipase with heat pretreatment (100°C for 15 min) was used as inactivated lipase. The hydrolysis products were confirmed by TLC analysis using silica gel 60 plate (Merck, Darmstadt, Germany) and the developing solvent of n-hexane/diethyl ether/acetic acid (80:30:1, v/v/v). Spots were detected by exposure to iodine vapor.

Oxidation of lipase-hydrolyzed fish oil in emulsion

After extraction, fish oil hydrolyzate (23 µmol) was again emulsified with 2.0 ml Tris-HCl buffer (10mM,pH7.4) containing 10mM TritonX-100 as described above. To 0.8ml emulsion, each 0.2 ml of Tris-HCl buffer (10 mM, pH 7.4) containing 500 µM metmyoglobin, or 500 µM $Fe(NO_3)_3$ and 5 mM ascorbic acid, or 50 mM AA-PH were added. In the case of AMVN-induced oxidation of the emulsion, the emulsion was prepared after mixing fish oil hydrolyzate (23 µmol) and (4 µmol). The emulsion was incubated at 37°C with shaking. In autoxidation experiment, the emulsion was incubated without addition of any initiators for 120 h.

Measurement of PV and thiobarbituric acid reactive substances (TBARS)

PV was assayed by the method of Burge and Aust (12) and expressed as meq/kg. The thiobarbituric acid (TBA) assay was done by the method of Uchiyama and Mihara (13) and the amount of TBARS was expressed as µM malonaldehyde.

Analysis of fatty acid composition of hydrolysis products

Hydrolysis products were separated on TLC plates as described above and then each spot was scraped off and extracted with chloroform and methanol (2:1, v/v). After evaporation, the residue was subjected to GLC analysis to determine their fatty acid compositions. The extracts were mixed with 5% HCl in methanol (1.0 ml) in a screw capped tube and heated at 70 °C for 3h. After the reaction was completed, methyl ester was extracted with n-hexane and then injected into the column of fused silica (DB-225, J&W Scientific Inc., Folsom, CA) using a Shimadzu GC-15A Gas Chromatograph (Kyoto, Japan). The operational column temperature was kept at 200 °C.

Results and Discussion

Lipase hydrolysis and its effect on metmyoglobin-induced lipid oxidation of fish oil emulsion

The effect of lipase hydrolysis on the lipid peroxidation of fish oil emulsion was monitored using metmyoglobin as an initiator. This hemoprotein is known to act as a powerful prooxidant in muscle foods (*14*). TLC analysis showed that fish oil was partly hydrolyzed by the treatment of porcine pancreatic lipase and released free fatty acids, diacylglycerols and monoacylglycerols as hydrolysis products (Fig. 1). The analysis also demonstrated complete loss of hydrolytic activity of the enzyme by heat-treatment at 100°C for 15 min. Table I shows that major fatty acids of fish oil triacylglycerols and its hydrolysis products as palmitic acid (16:0), oleic acid (18:1) eicosapentaenoic acid (20:5) and docosahexaenoic acid (22:6). In the case of non-treatment and the treatment with heat-inactivated lipase, TBARS increased gradually with time (Fig. 2). However, no increase was observed in the case of intact lipase-treated fish oil emulsion, indicating that lipase-treatment significantly elevated the resistance against metmyoglob'n-induced lipid peroxidation of fish oil emulsion.

Fig.3 shows the relationship between the increase of TBARS or PV, and the level of released free fatty acid by lipase-treatment of fish oil emulsion. It is apparent that increases of TBARS and PV were lowered with incubation time and were inversely related to the amount of free fatty acids released from fish oil triacylglycerols. It is therefore reasonable to assume that lipase-hydrolysis products are responsible for enhanced oxidative stability of fish oil emulsions.

Effect of lipase-hydrolysis on lipid peroxidation of fish oil emulsions with different initiators

Table II shows the effect of lipase-hydrolysis on the increase of TBARS of fish oil emulsion after incubation with different initiators of lipid peroxidation or autoxidation. In this table, AAPH and AMVN are water-soluble radical generator and lipid-soluble radical generator, respectively. In all oxidation systems, the treatment of lipase reduced the TBARS formation in the fish oil emulsion. It is therefore concluded that the effect of lipase-hydrolysis on oxidative stability of fish oil is independent of the initiation process of lipid peroxidation.

Lipid peroxidation proceeds *via* a radical chain reaction and consists of a chain-initiation reaction and a chain-propagation reaction (Fig. 4). Reactive oxygen species such as perhydroxyl radical (HOO), hydroxyl radical (HO), and peroxynitrite (ONOO), are possibly responsible for the initiation process of chain reaction, in which lipid alkyl radical (L) is formed by the abstraction of a bisallylic hydrogen from esterified or non-esterified PUFA. Chain-initiating radicals may

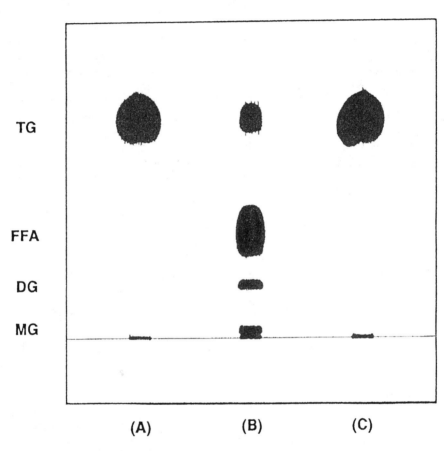

Figure 1. TLC of the extract of fish oil emulsion after incubation without lipase (A), with intact lipase (B) or with heat-inactivated lipase (C). The mixture was incubated at 37 °C for 60 min. TG, triacylglycerol; FFA, free fatty acid; DG, diacylglycerol; and MG, monoacylglycerol.

Table I. Fatty acid composition of lipase-hydrolyzate of fish oil

Fatty Acid	Triacylglycerol	Diacylglycerol	Monoacylglycerol	Free fatty acid
C14:0	7.7	13.3	8.4	9.2
C16:0	22.1	29.9	20.8	21.8
C16:1	8.4	10.5	8.6	10.2
C18:0	4.2	1.1	2.4	5.3
C18:1	15.1	6.5	11.4	23.4
C18:2	1.5	1.3	1.6	2.1
C18:3	2.6	2.4	3.3	1.8
C20:1	2.9	2.5	2.5	3.7
C20:5	16.1	20.3	20.3	9.1
C22:1	4.7	3.8	3.8	5.8
C22:6	11.1	14.1	14.1	2.9
Unknown	3.6	2.8	2.8	4.7

Table II. Effect of lipase hydrolysis on the increase of TBARS of fish oil emulsion induced by different initiators

Initiator/Treatment	None (μM)	Lipase (μM)	Deactivated lipase (μM)
Metmyoglobin	312 ± 22	51 ± 1	204 ± 3
FESO + ascorbic acid	306 ± 3	68 ± 2	248 ± 2
AAPH	176 ± 7	67 ± 5	151 ± 9
AMVN	245 ± 2	52 ± 1	176 ± 9
Autoxidation	326 ± 68	120 ± 4	220 ± 9

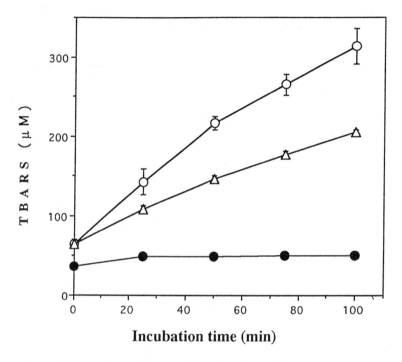

Figure 2. Effect of intact lipase or heat-inactivated lipase on metmyoglobin-induced lipid peroxidation of fish oil emulsion. (•) intact lipase-treatment; (Δ) heat-deactivated lipase treatment; (o) nontreated. Treated or nontreated fish oil emulsion was incubated with metmyoglobin at 37°C.

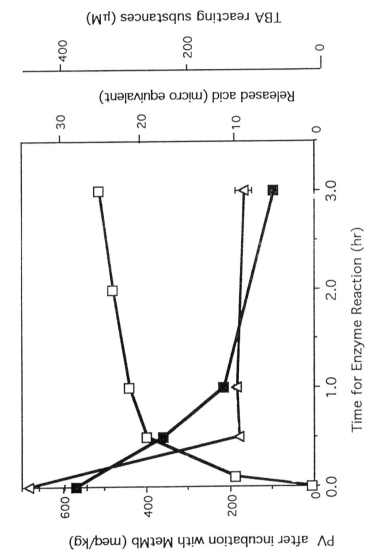

Figure 3. Relationship between lipase-hydrolysis and oxidative stability of fish oil emulsion. Lipase-hydrolyed fish oil was incubated in emulsion with metmyoglobin at 37 °C. (■) TBARS, (△) PV, (□) released free fatty acid.

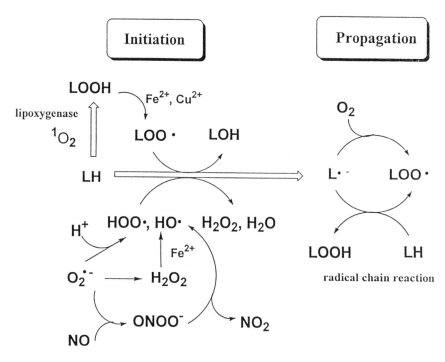

Figure 4. Possible mechanism of lipid peroxidation in food and biological systems.

be generated spontaneously from reactive oxygen species such as superoxide anion (O_2^-). Alternatively, lipid hydroperoxides (LOOH) - dependent lipid peroxidation happens when preformed LOOH is cleaved to produce reactive oxygen species responsible for the chain initiation. In both cases, propagation reaction proceeds *via* chain reaction in which lipid peroxyl radicals (LOO) act as mediators of LOOH accumulation. Thus, lipase hydrolysis seems to affect the rate of chain propagation reaction mediated by LOO in fish oil emulsion.

Effect of nonesterified EPA on lipid peroxidation of fish oil emulsion

Lowering of the oxidizability of fish oil emulsion is likely to be originated from the formation and accumulation of hydrolysis products in the emulsion. Thus, we selected free EPA and compared its oxidizabilty with that of fish oil by measuring TBARS, PV and oxidized fatty acid contents in the emulsion system (Fig. 5). Apparently, free EPA was more resistant to lipid peroxidation than fish oil, indicating that release of free fatty acids participates in the lowering of oxidizability of fish oil by lipase treatment. However, oxidizability of EPA methyl ester was much higher than that of free EPA as shown in Fig. 6. It is therefore concluded that generation of free carboxyl group in PUFA moiety is necessary for the lowering of oxidizability of fish oil emulsion. It is speculated that radical chain reaction easily proceeds in the nonpolar core portion of PUFA moiety of triacylglycerol assembly. Lipase hydrolysis releases free PUFA containing polar carboxyl group so that free PUFA in the assembly seems to move into polar surface resulting in the decrease of PUFA concentration in the core portion.

Furthermore, we found that the addition of free EPA suppressed lipid peroxidation of fish oil emulsion in a concentration-dependent manner (Fig. 7). In this figure, free EPA suppressed TBARS accumulation from fish oil emulsion with an increase of the molar ratio of EPA to fish oil, from 0.3:1 to 3:1. Therefore, it is clear that nonesterified PUFA possesses an activity to suppress lipid peroxidation of fish oil emulsion. The suppressive effect by EPA was compared with that of other fatty acids, namely 20:4, 18:2 and 10:0, to know the role of chain length and double bond in the effect of free EPA. Fig. 8 shows that the suppression of lipid peroxidation increases in the order of 10:0 < 18:2 < 20:4 < 20:5. It is therefore likely that longer chain length and more double bonds enhance the suppressive effect of free PUFA. Yazu et al. (*8*) have already pointed out that peroxyl radicals of free EPA in aqueous micelles diffuse from core portion to the micelle surface. Hence, free PUFA may intercept the lipid peroxyl radicals responsible for chain-propagation of triacylglycerols in core portion and resulting peroxyl radical of free PUFA may accelerate the chain-termination due to the transfer from core to the surface area. Moreover, intramolecular oxygenation from PUFA peroxy radical may also be responsible for the chain-termination because intermolecular oxygenation is required to promote lipid peroxidation in edible oils.

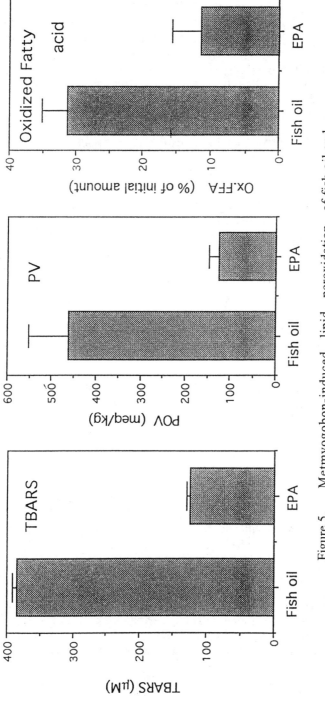

Figure 5. Metmyogobon-induced lipid peroxidation of fish oil and nonesterified EPA in emulsion system. Fish oil (23 µmol) or EPA (69 µmol) was dispersed in the emulsion (2.0 ml) and incubated with metmyoglobin at 37 °C for 100 min.

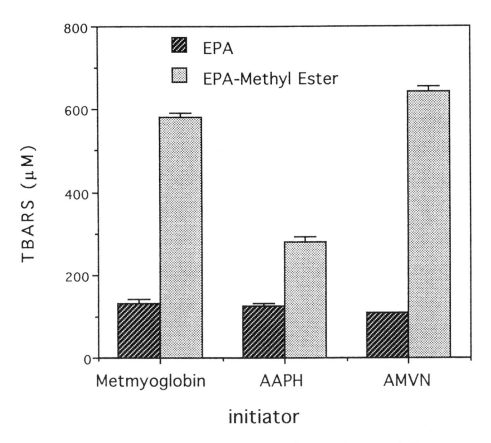

Figure 6. Lipid peroxidation of nonesterified EPA and EPA methyl ester in emulsion system. EPA (69 μmol) or EPA metal ester (69 μmol) was dispersed in the emulsion (2.0 ml) and the oxidation was initiated with metmyoglobin, AAPH or AMVN and incubated at 37 °C for 100 min.

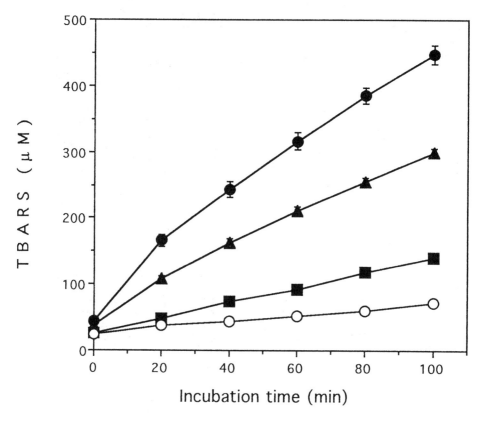

Figure 7. Effect of nonesterified EPA on metmyoglobin-induced lipid peroxidation of fish oil emulsion. EPA was added to the fish oil emulsion (fish oil; 23 µmol in 2.0 ml) and incubated with metmyoglobin at 37 °C. (●) no EPA; (▲) 7.7 µmol EPA; (■)23 µmol EPA; (O)69 µmol EPA.

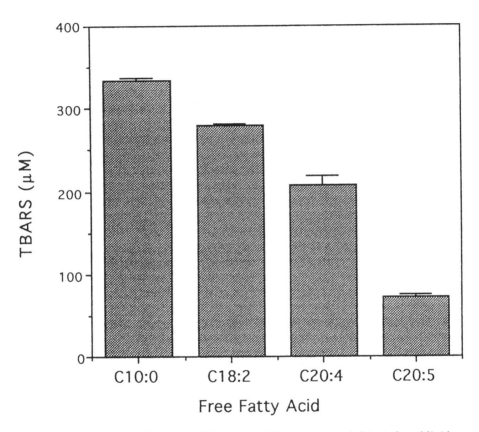

Figure 8. Effect of non-esterified fatty acid on metmyoglobin-induced lipid peroxidation of fish oil emulsion. Each fatty acid (69 μmol) was added to fish oil emulsion (fish oil 23 μmol in 2.0 ml) and incubated with metmyoglobin at 37 °C for 100 min.

Conclusion

Lipase hydrolysis improves oxidative stability of fish oil in emulsion systems. Polyunsaturated fatty acids in their free forms are resistant to lipid peroxidation in emulsions and can protect lipid peroxidation occurring in fish oil emulsions.

References

1. Addis, P.B. *Food Chem. Toxicol.* **1986**, *24*, 1021-1030.
2. Kaneda, T.; Miyazawa, T. *World Rev. Nutr. Diet.* **1987**, *50*, 186-214.
3. Kubow, S. *Free Radical Biol. Med.* **1992**, *12*, 63-81.
4. Cosgrove, J.P.; Church, D.F.; Pryor, WA. *Lipids* **1987**, *22*, 299-304.
5. Porter, N.A.; Caldwell, S.E.; Mills, K.A. *Lipids* **1995**, *30*, 277-290.
6. Bruna, E.; Petit, E.; Beljean-Leymarie M., Huynh, S.; Nouvelot, A. *Lipids* **1989**, *24*, 970-975.
7. Miyashita, K.; Nara E.; Ota, T. *Biosci. Biotech. Biochem.* **1993**, *57*, 1638-1640.
8. Yazu, K.; Yamamoto, Y.; Ukegawa, K.; Niki, E. *Lipids* **1996**, *31*, 337-340.
9. Hoshio, C.; Tagawa, Y.; Wada, S.; Oh, J-H.; Park, D-K.; Nagao, A.; Terao, J. Biosci. Biotech. *Biochem.* **1997**, *61*, 1634-1640.
10. Bligh, E.G.; Dyer, W.J. *Can. J Biochem. Physiol.* **1959**, *37*, 911.
11. Satouchi, K.; Matsushita, S. *Agric. Biol. Chem.* **1976**, *40*, 889-897.
12. Buege, J.A.; Aust, S.D. *Methods Enzymol.* **1978**, *52*, 302-310.
13. Uchiyama, M.; Mihara M. *Anal. Biochem.* **1978**, *86*, 271-278.
14. Love, J.D. *Food Technol.* **1983**, 117-129.

Chapter 15

Farmed Atlantic Salmon as Dietary Sources of Long-Chain Omega-3 Fatty Acids: Effect of High-Energy (High-Fat) Feeds on This Functional Food

Robert G. Ackman, Xueliang Xu, and Catherine A. McLeod

Canadian Institute of Fisheries Technology, DalTech, Dalhousie University, P.O. Box 1000, Halifax, Nova Scotia B3J 2X4, Canada

High energy diets are popular in today's salmon farming, but the lipid deposition in Atlantic salmon Salmo salar, particularly in steaks and fillets, and finally the effect of these diets on product quality, have not been examined. Over 12 months three commercial diets were fed, these being similar in all nutrients except the dietary fat levels (25, 29 and 31%). The muscle protein content was not modified by the three dietary treatments. Texture measurements and sensory panels were conducted on two specific sites in dorsal muscle steak cuts. A significant difference in muscle texture was found. There was a positive correlation between dietary fat and lipid deposition in the fish muscles. The total lipids in white muscle were, respectively, 1.27, 1.41 and 2.25 %, in the trimmed steaks, 2.97, 4.58 and 5.38%, and in the whole steaks 5.45, 7.64 and 9.76 %. The fatty acid composition analyses showed only a slight effect of diet on phospholipids rich in DHA. In the triacylglycerols the fatty acids reflected the dietary fatty acids with DHA and EPA at around 10% each. However, the DHA content slightly exceeded that of the EPA.

Aquaculture is now one of the world's largest fishing operations and functions on a world wide basis (*1*). Among the fish types farmed are salmonids, and among the various species the Atlantic salmon *Salmo salar* is now farmed in salt water in various oceans with suitable cool water temperatures. Among these are sites in Norway, Scotland, Iceland and the Faroe Islands, Chile, New Zealand, the USA and Canada.

Taste is now recognized as the most significant factor in food selection (*2*), closely followed by aroma, a factor not necessarily totally separated (*3*). How ever not far behind these variables is texture, a co-determinant since consumers evaluate food with the same systems although texture is primarily based on mastication. This releases non-volatile flavor and also aromas, and facilitates their physiological evaluation. Texture may be a slower phase of the evaluation than appearance, odor, or flavor, but is important with each class of food types (*2*). Although, fish muscle is usually regarded as very tender, except when toughened by extended frozen storage (*4,5*), texture has become of acute interest to the salmon farming industry of Atlantic Canada as competition has increased in the last decade. The most recognizable reason for this concern has its roots in the steady increase in the fat content of farmed salmon diets shown in Figure 1. The concepts behind this change reflected the increasing world price of fish meal and the relatively low price of fish oil. The latter is a traditional ingredient in the diets of salmonids since for them dietary long-chain omega-3 fatty acids are "essential" (*6-9*). Some reports of soft texture in Atlantic salmon were felt to be due to the extra fat deposited in the muscle from the use of these so-called "high energy" diets. Accordingly our project was designed to evaluate salmon texture on an instrumental food science basis as well as by sensory panels. Since fats were intimately involved this review will separate into two parts, the salmon growth and texture and the exploration of the long-chain polyunsaturated omega-3 fatty acids available to consumers from salmon both as a food and as an oil source.

Experimental

All fish were smolts obtained from the Department of Fisheries and Oceans, Halifax for the pilot study, or from a commercial operation for the second study. They were held in 2m tanks provided with running sand-filtered and aerated seawater. Feeding was ad libitum on a daily basis. A 12 h photoperiod was provided. Weighing was done by netting a dozen fish at random from each tank for representative weights. Diets (Table I) were distributed to pairs of tanks so that the pilot study had two diets fed in triplicate and the subsequent final study had three diets fed in duplicate.

Fish for research were placed in an ice-seawater mix for 30 min and after cutting the gill were then bled for 20 min and gutted before transport to the laboratory for analysis. Steaks were cut from just in front of the dorsal fin, the foremost for texture measurement and for protein and lipid analysis. The adjacent steak under the fin was used for sensory panel evaluations.

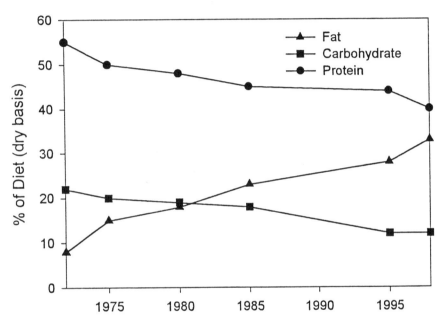

Figure 1. Percentages of major components in Atlantic salmon diets prepared by one firm in recent years. Data courtesy of W.D. Robertson (Connors Bros. Limited) and S.P. Lall (Institute for Marine Biosciences, National Research Council Canada)

Before texture measurement the steaks were held in ice for 24 h to allow the muscle to pass through rigor and were then allowed to warm to 17°C. At a selected central site in each loin a physical measurement of depression, rebound and recovery time was conducted with a TA-XT2 Texture Analyser (Texture Technologies Corp. Scarsdale, NY) following the principles set forth by Gill et al. (4) and Botta (10). The texture index (dr/di) is derived from deformation distance (di) and rebound distance (dr). Sensory panel evaluation was conducted by a trained panel using the triangle test (11). Loin white muscle of steaks was cut into small cubes, mixed and oven baked at 250°C in a covered -lass petri dish. Panelists were especially instructed in evaluation of mastication qualities (12).

All lipid extractions, including feeds, were conducted by the method of Bligh and Dyer (13). Fish muscle lipids were separated by streaking on thin-layer chromatography plates coated with silica gel, with development in hexane:diethyl ether:acetic acid (85:15:1, v/v/v). Two individual lipid classes (phospholipids, PL, and triacylclycerols, TAG) were recovered from the silica get by $CHCl_3$-MeOH extraction. The fatty acids of these lipids were converted to methyl esters with BF_3-MeOH (14). This analysis was conducted on an Omegawax-320 capillary column operated in a Perkin Elmer 8420 gas chromatograph. FID peak areas were converted to weight percent fatty acid (15).

Results and Discussion

The pilot study was run with two levels of fat in commercial fish diets (25% and 30%, Table 1). The fish muscle (whole steaks) differed in fat content (6.3 ± 0.6% and 8.3 ± 0.34%) after 12 months on diets, but there was no difference in either a mechanically measured texture index or sensory panel evaluation of texture (16). Accordingly the subsequent definitive study was modified to include three diets. Table I also gives the proximate composition of the second set of diets. The growth of these fish over one year on diets is given in Figure 2. The inflections in the growth lines presumably reflect water temperature changes. Although the supplies of commercial fish diets were purchased at different times over the feeding period, lipid analyses on each lot showed the fat contents to fluctuate only slightly, usually within ±1 % of the expected figure, and were not thought to be responsible for the change in relative growth curves where lines for the intermediate group fed diet B and the high fat diet group fed diet C finally converged. This is in fact thought to result from a reduction in growth rate in the high fat group for reasons discussed below, although water temperature fluctuations affecting, feed consumption could be applicable to all groups.

The necessity of aging steaks (Figure 3) to avoid an excessive fat figure from inclusion of the belly flaps and/or the fat deposit under the dorsal fin is described in detail by Sigurgisladóttir et al. (17) and Bell et al. (8). Table II provides the lipid contents for whole steaks, aged steaks, and the white muscle loin area where the

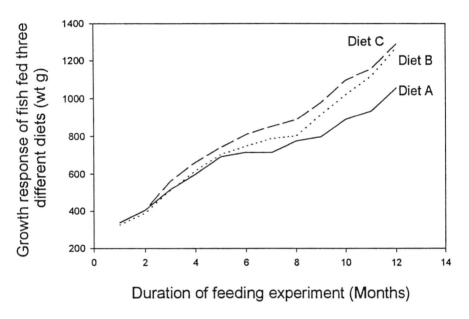

Figure 2. Actual weights of fish on the three diets for 12 months.

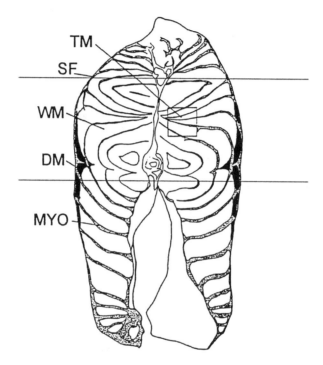

Figure 3. Parts of Atlantic salmon steak used in this study. WM is the white muscle including area for mechanical texture measurement. DM = dark muscle, SF = skin fat, MYO = myocommata (connective tissue). Texture measurements were made at the site TM, with minimal myocommata, shown for one side of WM by the square box. "Trimmed" portion of steak is between the two lines.

texture measurements were made, as well as of other body parts and the actual texture index results.

The basis of fat distribution in Atlantic salmon has recently been clarified by this laboratory (18,19) and by Aursand et al. (20). In brief there is in white muscle a basic muscle cell functional membrane composition of 0.6-0.8% lipid, primarily phospholipids (21,22). Bell et al. (8) show this in the form of the proportions of triacylglycerol and polar lipid plotted for a large number of Scottish farmed salmon. All other fat is in the form of adipocytes, literally bags of fat within a thin membrane and with the nucleus on the outside. The adipocytes are distributed along the connective tissue bands (myocommata in Figure 3) in the white muscle, but since the adipocytes are not very pigmented these appear to be the white streaks usually seen in raw salmon steaks, running through the "white" muscle, which is in fact usually well pigmented. The term "white" muscle is used to distinguish this type from the darker lateral line muscle common in all fish. This "dark" or "red" muscle has different lipid characteristics (23), but is relatively small in Atlantic salmon (Figure 3).

Table I. Proximate Compositions of diets[a] used in salmon feeding studies by actual analyses

	Pilot study		Final study		
	Low	High	A	B	C
Crude protein	44	43	46	45	45
Crude fat	25	30	25	29	31
Crude fibre	2	1	2	2	2
Ash	8	8	10	9	9
Digestible energy (Mj/kg)	17.8	19.3	19.7	20.5	21.1

[a]Two different suppliers

In this definitive study there were significant differences ($p<0.05$) in texture indices for the 12 fish examined from each of the three diets (Figure 4). The white muscle lipid contents were universally related to dietary fat in the fish from the three diets (Table 11). The muscle was also examined for the role of cathespsins A and B, and calpains, but no difference were detected that would explain the texture difference (S. White, unpublished data).

The sensory panelists were unable to satisfactorily detect difference in texture related to diet. This may have been partly due to the limited usefulness of this type of evaluation (24).

Table II. The fat content of whole steak, trimmed, white muscle and other quality parameters of Atlantic salmon

Diet and Fat content (%)	A (25%)	B (29%)	C (31%)
Fat contents (% wet weight) of			
Whole steaks (n=3)	5.23 ± 0.2^a	7.56 ± 0.01^b	9.70 ± 0.06^c
Trimmed steaks (n=3)	2.79 ± 0.01^a	4.58 ± 0.07^b	5.38 ± 0.01^c
White muscle loin (n=12)	1.27 ± 0.05^a	1.41 ± 0.05^a	2.25 ± 0.01^b
Total mesenteric tissues per fish (g) (n=12)	21.96 ± 2.49^a	27.79 ± 3.15^b	29.09 ± 5.11^c
Mesenteric lipid as % wet wt (n=3)	71.80 ± 1.38^a	69.98 ± 0.37^a	71.13 ± 1.84^a
Visceral index (n=12)	12.08 ± 0.56^a	11.89 ± 0.81^a	9.44 ± 0.30^b
Liver (g)/fish	23.51 ± 4.69^a	20.44 ± 2.66^{ab}	19.24 ± 3.86^b
Texture index (n=12)	36.00 ± 0.43^b	32.22 ± 0.43^b	30.00 ± 0.59^c

Different superscripts indicate significant differences across sample results (P<0.05 or P<0.01).

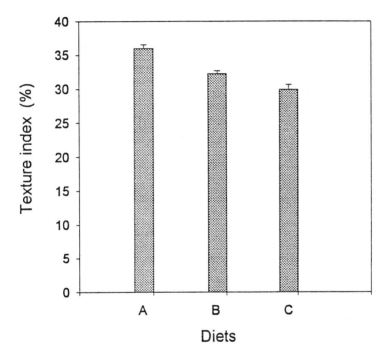

Figure 4. Texture indices in the loin area of experimental salmon based on three dietary treatments: A-LF (25.6%), B-MF (28.3%) and C-HF (30.9%) diets. Data are mean and standard deviation of 12 fish for each dietary treatment. There is a significant difference (P<0.05) in texture between all three diets.

Table III. Examples of lipid contents, and of polyunsaturated fatty acids in that lipid, comparing farmed Atlantic salmon with wild salmon.

	Icelandic			Scottish	
	Ocean-ranched[a]	Wild[b]	Farmed[b]	Wild[c,d]	Farmed
Lipid (w/w%)	4.8	5.0	18.6	3.5	10.1
18:4n-3	0.8	0.7	1.3	-	-
20:5n-3	4.7	4.5	3.8	5.9	5.8
22:5n-3	2.5	2.7	1.4	-	-
22:6n-3	11.3	13.2	5.1	12.9	11.6

[a]Jónsson et al. (Ref. 36)
[b]Sigurgisladóttir and Pálmadóttir (Ref. 37)
[c]Bell et al. (Ref. 8)
[d]TAG of wild fish lipids 72.5%, of farmed fish lipid, 89.2%

Subsequent to the sensory panelists evaluation of texture, the lipid distribution in the dorsal loin tissue revealed that by chance the site selected for texture measurements had a phenomenally low fat content ($\leq 2\%$). The sensory panelists thus received a mixture of pieces of muscle with different fat contents, a well-known problem with salmon muscle fillets but one to which we had not given sufficient attention in the case of steaks.

Omega-3 Fatty Acids

The concept of "wellness" from modify diets is now widely accepted in western society. One of the earliest medical reports in the use of fish oil fatty acids except as a source of vitamins A and D (cod liver oil), was that of Wome and Smith (25). Although handicapped by the absence of gas chromatography for fatty acid details, they prepared concentrates of fish oil fatty acids and studied their impact on serum cholesterol in patients. Any change in cholesterol was unlikely to have immediately benefitted the morale of the patients, but the reports of patients of improved agility and well-being were suggestive that the omega-3 fatty acids could have improved circulation, alleviated mild arthritis, or improved neural control. It is impossible to now evaluate these comments, but it is refreshing, to have a spontaneous health report attributed to fish oil fatty acids.

To promote fish as a healthy food a U.K. medical committee recommended (26) eating fish twice a week and further suggested that one of these meals be of "oily fish". This reflected the large-scale studies of Burr et at. published in 1989 in brief form (27) and later in more detail (28). The suitability of fish of various origins as sources of omega-3 fatty acids has been reviewed by Sargent (9) and the concept has been confirmed by numerous studies, firstly that of Kromhout et al. (29), and also by much more recent studies (30, 31). Epidemiology is subject to many variables including social status and life style (32). A frequent question relates to how much EPA and DHA is required daily to be useful for cardiovascular health in a "preventive" mode. In 1989 a total of about 300 mg seemed adequate (33), but exactly a decade later the recommendation has been doubled to 650 mg per day for a 2000 kcal diet or 0.3% of energy intake (34). Another recent recommendation was that people who do not eat fish should "consider obtaining 200 mg of very long-chain n-3 PUFA daily from other sources" (35).

In this context wild salmon provides a baseline for comparison with farmed fish (Table II). Icelandic ocean-ranched salmon are in effect wild fish and the omega-3 fatty acids total about 20% of fatty acids, although the fat content was relatively low (ca. 5%). The Scottish fatty acid data (8) is incomplete but if 18:4n-) and 22:5n-3 are added in at the likely 1% and 2%, respectively, the total omega-3 fatty acid content is again approximately 20% for the farmed fish. In all cases where the lipid exceeds 5% the triacylglycerols dominate the fatty acid composition. It seems safe to say that from a nutritional point of view the total of long-chain omega-3 fatty acids will be about 20% by weight of the fat content of farmed Atlantic salmon. Unfortunately, the fat content itself is variable and associated with fat levels in feed (8,20,38,39) and other factors such as starvation before slaughter (40) for farmed fish.

Our muscle lipid research data (Table IV), although based on relatively small and relatively low fat fish (Table II) confirm that there is a very good chance that in farmed salxnon DHA will exceed EPA in triacylglycerols. This is shown by careful examination of authentic salmon oil available in Europe in capsule form (41). With the addition of some phospholipid even more DHA will be provided in the amount consumed in our diets. This is possibly important in the context of preventing cardiac arrhythmia, a leading cause of "sudden" death (30,42,43).

The limited data in Tables III and IV suggest that there is an unusual accumulation of n-3 DPA (docosapentaenoic acid) in salmon lipids. In most fish oils this is about 1/10th of DHA (44) but in salmon triacylglycerols or oils it can be from 1/3 to 1/4th. This is also shown in an analysis of a salmon oil made from farmed salmon waste (Table IV). Primarily this type of oil is made from the viscera of farmed salmon (45) especially from the mesentery tissue (Table II), which is heavily laden with adipocytes (19). The n-3 DPA may be of great importance as it has been described as many times more effective than EPA in the epithelium of blood vessels (46).

Table IV. Distribution of omega-3 fatty acids (w/w%) in phospholipids and triacylglycerols in white muscle of experimental Canadian Atlantic salmon[a] compared to oil made from viscera of Canadian farmed Atlantic salmon[b].

	Phospholipids			Triacylglycerols			Visceral waste oil
	Low	Medium	High	Low	Medium	High	
18:4n-3	0.2	0.2	0.2	1.3	1.1	1.5	1.6
20:5n-3	8.4	8.3	9.0	7.8	7.5	10.1	7.5
22:5n-3	2.7	2.7	3.3	3.1	2.9	3.7	3.4
22:6n-3	44.0	40.2	42.0	11.0	9.8	11.0	7.6

[a]This study, are average of two fish
[b]Ackman, unpublished results

Overall it appears that 30% fat in farmed salmon diets may be almost a limiting factor in growth, accounting for the convergence of the two growth curves of Figure 2. Additional fat is simply deposited in the mesentery, and probably also in subdermal fats and elsewhere in the fish (20). If the Sprecher shunt (47) is unable to complete the EPA to DHA desaturation step with a Δ6 desaturase there may be a limit to the bioaccumulation of longer chain n-3 fatty acids. The fatty acids may then suffer from disruption of the balance between 20:5n-3 and 22:6n-3. A somewhat similar situation was noted in triacylglycerols from different tissues of another oily fish, the Atlantic mackerel *Scomber scombrus* (48). In that case the association of chain lengths (ie 22:5n-3 and 22:6n-3) was deemed a possible explanation.

The salmon family, including trout (49) has probably received more fatty acid research than almost any other fish species. However it is clear that we still do not fully understand the fatty metabolism in even this one species with its tremendous commercial implications.

Postscriptum

As recently as July 10, 1999, a leading Canadian newspaper ran an article on Pacific salmon (50). A highlighted section called "Tops in Nutrition" included a statement:

"But not all salmon is equal. Wild salmon eat other fish and sea vegetables, which gives them the high Omega-3 count. Farm fish eat soybean and canola and some ground fish, they are not high in Omega-3, but in Omega-6, which is less beneficial."

There is certainly totally misleading information in this in respect to wild salmon, which do not eat sea vegetables. There is instead a multi-step process of biofiltration and concentration of longer-chain omega-3 fatty acids in these fish (Figure 5). Omega-6 fatty acids are nearly all excluded by this enrichment process. For the farmed fish, soybean and canola meals as protein sources have been tested as partial replacement for fish meal (ground fish?), but the resulting fatty acids of salmon muscle are not necessarily "higher in omega-6", because vegetable meals are the residue after fat is extracted. In any case the salmon has a preference for accumulating omega-3 fatty acids in the muscle fats. The oxidation of any excess of omega-6 fatty acids could alter the spectrum of salmon flavors (51), as shown for poultry fed fish meal (52), but this remains to be tested.

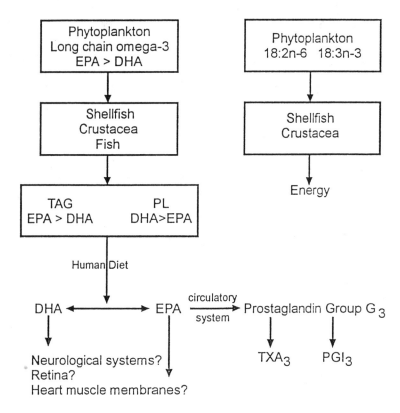

Figure 5. Role of marine phytoplankton fatty acids in the fats of seafoods.

References

1. Howgate, P. *Int. J. Food Sci. Technol.* **1998**, *33*, 99-125.
2. Glanz, K.; Basil, M.; Maibach, E.; Goldberg, J.; Snyder, D. *J. Am. Diet. Assoc.* **1998**, *98*, 1118-1126.
3. Taylor, A. J. *Crit. Rev. Food Sci. Nutr.* **1966**,*36*, 765-786.
4. Gill, T. A.; Keith, R. A.; Smith-Lall, B. *J. Food Sci.* **1979**, *44*, 661-667.
5. Haard, N. F. *Food Res. Int.* **1992**, *25*, 289-307.
6. Castell, J. D.; Sinnhuber, R. O.; Wales, J. H.; Lee, D. J. *J. Nutr.* **1972**, *102*, 77-85.
7. Castell, J. D.; Sinnhuber, R. O.; Lee, D. J.; Wales, J. H. *J. Nutr.* **1972**, *102*, 87-92.
8. Bell, J. G.; McEvoy, J.; Webster, J. L.; McGhee, F.; Millar, R. M.; Sargent J. R. *J. Agric. Food Chem.* **1998**, *46*, 119-127.
9. Sargent, J. R. *Brit. J. Nutr.* **1997**, *78*, Suppl. 1, S5-S13.
10. Botta, J. R. *J. Food Sci.* **1991**, *56*, 962-964, 968.
11. Poste, L. M.; Mackie, D. A.; Butler, G.; Larmond, E. Laboratory Methods for Sensory Analysis of foods. Agriculture Canada Publication 1864/E, Ottawa, 1991; 90 p.
12. Seurman, L.; Martinsen, C.; Little, A. In Proceedings World Symposium on Finfish Nutrition and fish feed Technology, Hamburg 20-23 June, 1978. Berlin. 1979; Vol. II, pp. 401-413.
13. Bligh, E. G.; Dyer, W. J. *Can. J. Biochem. Physiol.* **1959**, *37*, 911-917.
14. Ackman, R. G. *J. Am. Oil Chem. Soc.* **1998**, *75*, 541-545.
15. Ackman, R. G.; Ratnayake, W. M. N. In The Role of Fats in Human Nutrition; Crawford, M.; Veraroesen, A.J., Eds; Academic Press: London. 1989; 2nd. edn. pp. 441-514.
16. Ackman, R. G.; Gill, T. A.; Xu, X. L. In Quality Attributes of Muscle Foods; Xion, W. L.; Ho, C-T; Shahidi, F., Eds; Kluwer Academic / Plenum Publishers, 1999; pp. 61-71.
17. Sigurgisladóttir, S.; Torrissen, O.; Lie, O.; Thomassen, M.; Hafsteinsson, H. *Rev. Fish. Sci.* **1997**, *5*, 223-252.
18. Zhou, S.; Ackman, R. G.; Morrison, C. *Fish. Physiol. Biochem.* **1995**, *14*, 171-178.
19. Zhou, S.; Ackman, R. G.; Morrison, C. *Can. J. Fish. Aquatic Sci.* **1996**, *53*, 326-332.
20. Aursand, M., Bleivik, B., Rainuzzo, J.R., Jørgensen, L., and Mohr, V. *J. Sci. Food Agric.* **1994**, *64*, 239-248.
21. Polvi, S. M.; Ackman, R. G.; Lail, S. P.; Saunders, R. L. *J. Food Process Preserv.* **1991**, *15*, 167-181.
22. Polvi, S. M.; Ackman, R. G. *J. Agric. Food Chem,.***1992**, *40*, 1001-1007.
23. Ackman, R. G.; Eaton, C. A. *Can. Inst. Food Sci. Technol. J.* **1971**, *4*, 169-174.

24. Parrish, C. C.; McLeod, C. A.; Ackman, R. G. *J. Sci. Food Agric.* **1995**, *68*, 325-329.
25. Worne, H.E.; Smith, L. W. *Am. J. Med. Sci.* **1959**, *237*, 710-721.
26. U.K. Department of Health. Report on Health and Social Subjects, No. 46, Nutritional Aspects of Cardiovascular Disease. Report of the Cardiovascular Review Group, Committee on Medical Aspects of Food Policy, M4SO, London, 1994; 186 p.
27. Burr, M. L.; Gilbert, J. F.; Holliday, R. M.; Elwood, P. C.; Fehily, A. M.; Rogers, S.; Sweetnam, P. M.; Deadman, N. M. *Lancet* **1989**, Sept. 30, 757-761.
28. Burr, M. L. *Prog. Food Nutr. Sci.* **1989**,*13*,291-316.
29. Kromhout, D.; Bosschieter, E.B.; de Lezenne Coulander, C. *N. Engl. J. Med.* **1985**, *312*, 1205-1209.
30. Siscovick, D. S.; Raghunathan, T. E.; King, I.; Weinmann, S.; Wicklund, K. G.; Albright, J.; *JAAM* **1995**, *274*, 1363-1367.
31. Daviglus, M. L.; Stampler, J.; Orencia, A. J.; Dyer, A. R.; Liu, K.; Greenland, P.; Walsh, M. K.; Morris, D.; Shekelle, R. B. *N. Engl. J. Med.* **1997**, *336*, 1046-1053.
32. Johansson, L. R. K.; Solvoll, K.; Bjørneboe, G-E.; Drevon, C. A. *Eur. J. Clin. Nutr.* **1998**, *52*, 716-721.
33. Simopoulos, A. P. *J. Nutr.* **1989**, *119*, 521-528.
34. Simopoulos, A. P.; Leaf, A.; Salem, N., Jr. *ISSFAL Newsletter* **1999**, *6*, 14-16.
35. De Deckere, E. A. M.; Korver, O.; Verschuren, P. M.; Katan, M. B. *Eur. J. Clin. Nutr.* **1998**, *52*, 749-753.
36. Sigurgisladóttir, S.; Pálmadóttir, H. *J. Am. Oil Chem. Soc.* **1993**, 70, 1081-1087.
37. Jonsson, A., Pálmadóttir, Kristbergsson, K. *Internat. J. Food Sci. Tech.* **1997**, *32*, 547-551.
38. Sheehan, E. M.; O'Connor, T. P.; Sheehy, P. J. A.; Buckley, D. J.; Fitzgerald, R. *Irish J. Agric. Food Res.* **1996**, *35* ,37-42.
39. Bjerkeng, G.; Refstie, S.; Fjalestad, K. T.; Storebakken, T.; Rødbotten, M.; Roem, A. J. *Aquaculture* **1997**, *157*, 297-309.
40. Einen, O.; Waagan, B.; Thomassen, M.S. *Aquaculture* **1998**, *166*, 85-104.
41. Sagredos, A. N. *Fat Sci. Technol.* 93, 1991, 184-191.
42. Kang, J. X.; Leaf, A. *Circulation* **1996**, *94*, 1774-1780.
43. Sheard, N.F. *Nutr. Rev.* **1998**, *56*, 177-179.
44. Ackman, R. G. In Nutritional Evaluation of Long-Chain Fatty Acids in Fish Oils; Barlow, S. M.; Stansby, M. E., Ed.; Academic Press, London, 1982; pp 25-88.
45. Skara, T.; Cripps, S. In Seafood from Producer to Consumer, Integrated Approach to Quality; Luten, J.B.; Borresen; Oehlenschiager, Eds; Elsevier Science B.V., Amsterdam, 1997; pp 103-111.

46. Kanayasu-Toyoda, T.; Morita, I.; Murota, S-1. *Prostaglandins, Leukotrienes and Essential Fatty Acids* **1996**, *54*, 319-325.
47. Voss, A.; Reinhart, M.; Sprecher, H. *Biochim. Biophys. Acta* **1992**, *1127*, 33-40.
48. Ackman, R. G.; Orozco, V. R.; Ratnayake, W. M. N. *Fat Sci. Technol.* **1991**, *93*, 447-450.
49. Færgemand, J.; Rønsholdt, B.; Alsted, N.; Børresen, T. *Wat. Sci. Techol.* **1995**, *31*, 225-231.
50. Mallet, G. National Post, The National Post Company, Don Mills, ON, 1999, July 10, 20-21.
51. Farmer, L. J.; McConnell, J. M.; Graham, W. D. In Flavor and Lipid Chemistry of Seafoods; Shahidi, F.; Cadwallader, K. R.; ACS Symposium Series No. 674; American Chemical Society: Washington DC, 1997; pp 95-109.
52. O'Keefe, S. F.; Proudfoot, F. G.; Ackman, R. G. *Food Res. Int.* **1995**, *28*, 417-424.

Chapter 16

Highly Unsaturated Fatty Acids as Precursors of Fish Aroma

Toshiaki Ohshima, Janthira Kaewsrithong, Hiroyuki Utsunomiya, Hideki Ushio, and Chiaki Koizumi

Department of Food Science and Technology, Tokyo University of Fisheries, Konan 4, Minato-ku, Tokyo 108-8477, Japan

Certain fish species, including sweet smelt (*Plecoglossus altivelis*), rainbow smelt (*Osmerus eperlanus mordax*), shishamo smelt (*Spirinchus lanceolatus*), and Japanese smelt (*Hypomesus transpacificus*) are known as aromatic fish. These live fish has a characteristic cucumber-like or watermelon-like aroma that is completely a different characteristic from the so-called fishy smell. It is generally accepted that formation of volatile carbonyl compounds from unsaturated lipids via their oxidation contributes to the development of off-flavor. In case of aromatic fish, 2,6-nonadienal and 3,6-nonadien-1-ol are known as the key compounds that contribute to the development of the characteristic aroma. However, the origins of these C9 carbonyls remain obscure. We found that unusual amount of lipid hydroperoxides accumulated in the blood and liver of certain aromatic fish. It was considered that the hydroperoxides were generated as precursors of certain carbonyls from highly unsaturated fatty acids as a result of oxygenation by endogenous 12-lipoxygenase.

It is generally accepted that lipoxygenases are widely distributed in plants (*1-3*), mammalian (*4*) and fish (*5-10*) tissues. Generally, lipoxygenases catalyze oxygenation of highly unsaturated fatty acids (HUFAs) to form hydroperoxide geometrical isomers. Thus, lipoxygenases are categorized into several types depending on the site specificity to fatty acid oxygenation. It is found that there are two kinds of soybean lipoxygenases with different site specificities, 9- and 13-lypoxygenases, in four isozymes when linoleic acid is used as a substrate (*2*). In mammalian platelet (*11,12*), lung (*13,14*) and skin (*15*), 12-lipoxygenase, which is reactive to ward arachidonic acid, has been identified. 15-Lipoxygenase reacts with arachidonic acid in human reticulocyte (*16*) and eosinophilis (*17*). The gill and skin tissues of many species of fish contain 12- and/or 15-lipoxygenase activities when eicosapentaenoic acid (EPA) and docosahexaenoic acid (DHA) are used as substrates (*18-20*). The hydroperoxides are further converted by endogenous lyase (*20*) and/or by classical chemical cleavage reactions to several secondary volatile compounds, including relatively short chain alcohols, aldehydes, and ketones (*4*).

Fish and shellfish tissues are generally rich in HUFAs. Thus, n-3 HUFAs including EPA and DHA could be the precursors of the biogeneration of volatile aroma compounds in fresh seafoods (*5,21,22*). Site-specific lipoxygenase oxygenates EPA to generate geometrical isomers of fatty acid hydroperoxides.

Certain species of fish are known to develop a characteristic aroma at their live state. Josephson and Lindsay (*20*) and Josephson *et al.* (*21*) characterized the aroma profiles of emerald shiner (*Notropis artherinoides*) and proposed a mechanism for enzymatic conversion of HUFAs to volatile aroma compounds. Zhang *et al.* (*23*) revealed the composition of volatiles of sweet smelt (*Plecogrossus aureus*) and postulated some relation of lipoxygenase-like enzymes to the development of the characteristic cucumber-like aroma of the fish.

There are several interesting fish species that develop characteristic aromas. Japanese smelt (*Hypomesus transpacificus*) is a freshwater fish and has a slight green aroma. Rainbow smelt (*Osmerus eperlanus mordax*) inhabits both in the freshwater and in the seawater and has an intense cucumber-like aroma. This aroma usually transfers to other fish species with no aroma, and therefore most fishermen try to avoid catching rainbow smelt since most consumers do not like this intense cucumber-like aroma. Sweet smelt (*Plecoglossus altivelis*) with a watermelon-like aroma inhabits in Japan as well as in the south parts of Korea and China. Aroma of shishamo smelt, however, (*Spirinchus lanceolatus*) is also slightly green similar to the Japanese smelt. These does not have the so-called fishy smell, usually found in fresh fish such as rainbow trout. In the present study, contribution of HUFAs in fish and oyster to the development of characteristic aromas were studied by determining the levels of hydroperoxide as precursors of volatile compounds in the tissues of aromatic fish.

Materials and Methods

Samples

Live Pacific oysters (*Crassostrea gigas*), cultured in Hiroshima Prefecture as well as in Miyagi Prefecture, were hand-shucked and frozen quickly to −50 °C to be transferred into the laboratory and subjected to analyses. Commercially available cultured fish, including amberjack, flounder, red sea bream, and yellow tail were donated by Nippon Suisan Kaisha, Ltd. (Tokyo). Rainbow trout was cultured at the Tokyo University of Fisheries. Rainbow smelt and Japanese smelt were caught at the Saroma lake and its branch stream, respectively, in Hokkaido Prefecture, Japan. Shishamo smelt was caught at the Tokachi River in Hokkaido Prefecture. Some specimens of sweet smelt were obtained from hatcheries in Kanagawa Prefecture, Japan.

Lipid extraction and Fractionation

Total lipids (TL) were extracted with a mixture of chloroform and methanol (1:2, v/v) according to the procedure of Bligh and Dyer (*24*) from the total wet tissues of the shucked oyster. An aliquot of TL (about 100mg) was separated into PL and NL, using silicic acid cartridges (25mm x 10mm id, Sep-Pak silica, Waters Associates, Milford, MA) according to the method of Juaneda and Rocquelin (*25*).

Determinations of Fatty Acid Compositions of Total Lipids (TL), Phospholipids (PL) and Neutral Lipids (NL)

An aliquot of the lipid sample was saponified with ethanolic potassium hydroxide and subsequently methyl-esterified by 14% BF_3 in methanol to obtain the corresponding fatty acid methyl esters (FAME). Quantitative analysis of the FAME was carried out by gas-liquid chromatography (GLC) using a Shimadzu GC 15A instrument equipped with a Supelcowax-10 fused silica open tubular column (0.25 mm i.d. x 30 m, 0.25 μm in film thickness, Sigma Aldrich Group, Tokyo, Japan) and a flame ionization detector. The column oven temperature was programmed from 170 to 230 °C at a rate of 1 °C/min. Helium was used as a carrier gas at a column inlet pressure of 2 kg/cm^2 and a split ratio of 1:50.

Determination of Lipid Class Compositions

Quantitative determination of lipid classes was carried out for each lipid sample using Chromarod S-III and an Iatroscan MK-5 TLC-FID analyzer (Dia-Iatron Co., Tokyo, Japan). Briefly, an aliquot of the TL in chloroform was spotted on the Chromarod S-III and developed with hexane/diethyl ether/formic acid (90:10:1, v/v/v) to separate non-polar lipids, according to the method of Ohshima and Ackman (26), and acetone and subsequently chloroform/methanol/water (65:35:4, v/v/v) to separate polar lipid classes as described by Ratnayake and Ackman (27). The developed Chromarod S-III was evaporated for 2 min in an oven set at 110 °C and subjected to scanning at a scan speed of 40s, a hydrogen flow rate of 150 mL/min, and an air flow rate of 2,000 mL/min.

Analysis of Volatiles

The total wet tissues of hand-shucked oyster (80 g) were homogenized in a Waring Blender. The homogenate was transferred to a three-necked round bottom flask, and a stream of nitrogen gas was introduced into the flask to evaporate the volatiles at a flow rate of 100 mL/min with gentle stirring of the sample homogenate. The volatiles were transferred and collected onto 300 mg of Tenax TA (60-80 mesh, GL Science, Tokyo, Japan) packed in a glass tubing (4 mm i.d. x 11.5 mm) for 3 h at room temperature (20 °C).

The volatiles were desorbed from the Tenax TA by heating at 250 °C for 5 min with a Thermal Desorption Unit Model 890 (Dynatherm Analytical Instruments, Inc., Kelton, PA), and transferred to a Supelcowax-10 fused silica open tubular column (0.32 mm i.d. x 60 m, 0.25 µm in film thickness) for gas chromatography/mass spectrometry. The column oven temperature was held at 50 °C for 5 min and then programmed to 230 °C at a rate of 2 °C/min. Mass spectrometric analysis of the volatile compounds was carried out with a Shimadzu GC-MS 9020 DG instrument with an electron impact ionization source to which the outlet of the open tubular column was connected directly. The accelerating voltage and electron beam energy were set at 3 kV and 70 eV, respectively; the ion source temperature was at 250 °C. The data acquisition and processing were carried out with an on-line Shimadzu 1100 data system.

Determination of Hydroperoxides

Heparinated total blood was collected from the vein in the tail part of live fish. Total lipid was extracted from plasma, red blood cells (RBC) or liver with a mixture of chloroform and methanol containing butylated hydroxyanisole (BHA). 1-Palmitoyl-2-[12-(7-nitro-2-1,3-benzoxadiazole-4-yl)amino[dodecanoyl]-*sn*-glycero-3-phosphocholine (NBD-labeled PC, Avanti Polar-Lipids, Alabaster, AL) was added as an internal standard to the sample in the lipid extraction process step. The levels of phosphatidylcholine (PC) hydroperoxide of plasma, RBC and livers were determined by HPLC with a diphenyl-1-pyrenylphosphine (DPPP) fluorescent post-column detection system (*28,29*). A normal-phase silica column was used to separate hydroperoxy PC using a mixture of chloroform/methanol/water (50:8:2, v/v/v) as a mobile phase at a flow rate of 0.6 mL/min. The separated PC hydroperoxides react with DPPP to form DPPP oxide, which was excited at 352 nm and emission was monitored at 380 nm. The NBD-labeled phosphatidylcholine was detected at 534 nm with excitation at 460 nm.

In order to compare the hydroperoxide levels of several fish species on the same basis, the number of RBC were determined. The hydroperoxide levels were then expressed based on fmol per 10^5 cells.

Results

Fatty Acid Compositions of the Oyster Wet Tissues and Fish Blood and Liver

Table I compares major fatty acids of total lipids extracted from total wet tissues of the Pacific and European oysters. The total lipids consisted of 70% non-polar and 30% polar lipids. The non-polar lipids contained 80% triacylglycerols and 20% sterols. As expected from the lipid class compositions (data not shown), the percentage of EPA was high, amounting to 23% in the Pacific oyster and 16% in the European oyster. DHA was another prominent fatty acid, accounting for over 10% of the total. The level of n-6 unsaturated fatty acids was lower than that of n-3 unsaturated fatty acids.

Table II summarizes fatty acid compositions of plasma and red blood cells of sweet smelt. DHA was the most prominent HUFA and a small amount of EPA and arachidonic acid were also present. The total content of n-3 unsaturated fatty acids was higher than that of n-6 unsaturated fatty acids, because most of fatty acids are derived from cell membrane phospholipids. These particular fatty acid profiles show that the lipids of oyster tissues and sweet smelt blood are rich in HUFAs such as EPA and DHA.

Table I. Prominent fatty acid compositions of total lipids in the Pacific (Crassostrea gigas) and the European (Ostrea edulis) oysters.

Fatty acid	C. gigas	O. edulis
16:0	16.73	14.73
18:0DMA	3.55	8.30
18:0	3.11	4.78
16:1n-7	3.79	2.93
18:1n-9	2.25	2.66
18:1n-7	6.67	2.42
20:1n-11	1.19	1.38
20:1n-9	0.27	0.43
20:1n-7	3.08	4.75
18:4n-3	3.04	2.64
20:4n-6	1.29	2.85
20:5n-3	22.77	16.35
22:2NMID	2.25	4.19
21:5n-3	1.26	0.70
22:6n-3	10.37	13.61
Σn-6	4.00	6.22
Σn-3	39.77	36.66

NOTE: Units are weight percent.
DMA, Dimethyl acetal; NMID non-methylene interrupted diene.

Compositions of Volatiles Compounds from the Total Wet Tissues of Oyster

Table III summarizes the volatile compounds identified in the Pacific and European oysters. The compositions of some compounds showed a difference between the two oyster species. In the Pacific oyster, 1-penten-3-ol, 1-hepten-3-ol and 1,5-octadiene-3-ol were found in higher levels compared to those of the European oyster. On the contrary, 5-methyl-3-heptanone and 2,5-octadiene-1-ol were found in higher amounts in the European oyster. It is well accepted that (E,E)-2,6-nonadienal and 3,6-nonadien-1-ol and (E)-2-nonenal have the aroma note of clean cucumber. 1,5-Octadiene-3-ol shows a melon-like aroma note and 2,5-octadiene-3-ol has a fresh fish undernote. Josephson et al. (5) determined volatile compounds of the Atlantic and Pacific oysters and found that the levels of 1,5-octadiene-3-ol, 1-octen-3-ol, 2,5-octadien-1-ol, and 3,6-nonadien-1-ol were higher in the Pacific oyster. 3,6-Nonadien-1-ol was not found in the Atlantic oyster. In the present study, 3,6-nonadien-1-ol was found only in the Pacific oyster and was absent in the European oyster.

Table II. Prominent fatty acid compositions of total lipids in the sweet smelt (*Plecoglossus altivelis*) plasma and red blood cell.

Fatty acid	Plasma	Red Blood Cell
14:0	1.20	1.11
16:0	28.73	30.96
18:0	3.03	1.47
16:1n-7	3.79	2.93
18:1n-9	8.43	5.86
18:1n-7	1.46	1.44
18:2n-6	1.76	2.12
20:4n-6	1.29	2.49
20:5n-3	3.25	5.81
22:4n-6	0.58	1.11
22:4n-3	0.52	1.06
22:5n-3	3.51	2.79
22:6n-3	38.85	32.04
Σn-6	5.37	7.15
Σn-3	46.91	42.83

NOTE: Units are weight percent.
SOURCE: Adapted from ref. 29.

Table III. Prominent volatile compounds identified in the Pacific (*Crassostrea gigas*) and the European (*Ostrea edulis*) oysters.

Compound	C. gigas	O. edulis
Pentanal	6.2	15.4
1-Penten-3-ol	11.6	5.1
4-Methyl hexanal	tr	0.3
5-Hexen-3-one	tr	0.7
5-Methyl-3-heptanone	9.2	19.0
1-Hepten-3-ol	11.1	4.8
1-Octen-3-ol	6.0	7.4
1-Methyl propanoate-1-butanol	0.7	0.3
1,5-Octadien-3-ol	25.9	9.1
(*E,E*)-2,4-Heptadienal	tr	0.7
(*E,Z*)-2,6-Nonadienal	-	0.2
2-Undecanone	0.5	0.4
(*Z*)-2-Octen-1-ol	0.6	6.4
2,5-Octadien-1-ol	2.7	15.1
3,6-Nonadien-1-ol	2.0	-

NOTE: Units are weight percent.

Hydropeoxide Levels in Fish Blood and Liver

The levels of PC hydroperoxides in non-aromatic fish ranged between 2.9 and 5.1 nmol/mL of plasma. Rainbow trout plasma contained lesser levels of hydroperoxides, amounting to about 1.2 nmol/mL. On the contrary, the levels of hydroperoxides in the aromatic fish plasma were extremely high. Especially, sweet smelt plasma contained the highest amount of hydroperoxide, amounting to 23 nmol/mL sample.

The levels of PC hydroperoxides in the RBC were low in the non-aromatic fish, ranging between 24 and 33 fmol/10^5 RBC. For the non-aromatic fish, there was no significant difference in the levels between marine fish RBC and fresh water fish RBC. On the contrary, a large amount of PC hydroperoxide accumulated in all of the aromatic fish used in this study, ranging between 122 and 354 fmol/10^5 RBC. Especially, the levels of PC hydroperoxides in sweet smelt and the Japanese smelt RBC were remarkably high.

The sweet smelt livers contained a very large amount of hydroperoxides compared to the non-aromatic fish livers (Table IV).

Table IV. Contents of phosphatidylcholine hydroperoxide in several fish plasma, red blood cell, and livers.

Fish (Scientific name)	Plasma nmol/mL	Red Blood Cell fmol/10^5 cell	Liver nmol/g
Yellowtail (*Seriola quinqueradiata*)	4.7±1.8	25±2.7	77.4±20.2
Amberjack (*Seriola dumerili*)	2.9±0.95	24±4.9	-
Flounder (*Paralichthys olivaceus*)	5.1±0.45	33±3.0	7.89±0.68
Sea bream (*Pagrus major*)	3.6±1.1	32±3.5	118±48.8
Rainbow trout (*Oncorhynchus mykiss*)	1.2±0.21	25±3.1	118±17.7
Japanese Smelt (*Hypomesus transpacificus*)	21±2.0	354±29	-
Rainbow smelt (*Osmerus mordax*)	10±0.76	178±8.7	-
Shishamo smelt (*Spirinchus lanceolatus*)	10±0.59	122±5.3	-
Sweet smelt (*Plecogrossus altivelis*)	23±2.8	202±27	670±93.0

-, not analyzed
SOURCE: Adapted from ref. 29.

Discussion

Among the several biogenerated volatile compounds from fresh seafood, each compound has its own characteristic aroma note. 2,4,7-Decatrienal contributes to the oxidized fish oil smell. Contrary to this, both 2-nonenal and 2,6-nonadienal have a cucumber rind like aroma. Also, 3,6-nonadien-1-ol has the aroma note of clean cucumber. Octadiene-3-ol has a seaweed-like aroma. It should be notified that the threshold of each volatile compound is significantly different. This means that we can distinguish the aroma of the compounds with low threshold even though their actual content might be very low. For example, one can identify cucumber-like aroma in foods even though the contents of both nonenal and nonadienal are often quite low because the threshold of these compounds is low, being 0.08 and 0.01 ppb, respectively (*30*). Josephson *et al.* (*5*) found that 3,6-nonadien-1-ol did not exist in the Atlantic oyster and postulated that this was probably due to lack of lipoxygenase activity in the Atlantic oyster. In the present study, only the Pacific oyster contained 3,6-octadien-1-ol. Other volatiles such as 1,5-octadien-3-ol, 3,6-nonadien-1-ol and 1-penten-3-ol are biogenerated by oxygenation of n-3 HUFAs by 12-lipoxygenase. It is postulated, therefore, that the Pacific oyster was high in 12-lipoxygenase activity compared to the Atlantic oyster.

There are many factors that affect both the formation and decomposition of hydroperoxides in biological tissues. Superoxide dismutase, glutathione reductase, and true antioxidants such as α-tocopherol in fish tissues will act as suppressors against formation of oxidation products. On the contrary, lipoxygenase, heme protein as well as trace transition metals in the tissues will accelerate lipid peroxidation and their further decomposition to volatile compounds. Therefore, the level of hydroperoxides as precursors of aroma compounds in fish tissue is controlled by the balance of antioxidative factors and prooxidative factors.

Josephson and Lindsay (*20*) have presented a mechanism for the biogeneration of volatile aroma compounds from n-3 HUFAs in fresh fish. We would propose another possible mechanism of formation of nonenal, which contributes to the cucumber-like aroma of aromatic fish. The arachidonic acid compositions in the aromatic fish usually accounts for around 2%. This suggests that biogenerated amount of nonenal must be low, however, the threshold of nonenal determined in water is only 0.08 ppb (*30*). Therefore, nonenal should contribute to development of the characteristic aroma of aromatic fish.

In conclusion, relationship between accumulated hydroperoxides and characteristic aroma development may be explained as shown in Fig. I. Highly unsaturated fatty acids in aromatic fish are oxygenated to positional isomers of hydroperoxides by 12- and 15-lipoxygenases. It is known that 15-lipoxygenase

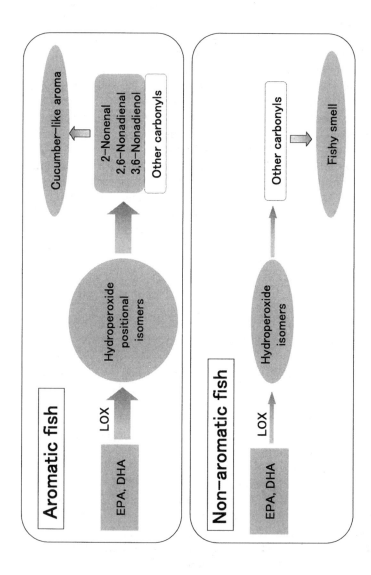

Figure 1. Generation of cucumber-like arom in aromatic fish.
The level of hydroperoxide is higher in aromatic fish tissues. Highly unsaturated and position-specific hydroperoxides play an important role as precursors of aroma compounds which are subsequently formed by enzymatic and non-enzymatic degradation.

and 12-lipoxygenase exist in also non-aromatic fish such as rainbow trout (8,19). In aromatic fish, however, lipoxygenase activity might be higher than that of non-aromatic fish, because the level of hydroperoxide was higher. Therefore, highly unsaturated and position-specific hydroperoxides play an important role as precursors of aroma compounds which are formed by both enzymatic and non-enzymatic degradation.

References

1. Galliard, T.; and Phillips, D. R. *Biochem. Biophys. Acta*, **1976**, *431*, 278-287.
2. Grechkin, A. *Prog. Lipid Res.* **1998**, *37*, 317-352.
3. Perez, A. G.; Sanz, C.; Olias, R.; Olais, J. M.. *J. Agric. Food. Chem.* **1999**, *47*, 249-253.
4. Frankel, E. N. **1998**. *Lipid Oxidation*. The Oily Press, Scotland, 249-291.
5. Josephson, D. B.; Lindsay, R. C.; Stuiber, D. A. *J. Food Sci.* **1985**, *50*, 5-9.
6. German, J. B.; Bruckner, G. G.; Kinsella, J. E. *Biochem. Biophys. Acta*, **1986**, *875*, 12-20.
7. Josephson, D. B.; Lindsay, R. C.; Stuiber, D. A. *J. Food Sci.* **1987**, *52*, 596-600.
8. Hsieh, R. J.; Kinsella, J. E. *J. Agric. Food. Chem.* **1989**, *37*, 279-285.
9. Mohri, S.; Cho, S.; Endo, Y.; Fujimoto, K. *J. Agric. Food Chem.* **1992**, *40*, 573-576.
10. Harris, P.; Tall, J. *J. Food Sci.* **1994**, *59*, 504-516.
11. Hamberg, M.; Samelsson, B. *Proc. Natl. Acad. Sci. USA* **1974**, *71*, 3400-3404.
12. Nutgeren, D.H. *Biochim. Biophys. Acta* **1975**, *380*, 299-307.
13. Hamberg, M. *Biochim. Biophys. Acta* **1976**, *431*, 651-654.
14. Pace-Asciak, C.R.; Mizuno, K.;Yamamoto, S. *Biochim. Biophys. Acta* **1981**, *665*, 352-354.
15. Ruzicka, T.; Vitto, A.; Printz, M.P. *Biochim. Biophys. Acta* **1983**, *751*, 368-374.
16. Schewe, T.; Papoport, S.M.; Kuhun, H. *Adv. Enzymol.* **1986**, *58*, 191-276.
17. Turk, J.; Maas, R.L.; Brash, A.R.; Robert II, L.J.; Oates, J.A. *J. Biol. Chem.* **1982**, *257*, 7068-7076.
18. German, J. B.; Kinsella, J. E. *J. Agric. Food Chem.* **1985**, *33*, 680-683.
19. German, J. B.; Creveling, R. K. *J. Agric. Food Chem.* **1990**, *38*, 2144-2147.
20. Josephson, D. B.; Lindsay, R. C. *Biogeneation of Aromas*. American Chemical Society, Washington, D.C. **1986**, pp 201-219.

21. Josephson, D. B.; Lindsay, R. C.; Stuiber, D. A. *J. Agric. Food Chem.* **1984**, *32*, 1347-1352.
22. Hsieh, R. J. *Lipids in Food Flavors.* American Chemical Society, Washington, D.C. **1986**, pp 30-48.
23. Zhang, C.; Hirano, T.; Suzuki, T.; Shirai, T. *Nippon Suisan Gakkaishi*, **1992**, *58*, 559-565.
24. Bligh, E. G.; Dyer, W. J. *Can. J. Biochem. Physiol.* **1959**, *37*, 911-917.
25. Juaneda, P.; Rocquelin, G. *Lipids* **1985**, *20*, 40-41.
26. Ohshima, T.; Ackman, R. *J. Planar Chromatogr.* **1991**, *4*, 27-34
27. Ratnayake, W.M.N.; Ackman, R.G. *Can. Inst. Food Sci. Technol. J.* **1985**, *18*, 284-289.
28. Ohshima, T.; Hopia, A.; German, J. B.; Frankel, E. N. *Lipids* **1996**, *31*, 1091-1096.
29. Kaewsrithong, J.; Qiau, D.-F.; Ohshima, T.; Ushio, H.; Yamanaka, H.; Koizumi, C. *Fisheries Sci.* in press.
30. Tressl, R.; Bahri, D.; Engel, K.H. *Lipid oxidation in fruits and vegetables.* American Chemical Society, Washington, D.C. **1986**, p16.

Chapter 17

Identification of Potent Odorants in Seal Blubber Oil by Direct Thermal Desorption–Gas Chromatography–Olfactometry

Keith R. Cadwallader[1,3] and Fereidoon Shahidi[2]

[1]Department of Food Science and Technology, Mississippi Agricultural and Forestry Experiment Station, Mississippi State University, Box 9805, Mississippi State, MS 39762
[2]Department of Biochemistry, Memorial University of Newfoundland, St. John's, Newfoundland A1B 3X9, Canada
[3]Present address: Department of Food Science and Human Nutrition, University of Illinois at Urbana-Champaign, 202 Ag. Bioprocess Lab, 1302 West Pennsylvania Avenue, Urbana, IL 61801

Volatile components of crude and refined-bleached-deodorized seal blubber oils were analyzed by direct thermal desorption-cryogenic trapping-gas chromatography olfactometry (DTD-GCO). Sample dilution analysis (SDA) was used in conjuction with DTD-GCO to indicate predominant odor-active compounds in the oils. A total of 27 odorants were detected consisting of 15 aldehydes, five ketones, two alcohols, and five unknown compounds. More (27) odorants were identified in the crude oil than in the refined oil (17). In general, odorants were present at higher odor-potency in the crude oil. Predominant odorants included (Z)-1,5-octadien-3-one (*metallic*), (E,E,Z)-2,4,7-decatrienal (*fatty, fishy*), (Z)-3-hexenal (*green, cut-leaf*), (E,Z)-2,6-nonadienal (*cucumber*), and an unknown (*plastic, water bottle*). DTD-GCO and SDA provided a simple, rapid, and sensitive approach for the study of the occurrence and removal of off-odors in seal blubber oils.

Omega-3 polyunsaturated fatty acids (PUFAs), in the form of fish and algal oils, are popular dietary supplements. Likewise, the oil derived from seal blubber contains appreciable amounts of long chain ϖ-3 PUFAs, namely eicosapentaenoic acid (EPA) and docosahexaenoic acid (DHA), and is, therefore, also a potentially valuable nutraceutical product (*1*). However, seal blubber oil, like other marine oils, is highly unstable toward oxidative deterioration. Potent off-odors can rapidly develop during processing and storage causing quality deterioration (*2,3*). Furthermore, hydroperoxides, free radicals, and aldehydes formed during oxidation may have negative health consequences (*4*). Volatile components of marine oils and of their constituent fatty acids have been reported (*3,5-11*). However, comparatively little research has been published on the volatile compounds that cause characteristic off-odors in the oils (*8,9*). Knowledge of the identities of these odor-impact compounds could aid in the development of better analytical strategies to assess the effectiveness of purification/deodorization processes and antioxidant systems in retarding oxidation of marine oils. Use of instrumental-sensory based techniques, such as gas chromatography-olfactometry (GCO), is an excellent approach for the determination of the potent odorants in marine oils.

Gas chromatography-olfactometry (GC), in which the human nose is used as a GC detector, has been in existence for nearly four decades (*12*). It was recognized early on that the combination of GC with olfactometry was a potentially powerful technique for identifying and characterizing the aroma-active components in a complex volatile mixture (*13,14*). The past two decades have experienced tremendous growth in the development and application of GCO in aroma research (*15*). Numerous GCO techniques are currently in use, but the most popular, and possibly most effective methods, are the so-called dilution methods, which include aroma extract dilution analysis (AEDA)(*16*) and CharmAnalysis^{TM}(*17*). In both techniques a dilution series of an aroma extract is evaluated by GCO. For example, in AEDA the highest dilution at which a compound is detected by GCO is defined as its flavor dilution factor, which is directly proportional to the odor activity value of the compound. A major limitation of dilution techniques is that they do not account for highly volatile compounds that are lost during solvent extraction and workup procedures. Techniques such as GCO of decreasing headspace volumes (GCO-H) (*18*) and GCO of dynamic headspace volumes (GCO-DHS) (*19*) have been used along with AEDA to provide a more complete evaluation of food aroma components.

For the most part, the analysis of the volatile composition of edible oils has been accomplished by gas chromatographic-based methods. Static or dynamic headspace-GC have been the most popular techniques (*20-25*). However, in general, headspace methods are only suitable for the analysis of components of high and intermediate volatility. A good alternative to static and dynamic headspace techniques is direct thermal desorption (DTD). In DTD, a purge gas is used to extract and transfer the volatile constituents from a heated sample directly to a cryofusing trap at the head of a GC column (*26,27*). Potential advantages of DTD include ease of sample preparation, need for only a small quantity of sample, high sample throughput, and efficient isolation of volatile constituents of widely varying volatilities (*27*). DTD-GC is limited to analysis of low moisture foods and has been successfully employed in the analysis of edible oils and fats (*7,28-30*). A schematic of how DTD-GC is carried for the analysis of an oil sample is shown in Figure 1. In many cases, the results of DTD-GC compared well with sensory flavor scores of taste panels on the same products (*31,32*). Some potential disadvantages of DTD-

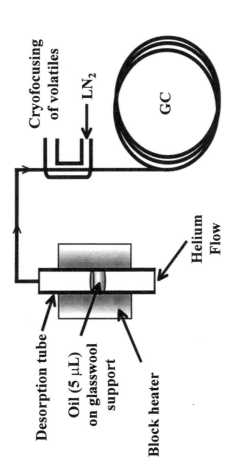

Figure 1. Apparatus for direct thermal desorption
Figure 1. Appareil pour désorption thermique directe

GC are sample carryover and decomposition of thermally labile components leading to the formation of artifacts and/or loss of thermally labile aroma components.

As previouly stated, direct thermal desorption (DTD) combined with cryogenic trapping and GCO allows for the evaluation of essentially the whole range of sample volatiles, from those of very high volatility to the semivolatiles. It is possible to conduct GCO 'dilution analysis' using this approach, in which stepwise dilutions (or decreased amounts or volumes) of a dry or semidry sample (e.g. an edible oil or dried herb or spice) are analyzed by DTD-GCO. Results of DTD-GCO-sample dilution analysis (SDA) allow for ranking of odorants based on lowest volume (or mass) of sample required for their detection by DTD-GCO. The aim of this study was to evaluate the use of DTD-GCO and SDA for the determination and comparison of predominant odorants in crude and refined-bleached-deodorized seal blubber oils.

Materials & Methods

Materials.

The seal blubber oil samples used in this study were prepared as previously described (*1*). A general scheme used for the preparation of crude and refined-bleached-deodorized oils is presented in Figure 2. Samples were stored under nitrogen at −20°C in amber vials equipped with teflon-line caps.

Reference compounds listed in Tables 1 and 2 were obtained from commercial sources. Compounds no. 11 (*33*) and 12 (*34*) were synthesized using published procedures.

Direct Thermal Desorption-Gas Chromatography-Olfactometry (DTD-GCO) and Sample Dilution Analysis (SDA).

A Tekmar™ 3000 Purge and Trap Concentrator/Cryofucusing Module (Tekmar Co., Cincinnati, OH) coupled to an HP5890 Series II GC (Hewlett-Packard Co., Palo Alto, CA) was used for DTD-GCO. The GC was equipped with a flame ionization detector (FID) and sniffer port. The Tekmar unit was configured to operate in the thermal desorption mode. A 30-cm length of 0.3-cm i.d. stainless steel glass-lined tubing (sample tube) was loosely packed with volatile-free salinized glass wool to permit diffusion of oil throughout the packing. Clearance of about 12.3 cm was allowed at the bottom of the liner and 2.5 cm at the top. The sample tube and glass wool packing were cleaned by baking at 225°C under a constant flow (40 mL/min) of helium.

For analysis, an oil sample (5 µL of neat oil or a 5 µL aliquot of a 1:10, 1:100, or 1:1000 serial dilution (e.g. 1:10 v/v, one part oil + nine parts of dichloromethane) was injected onto the top of the glass wool packing using a series 701 syringe (Hamilton Co. Reno, NV). The sampling tube was connected to the trap heater of

225

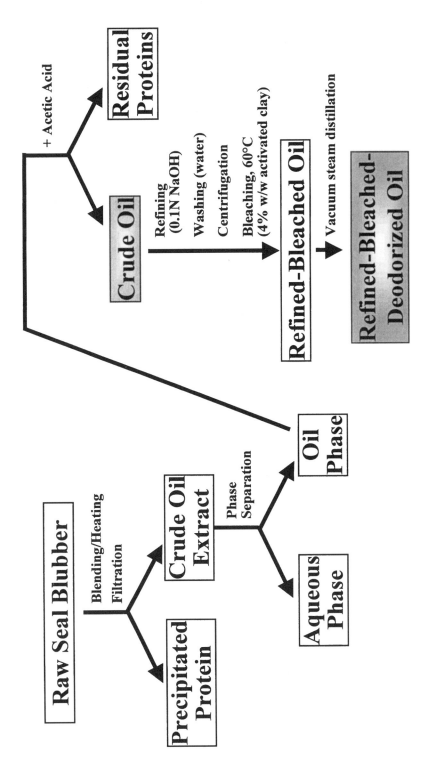

Figure 2. Generalized scheme for the refining of seal blubber oil.

the Tekmar unit. The volatiles were purged (with helium) from the oil sample at 180°C for 10 min and then subsequently cryofocused (-150°C) onto a 15-cm section of 0.53 mm i.d. deactivated fused silica capillary column. Transfer lines and valves were maintained at a temperature of 200°C. Helium flows during thermal desorption of sample tube (20 mL/min) and cryofocusing trap (1 mL/min) were controlled by the split/splitless electronic pressure control pneumatics of the GC as previously described (35). Cryofocused volatiles were thermally desorbed (180°C for 1 min) directly into the analytical GC column. Between each analysis, the DTD system was thoroughly baked out (225°C for 10 min) after installation of a clean sample tube. The above DTD temperature settings were chosen based on the literature, as this temperature is between that used previously for blended oils (170°C) (31) and fish oils (200°C)(7). GC separations were performed on a DB-WAX column (30 m length x 0.32 mm i.d. x 0.25 μm film thickness; J&W Scientific, Folson, CA). GC oven temperature was programmed from 40°C to 200°C at a rate of 10°C/min, with initial and final hold times of 5 and 30 min, respectively. FID and sniff port temperatures were maintained at 250°C.

Each sample dilution was evaluated by two trained panelists who were instructed to record odor intensity (scale from 0 = no odor, to 7 = very strong), retention time, and odor properties of each odorant perceived during GCO. Results are expressed as dilution volumes (DVs) on DB-WAX column. A detection volume is the lowest volume of sample required for detection of an odorant by DTD-GCO. Sample dilution factors were calculated for each odorant by dividing the largest sample volume analyzed (i.e. 5 μL) by the detection volume for that compound.

Direct Thermal Desorption-Gas Chromatography-Mass Spectrometry (DTD-GC-MS).

A Tekmar™ 3000 Purge and Trap Concentrator/Cryofucusing Module (Tekmar Co., Cincinnati, OH) coupled to an HP5890 Series II GC/HP5972 mass selective detector (Hewlett-Packard Co.) was used for DTD-GC-MS. DTD conditions were the same as those described above. Spiked oil sample was prepared by adding 5 μL (4.28 μg) of an internal standard solution (2-methyl-3-heptanone in methanol) to 2 g of oil, followed by 5 min of vigorous shaking and settling (1 h) prior to analysis. A 5 μL aliquot was then analyzed using DTD as described above.

GC separations were performed on a DB-WAX column (60 m length x 0.25 mm i.d. x 0.25 μm film thickness; J&W Scientific). GC oven temperature was programmed from 40°C to 200°C at a rate of 3°C/min, with initial and final hold times of 5 and 60 min, respectively. MSD conditions were as follows: capillary direct interface temperature, 280°C; ionization energy, 70 eV; mass range, 33-350 a.m.u.; electron multiplier (EM) voltage, 200 V above autotune; scan rate, 2.2 scans/s. Compound identifications were based on comparison of retention indices (RI), mass spectra, and odor properties of unknowns with those of standard reference compounds analyzed under identical experimental conditions. Tentative identifications were based on matching RI values and odor properties of unknowns with those of reference compounds. Concentrations of compounds were estimated based on area ratios relative to the internal standard and are not corrected for recovery and mass spectral reponse factors.

Results and Discussion

Volatile Profiles

Typical total ion chromatograms (TICs) of volatile constituents of crude and refined-deodorized seal blubber oils are shown in Figure 3. DTD-GC-MS analysis required only a 5 μL aliquot of oil sample, indicating the method is highly sensitive. Furthermore, the TICs demonstrate that the technique is capable of efficient isolation of components of widely varying volatilities, from the highly volatile constituents (RI < 1000) to the semivolatile compounds such as the 2,4,7-decatrienal isomers. The crude and refined-bleached-deodorized oils are readily distinguishable by their volatile profiles (Figure 3). As expected, there was a general reduction in the number and abundance of volatiles in the oil after refining, bleaching, and deodorization processes. Based on these results, DTD-GC-MS may serve as a good screening tool to monitor the effectiveness of refining, bleaching, and deodorization processes in the production of seal blubber oil.

Odor-Active Components

In addition to depicting the volatile profiles of the crude and refined-bleached deodorized oils, Figure 3 also indicates the prodominant odorants in these oils. The most intense odorants were assigned the highest sample dilution factor (SD-factors) of 1000. These compounds were detected at higher odor intensities in the crude oil than the refined-bleached deodorized oil. The SD-factors described here are analogous to FD-factors used in AEDA (*16*) and are equal to the ratio of the initial sample volume (5 μL) to the lowest sample volume required for detection (detection volume, DV) of an odorant by GCO. However, in sample dilution analysis DVs may be superior to SD-factors for ranking odor intesities, since the former gives a measure of how much sample is required for detection of the odorant by GCO. A complete listing of the odor-active components detected in crude and refined-bleached-deodorized oils is given in Table 1. A combined total of 27 odorants were detected with DVs ≤ 5000 nL. All 27 were detected in the crude oil, with only 17 being detected in the refined-bleached-deodorized oil.

Based on their low DVs and high odor intensity ratings, the most intense odorants were (Z)-1,5-octadien-3-one (no. 11; *metallic*), (E,E,Z)-2,4,7-decatrienal (no. 25; *fatty, fishy*) and an unknown having a *plastic, water bottle* note (no. 6). All three compounds had DVs equal to 5 nL in the crude oil and 50-500 nL in the refined-bleached-deodorized oil. Among these compounds, no. 25 was assigned the

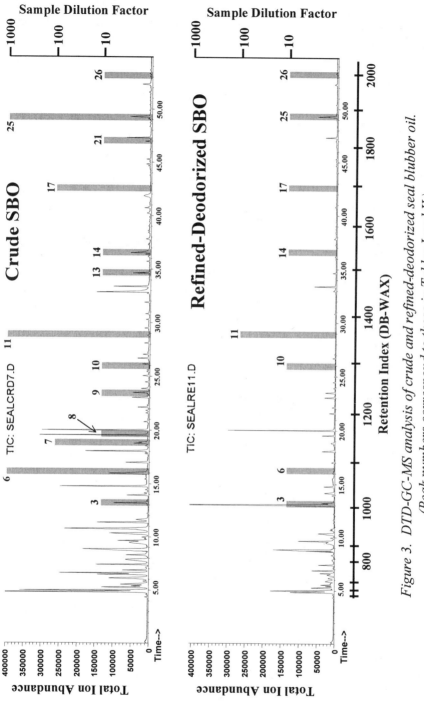

Figure 3. DTD-GC-MS analysis of crude and refined-deodorized seal blubber oil. (Peak numbers correspond to those in Tables I and II.)

Table I. Predominant odor-active components of crude and refined-deodorized seal blubber oil.

No.[a]	Compound	RI[b]	Odor property	Det. Vol. (nL) (Odor Intensity)[c] Crude	Refined
1	Acetaldehyde	<800	pungent, yogurt	5000 (2.5)	ND[d]
2	2,3-Butanedione	965	buttery, cream cheese	5000 (2.0)	5000 (0.5)
3	1-Penten-3-one	1014	plastic, water bottle	500 (2.5)	500 (1.5)
4	2,3-Pentanedione	1049	buttery, cream cheese	5000 (1.5)	ND
5	Hexanal	1076	green, cut-grass	5000 (2.0)	ND
6	Unknown	1088	plastic, water bottle	5 (2.5)	500 (2.5)
7	(Z)-3-Hexenal	1134	green, cut-leaf	50 (3.0)	5000 (2.5)
8	1-Penten-3-ol	1150	plastic, water bottle	500 (1.0)	ND
9	(Z)-4-Heptenal	1233	rancid, crabby	500 (2.5)	5000 (1.5)
10	1-Octen-3-one	1294	mushroom, earthy	500 (2.0)	500 (1.5)
11	(Z)-1,5-Octadien-3-one[e]	1362	Metallic	5 (3.5)	50 (3.0)
12	(E,E)-2,4-Heptadienal	1487	stale, rancid	500 (2.5)	5000 (2.0)
13	(Z,Z)-3,6-Nonadienal[e]	1491	fresh, watermelon	5000 (2.5)	5000 (2.0)
14	(E)-2-Nonenal	1530	stale, hay	500 (2.5)	500 (2.0)
15	(E,Z)-2,6-Nonadienal	1582	Cucumber	50 (3.0)	500 (3.0)
16	Butyric acid	1610	vomit, sour milk	5000 (2.0)	ND
17	3-Methylbutyric acid	1648	sweaty, smelly feet	5000 (2.0)	ND
18	Unknown	1708	stale, metallic	5000 (1.5)	5000 (2.0)
19	(E)-2-Undecenal	1749	green, cilantro	5000 (2.0)	5000 (0.5)
20	(E,Z)-2,4-Decadienal	1763	fatty, fried	5000 (1.0)	ND

Continued on next page.

Table I. continued

No.[a]	Compound	RI[b]	Odor property	Det. Vol. (nL) (Odor Intensity)[c] Crude	Refined
21	(E,E)-2,4-Decadienal	1810	fatty, fried	500 (2.5)	5000 (2.0)
22	(E,Z,Z)-2,4,7-Decatrienal*	1820	metallic, fatty	5000 (1.5)	ND
23	Unknown	1852	fatty, fish oil	5000 (2.5)	ND
24	Unknown	1870	fatty, fried, fishy	5000 (2.0)	5000 (2.0)
25	(E,E,Z)-2,4,7-Decatrienal*	1877	fatty, fishy	5 (4.0)	50 (3.5)
26	(E,E,E)-2,4,7-Decatrienal*	2000	stale, fatty	500 (2.0)	500 (1.5)
27	Unknown	2086	fatty, soapy	5000 (2.5)	ND

[a]Numbers correspond to those in figure 3 and Table II. [b]Retention index on DB-WAX column. [c]Detection volume, lowest sample volume required for detection by DTC-GCO. Numbers in parentheses represent average odor intensity ratings perceived during DTD-GCO analysis of 5 µL oil samples. [d]Not detected. [e]Compound tentatively identified by comparison of odor properties and RI with those of reference standard. *Compound tentatively identified based on literature data.

highest odor intensity rating in both oils, indicating that it was possibly the most intense odorant in the crude and refined-bleached-deodorized oils. Two additional potent odorants (DVs = 50 nL in crude oil) were identified as (Z)-3-hexenal (no. 7; *green, cut-leaf*) and (E,Z)-2,6-nonadienal (no. 15, *cucumber*). These compounds had DVs equal to 5000 nL and 500 nL, respectively, in the refined-beahed-deodorized oil. Compounds having DVs equal to 500 in the crude oil included 1-penten-3-one (no. 3; *plastic, water bottle*), 1-penten-3-ol (no. 8; *plastic, water bottle*), (Z)-4-heptenal (no. 9; *rancid, crabby*), 1-octen-3-one (no. 10; *mushroom, earthy*), (E,E)-2,4-heptadienal (no. 12; *stale, rancid*), (E)-2-nonenal (no. 14; *stale, hay*), (E,E)-2,4-decadienal (no. 21, *fatty, fried*) and (E,E,E)-2,4,7-decadienal (no. 26; *stale, fatty*). Except for no. 26, the DVs for these compounds were lower in the crude oil. Odorants found at low intensities included aldehydes (nos. 1, 5, 13, 19, 20, 22), ketones (nos. 2 and 4), short-chain volatile acids (nos. 16 and 17) and several unknowns (nos. 18, 23, 24, and 27).

Relative concentrations of selected volatile compounds are listed in Table 2. Concentrations are based on peak area comparisons to the internal standard spiked into each oil at a level of 2.1 ug/g. The relative response of the internal standard compared with the other volatile components of the oils is depicted in Figure 3 also. Aldehydes were the most abundant components of both oils, followed by ketones and alcohols. All compounds were in lower abundance in the refined-bleached-deodorized oil. Large decreases were observed for the very volatile components,

Table II. Relative concentrations of selected volatile components of crude and refined-deodorized seal blubber oil

No.[a]	Compound	RI[b]	Cnc (ng/g)[c]	
			Crude	Refined
2	2,3-Butanedione	975	1600 (± 18%)	36.0 (± 35%)
3	1-Penten-3-one	1014	884 (± 8.4%)	388 (± 19%)
4	2,3-Pentanedione	1049	2100 (± 0.5%)	312 (± 34%)
5	Hexanal	1076	1770 (± 9.1%)	121 (± 53%)
7	(Z)-3-Hexenal	1134	460 (± 9.8%)	18.3 (± 77%)
8	1-Penten-3-ol	1150	1990 (± 19%)	347 (± 41%)
9	(Z)-4-Heptenal	1233	67.0 (± 18%)	- -[d]
12	(E,E)-2,4-Heptadienal	1487	365 (± 6.2%)	86.2 (± 26%)
	(E,E)-3,5-Octadien-2-one	1513	86.6 (± 31%)	- -
14	(Z)-2-Nonenal	1530	119 (± 23%)	9.6 (± 26%)
17	(E,Z)-2,6-Nonadienal	1582	80.2 (± 15%)	27.9 (± 93%)
19	(E)-2-Undecenal	1749	101 (± 11%)	14.4 (± 53%)
20	(E,Z)-2,4-Decadienal	1763	79.4 (± 10%)	31.9 (± 53%)
21	(E,E)-2,4-Decadienal	1810	130 (± 12%)	32.9 (± 30%)
22	(E,Z,Z)-2,4,7-Decatrienal*	1820	508 (± 2.1%)	306 (± 23%)
25	(E,E,Z)-2,4,7-Decatrienal*	1877	121 (± 55%)	107 (± 34%)
26	(E,E,E)-2,4,7-Decatrienal*	2000	40.9 (±14%)	39.2 (± 43%)

[a]Numbers correspond to those in figure 3 and Table I. [b]Retention index on DB-WAX column. [c]Average relative concentration. Numbers in parentheses represent percent relative standard deviations (n = 2). [d]Not detected. *Compound tentatively identified based on literature data.

such as nos. 2-5, 7, 8; whereas, semivolatile compounds (nos. 19-22, 25, 26) exhibited smaller declines. Despite their high odor intensities, quantities for compounds nos. 10, 11, 13 and 15 were not estimated by DTD-GC-MS due to very low abundance. The predominance of these compounds (especially no. 11) in the odor profile of the seal oils is not surprising since these compounds have extremely low detection thresholds. For example, the odor detection thresholds (in vegetable oil) for compounds nos. 10 and 11 were reported to be 10, and 0.45 and ng/g, respectively (36).

The volatile compounds listed in Tables I and II are typical of those previously encountered in fish and fish oils (3,6-9,38). These volatile aldehydes, ketones and alcohols can be formed via autoxidation or lipoxygenase action from the abundant polyunsaturated fatty acids found in seal oil (PUFAs) (3,9,37-40). Results of the present study agree with previous researchers who reported (E,Z,Z)- and (E,E,Z)-2,4,7-decatrienal as predominant odorants in fish oils (3,6,9,38). Autoxidation PUFAs in marine oils leads to the formation of the 2,4,7-decatrienals (3,6,9,38,39).

In addition to the 2,4,7-decatrienal isomers, several other compounds (e.g. nos. 1, 2, 4, 5, 12, 14, 21, and 22) were likely formed as a result of autoxidation of PUFAs (3,9,37-40). On the other hand, (Z)-1,5-octadien-3-one (no. 11), as well as compounds nos. 3, 7, 8, 10, 13, and 15, could have arisen via breakdown of site-specific hydroperoxides formed as a result of lipoxygenase action on the PUFAs (41). Formation of (Z)-4-heptenal (no. 9) may have occurred by further breakdown of (E,Z)-2,6-nonadienal (no. 13) through retro-aldol condensation (42). (Z)-4-Heptenal has been reported to contribute to the off-odors of rancid fish oils (3,9) and may modify or enhance the burnt/fishy odor of 2,4,7-decatrienals (9).

Conclusions

The results of this study demonstrate that the combined use of DTD-GCO and SDA is a powerful tool for the identification of potent odorants in seal blubber oil. The technique allows for the simultaneous sampling and evaluation of components of widely varying volatilities. It is a simple and rapid technique that requires only a small sample size and minimal sample preparation. Furthermore, with DTD-GC-MS there is no interference from solvents or solvent impurity peaks, which allows for the quantitative analysis of highly volatile consitituents. DTD has some potential limitations in the analysis of food lipids, such as breakdown of thermally labile hydroperoxides (22,27,30). Of primary importance when analyzing edible fats and oils with DTD-GCO is that data analysis be conducted with special consideration of potential aritacts. Decomposition of hydroperoxides can give rise to a number of breakdown products, many of which are responsible for the typical off-flavors encountered in rancid fats. Nevertheless, the findings of this study clearly demonstrate the potential of DTD-GCO-SDA and DTD-GC-MS for monitoring the volatile profiles as an indicator of the effectiveness of the refining-deodorization processes in marine oil production.

Literature Cited

1. Shahidi, F.; Synowiecki, J.; Amarowicz, R.; Wanasundara, U. 1994. In *Lipids in Food Flavors*; Ho, C.-T.; Hartman, T.G. , Eds.; ACS Symposium Series 558; American Chemical Society: Washington, DC, 1994; pp. 233-243.
2. Stansby, M. *J. Am. Oil Chem. Soc.* **1974**, *48*, 820-823.
3. Lin, C.F. In *Lipids in Food Flavors*; Ho, C.-T.; Hartman, T.G. , Eds.; ACS Symposium Series 558; American Chemical Society: Washington, DC, 1994; pp. 208-232.
4. Sanders, T. *Food Sci. Technol.* **1987**, *1*, 162-164.
5. Chipault, J.R. *Hormel Inst. Univ. Minn. Ann. Rpt.* **1956**, pp. 61-65.
6. Ke, P.J.; Ackman, R.G.; Linke, B.A. *J. Am. Oil Chem. Soc.* **1975**, *52*, 349-353.
7. St. Angelo, A.J.; Dupuy, H.P.; Flick Jr., G.J. *J. Food Qual.* **1987**, *10*, 393-405.
8. Hsieh, T.C.; Williams, S.S.; Vejaphan, W.; Meyers, S.P. *J. Am. Oil. Chem. Soc.* **1989**, *66*, 114-117.
9. Karahadian, C.; Lindsay, R.C. *J. Am. Oil Chem. Soc.* **1989**, *66*, 953-959.
10. Rorbaek, K.; Jensen, B. *J. Am. Oil Chem. Soc.* **1997**, *74*, 1607-1609.
11. Endo, Y.; Aoyagi, T.; Fujimoto, K. 1998. *Nihon Yukagakkaishi* **1998**, *47*, 873-878.
12. Acree, T. *Analytical Chemistry News & Features*, **1997**, *March 1*, pp. 170A-175A.
13. Fuller, G.H.; Steltenkamp, G.A.; Tisserand, G.A. *Annals. N.Y. Acad. Sci.* **1964**, *116*, 711-724.
14. Acree, T.; Butts, R.M.; Nelson, R.R.; Lee, C.Y. *Anal. Chem.* **1976**, *48*, 1821-1822.
15. Blank, I. In *Techniques for Analyzing Food Aroma*; Marsili, R., Ed.; Marcel Dekker, Inc.: New York, 1997; pp. 293-329.
16. Grosch, W. *Trends in Food Science & Technology* **1993**, *4*, 68-73.
17. Acree, T. In *Flavor Measurement*. Ho, C.-T.; Manley, C.H., Eds.; Marcel Dekker, Inc.: New York, 1993; pp. 77-94.
18. Milo, C.; Grosch, W. *J. Agric. Food Chem.* **1995**, *43*, 459-462.
19. Cadwallader, K.R.; Baek, H.H. In *Proceedings of the Ninth International Flavor Conference, George Charalambous Memorial Symposium*; Mussinan, C.; Contis, E.; Ho, C.-T.; Parliment, T.; Spanier, A.; Shahidi, F., Eds.; Elsevier Science B.V.: Amsterdam, 1997; pp. 271-279.
20. Snyder, J.M.; Frankel, E.N.; Selke, E. *J. Am. Oil Chem. Soc.* **1985**, *62*, 1675-1679.
21. Snyder, J.M.; Mounts, T.L. *J. Am. Oil Chem. Soc.* **1990**, *67*, 800-803.
22. Lee, I.; Fatemi, S.H.; Hammond, E.G.; White, P.J. *J. Am. Oil Chem. Soc.* **1995**, *72*, 539-546.
23. Overton, S.V.; Manura, J.J. *J. Agric. Food Chem.* **1995**, *43*, 1314-1320.
24. Boyd, L.C.; Nwosu, V.C.; Young, C.L.; MacMillian, L. *J. Food Lipids* **1998**, *5*, 269-282.
25. Ulberth, F. *Z. Lebensm. Unters. Forsch. A* **1998**, *206*, 305-307.

26. Grimm, C.C; Lloyd, S.W.; Miller, J.A.; Spanier, A.M. In Techniques of Analyzing Food Aroma; Marsilli, R., Ed.; Marcel Dekker, Inc.: New York, 1997, pp. 59-79.
27. Hartman, T.G.; Lech, J.; Karmas, K.; Salinas, J.; Rosen, R.T.; Ho, C.-T. In *Flavor Measurement*. Ho, C.-T.; Manley, C.H., Eds.; Marcel Dekker, Inc.: New York, 1993; pp. 37-60.
28. Dupuy, H.P.; Flick, G.J.; Bailey, M.E.; St. Angelo, A.J.; Legendre, M.G.; Sumrell, G. *J. Am. Oil Chem. Soc.* **1985**, *62*, 1690-1693.
29. Suzuki, J.; Bailey, M.E. *J. Agric. Food Chem.* **1985**, *33*, 343-347.
30. Snyder, J.M.; Frankel, E.N.; Selke, E.; Warner, K. *J. Am. Oil Chem. Soc.* **1988**, *65*, 1617-1620.
31. Dupuy, H.P.; Rayner, E.T.; Wadsworth, J.I. *J. Am. Oil. Chem. Soc.* **1976**, *53*, 628-631.
32. Dupuy, H.P.; Rayner, E.T.; Wadsworth, J.I.; Legendre, M.G. *J. Am. Oil Chem. Soc.* **1977**, *54*, 445-449.
33. Ullrich, F.; Grosch, W. *J. Am. Oil Chem. Soc.* **1988**, *65*, 1313-1317.
34. Milo, C.; Grosch, W. *J. Agric. Food Chem.* **1993**, *41*, 2076-2081.
35. Cadwallader, K.R.; Howard, C.L. In *Flavor Analysis: Developments in Isolation and Characterization*. Mussinan, C.J.; Morello, M.J., Eds.; ACS Symposium Series 705; American Chemical Society: Washington, DC, 1998; pp. 343-358.
36. Guth, H.; Grosch, W. *Lebensm.-Wiss. u.-Technol.* **1990**, *23*, 513-522.
37. Seals, R.G.; Hammond, E.G. *J. Am. Oil Chem. Soc.* **1970**, *43*, 278-280.
38. Meijboom, P.W.; Stroink, J.B.A. *J. Am. Oil Chem. Soc.* **1972**, *49*, 555-558.
39. Karahadian, C.; Lindsay, R.C. In *Flavor Chemistry: Trends and Developments*; Teranishi, R.; Buttery, R.G.; Shahidi, F., Eds.; ACS Symposium Series 388; American Chemical Society: Washington, DC; 1989, 60-75.
40. St. Angelo, A.J.; *Crit. Rev. Food Sci. Nutr.* **1996**, *36*, 175-224.
41. Josephson, D.B.; Lindsay, R.C.; Stuiber, D.A. *J. Food Sci.* **1987**, *52*, 596-600.
42. Josephson, D.B.; Lindsay, R.C. *J. Am. Oil Chem. Soc.* **1987**, *64*, 132-138.

Chapter 18

Errors in the Identification by Gas–Liquid Chromatography of Conjugated Linoleic Acids in Seafoods

Robert G. Ackman

Canadian Institute of Fisheries Technology, DalTech, Dalhousie University, P.O. Box 1000, Halifax, Nova Scotia B3J 2X4, Canada

Several papers have appeared in which cis-9,trans-11-octadecadienoic acid (cis-9, trans-11 conjugated linoleic acid, CLA) has been stated to be present in raw seafoods. The trans-10-cis-12 isomer was listed as not detected in these reports. The errors lie in the presence of two fatty acids, 18:4n-3 and 18:4n-1, in most seafoods. When the methyl esters of fatty acids are examined on polyglycol- based capillary GLC columns the cis-9,trans-11 CLA isomer usually coincides with the natural 18:4n-3 and the trans-10,cis-12 CLA isomer with the 18:4n- 1. Thus 18:4n-3 was reported as cis-9,trans-11CLA, but relative to 18:4n-3 the 18:4n-1 is usually very small, and would not be reported as trans-10,cis-12 CLA. More polar liquid phases present a different set of identification challenges. The increasing use of vegetable proteins and oils in feeds for fish farming could create new problems and misunderstandings in respect to CLA in marine-based foods. Analyses of deep-fried fish and seafood dishes with substantial pastry or vegetable components could also be confusing. Fortunately, silver nitrite thin layer chromatography ($AgNO_3$-TLC) is a solution to this problem.

Conjugated linoleic acids (CLA) are currently a subject of intense interest among nutritionists, biochemists and medical researchers (*1*). In brief, one of the two cis ethylenic bonds of cis-9,cis-12-linoleic (octadecadienoic) acid migrates to a position adjacent to the other and in the process will be converted about 75 percent of the time to the lower energy trans configuration (*2*). In one common enzyme system the direction of this reaction is position-specific (*3*) and for that reason the dominant isomer found in ruminant fats will be the cis-9,trans-11-18:2. If chemical isomerization is promoted by alkali catalysis the two most likely CLA isomers (cis-9,trans-11 and trans-10-cis-12) will be found in equal proportions (*4*). The reaction conditions to prevent an excessive number of minor components were described in a thorough investigation by Mounts et al. in 1970 (*5*), and basically suggest optimum isomerization conditions of 90 °C or less (*6*).

In the last decade CLA has achieved some notoriety and this has led to the recording of several sets of analytical conditions for the GLC (gas-liquid chromatography) analyses of frying oils (*7*) or of dairy products (*8*). The most recent publication, that of Jung, and Ha (*9*) records the result of non-selective and selective hydrogenation conditions on the formation of CLA isomers. Table I gives the ECL (equivalent chain lengths) on SUPELCOWAX-10 for seven peaks accounting for over a dozen possible isomers. The two 18:2 components commonly expected are identifiable but the unusual feature of Table I, observed under non-selective conditions, was the very large GC peak (No.7) for three trans, trans isomers. This superfluity of isomers is accounted for by the relatively harsh reaction conditions, and the catalyst of hydrogenation. Much gentler conditions are preferred as noted above.

Highly polar liquid phases have been used in recent CLA studies and the elution order of CLA isomers on CP-Sil 88 has been established (Table II). An alternative cyanosilicone phase is SP-2560, used by Kramer et al. (*10*) to supplement the earlier work of Shantha et al. (*11*) on the loss of CLA during esterifications. Both groups carefully checked the production of artifacts or loss of CLA during various esterification procedures. Acid-catalysed methylation resulted in "the loss of 12% total CLA, 42% recovery of Δ-9c, 11-18:2, and a four-fold increase in Δ-9t,11t-18:2", as well as the introduction of methoxy artifacts. However these reports do not deal with the possible coincidence of CLA with 18:4n-3 or other fatty acids as discussed below for one case of the analysis of marine food lipids by GLC.

In a study of the uptake of CLA by fish (*13*) the GLC column was Omegawax 320 (a polyglycol similar to SUPELCOWAX-10) and six CLA peaks could be detected in an alkali-catalyzed preparation of CLA from safflower seed oil. The analysis was facilitated by $AgNO_3$-TLC (silver nitrate-thin layer chromatography) clean-up of the methyl esters of the fatty acids of the fish lipids with development in benzene. The CLA band will lie below that for saturated fatty acids and above that for cis-monoethylenic fatty acids.

Table I. Equivalent chain length (ECL) of CLA isomers (methyl esters) on a SUPELCOWAX 10 fused-silica capillary column (60 m x 0.25 mm, 0.25 film thickness)

Peak no.	CLA isomer[a]	From Ref. 9	From Ref. 12
1	c-9,t-11/c-8,t-10	19.51	19.49
2	c-10,t-12/t-9,c-11	19.56	19.53
3	t-10, c-12	19.63	19.62
4	c-11,t-13 or t-11, c-13	19.71	19.67
5	c-9, c-11/c-8, c-10	19.84	19.80
6	c-10, c-12/c-11, c-13	19.88	18.82
7	t-9, t-11/t-10, t-12/t-8, t-10	20.16	20.01

aPeak identification was based on the previously reported in ref. 12 and ref. 8.

It was involvement in this fish-oriented research that led to our surprise at the table published in 1998 by O'Shea et al. (15), listing seafood as having, 0.5 mg, CLA/g of fat, with no detectable cis-9,trans- 11 isomer. The reference for this data was to a definitive paper on CLA in foods by Chin et al. (16), when two tables listed 19 seafoods (also including lake trout) with totals of <0.1-0.8 mg CLA/g fat. It was noted by Chin et al. (16) that the totals were determined by HPLC and that GLC failed to show the cis-9,trans-11-18:2 isomer in seafood "due to interfering substances" eluting with the cis-9,trans-11isomer from the GC capillary column. The colunm in question was the polyglycol SUPELCOWAX-10.

This choice of liquid phase for GLC clarified the origin of the interference problem. Two C18 polyunsaturated fatty acids are commonly found in lipids of marine organisms, but are otherwise rare. They are 18:4n-3 (stearidonic acid) and an even more unusual 18:4n-1 (17). The former is found in terrestrial plants, but it is also produced in marine algae. Theoretically it could also be produced in animals by a Δ-6 desaturase acting on 18:3n-3. The 18:4n-1 arises by the addition of an acetate unit in fish to a dietary 16:4n- I fatty acid, also an algal fatty acid and part of a series with n-1, n-4 and n-7 bonds (18). The 18:4 n-1 peak is usually about 1/10th of the adjacent 18:4n-3. In Figure 1 it is not labeled, but is the small bump between the 18:4n-3 and 20:0 peaks. When a mixture of esters of the two synthetic CLA isomers produced from linoleic acid or from a vegetable oil is co-injected with esters of a marine oil or lipid the cis-9,trans-11 and trans-10, cis- 12 CLA peaks straddle the 18:4n-3 peak. Two minor CLA isomers fall into the 20: 1 n- 11and 20:1 n-9 positions. The 18:4n-3 has always had an ECL value of approximately 19.50 on polyglycol columns, including FFAP (free fatty acid

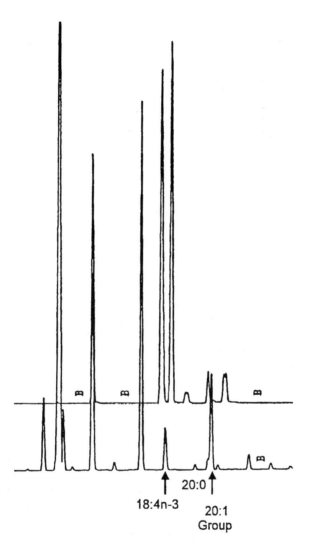

Figure 1. Analyses of methyl esters of a fish lipid (below), and of a mixture of CLA Prepared from linoleic acid with alkali (above), on an Omegawax-320 GC column. In order from the left unmarked peaks are 18:0, 18:1n-9, 18:1n-7, 18:2n-6, 18:3n-6 and 18:3n-3. Note coincidence of 18:2n-6 in the two GLC analyses.

Table II. Expected gas chromatographic elution order of positional and geometric CLA fatty acid methyl ester, or dimethyloxazoline, isomers on a 100 m CP-Sil 88 capillary column.[a,b]

cis/trans-18:2[c]	cis,cis-18:2[d]	trans,trans-18:2[e]
7c,9t	(7c,9c)[f]	12t, 14t
(6t,8c)	8c, 10c	11t, 13t
(8c, 10t)	9c, 11c	10t, 12t
7t,9c	10c, 12c	9t, 11t
9c, 11t	11c,13c	8t, 10t
8t, 10c	12c, 14c	7t, 9t
10c, 12t		
9t, 11c		
11c, 13t		
10t, 12c		
12c, 14t		

[a]Adapted from ref. *14*.
[b]The elution order was of all the cis,trans, followed by all the cis,cis, followed by all of the trans,trans CLA positional isomers. Many CLA isomers overlapped within eacgeometric group.
[c] The observed elution time of cis,trans CLA isomers increased as the Δ value of the cis double bond increased in the molecule. For a pair of cis,trans isomers in which the cis double bond has the same Δ value, the isomer with the lower Δ trans value eluted first. Therefore, it followed that for the same positional isomer, the cis,trans eluted before the trans,cis geometric isomer.
[d]The observed elution time of cis,cis CLA isomers increased with increased Δ values
[e]The observed elution time of trans,trans CLA isomers increased with decreased Δ values.
[f]CLA isomers shown in parentheses were predicted.

phase) as recorded by Ackman and Ratnayake (*19*). In fact it is the two factors of a high degree of reproducibility of ECL values between laboratories, even if polyglycol columns of different manufacturers are used, and the freedom from chain length overlap in the C16/C18, C18/C20 and C20/C22 re-ions, that recommend the use of this liquid phase for biomedical research (*20*).

In a project involving monkeys fed a control diet of a simulated Canadian food fatty acid mixture based on lard : corn oil in a 2:1 ratio (21) two putative CLA peaks were found when the depot fat was investigated by capillary open-tubular gas chromatography. These were traced to the diet and the GLC positions were shown to be approximately between the peaks for 18:3n-3 and 20:ln-11 on SILAR-5CP, a low polarity cyanosilicone liquid phase then in use for GLC. On the more polar SILAR-7 CP the two CLA peaks straddled the 20: 1 n-9 peak commonly observed in canola (low erucic acid rapeseed) oil. In that investigation several refined vegetable oils were shown to contain CLA in measurable amounts, as also reported by Chin et al. (16). Although these particular cyanosilicone phases are no longer much used they show the risks of using high polarity liquid phases. For example Segredos (22) used columns coated with a popular cyanosilicone liquid phase (CP-Sil 88) or the less well-known RTX 2330. He marked a peak between 18:ln-7 and 18:2n-6 as "18:2ω6ττ" whereas in fact it is most probably 16:4n- 1. The " 18:4n-3" on this column is correctly shown to fall after "20:1ω9" and before a "20:2" peak. The chances of finding and quantitating specific CLA isomers such as the cis-9,trans-11 isomer in this complex region of a GLC chart are not good, compared to the clean area between 18:3n-3 and 20:0 obtained with polyglycols as liquid phases. It is therefore probable that if the sample is of marine lipid origin, 18:4n-3 and the accompanying 18:4n-1 could interfere in GLC of CLA on polyglycol capillary columns unless a preliminary clean-up of the CLA is executed.

Chin et al. (16) remarked on finding both cis-9,trans- 11 and trans-10, cis- 12 acids in several plant oils, including similar CLA levels in their laboratory extract of corn. Some fish are omnivores, others herbivores, and some aquaculture fish products, for example the U.S. channel catfish *Ictalurus punctatus*, may show both CLA and 18:4n-3if given both fish meal and plant products rich in linoleic acid as part of the diet. As an example full fat soybean meal has been tested in diets for Atlantic salmon *Salmo salar* (23). The salmon industry prefers fish meal as a protein source, but has some flexibility in the use of supplementary oils other than fish oil. Canola oil, with about 20% linoleic acid, has been used as a fat supplement in trials for rearing Atlantic salmon (24,25). The Atlantic salmon has a strong and distinctive flavor (25-27), but there are concerns over adding too much vegetable materials to the diet (28). Farmer et al. (27) show as much as 6.15% 18:2n-6 in one sample of ocean pen-reared salmon muscle, several times the normal value of 1-2%. Obviously, the use of vegetable products is increasing so care must be taken to assure that CLA is a not reported as a metabolic component of component of fish lipids, especially in marine fish. However, no recent analyses of fish lipids, specifically for CLA, seem to be available except for the publication by Choi et al. (13)

REFERENCES

1. Haumann, B.F. *INFORM* **1996**, *7*,152-159.
2. Litchfield, C.; Lord, J.E.; Isbell, A.F.; Reiser, R. *J. Am. Oil Chem. Soc.* **1963**, *40*, 553-557.
3. Gurr, M. I.; Harwood, J. L. Lipid Biochemistry; Chapman and Hall: London, 1991; 4th edn, 406 pp.
4. Christie, W.W.; Dobson, G.; Gunstone, F. D. *J. Am. Oil Chem. Soc.* **1997**, *74*, 1201.
5. Mounts, T. L.; Dutton, H. J.; Glover, D. *Lipids* **1970**, *5*, 997-1005.
6. Ackman, R.G. *J. Am. Oil Chem. Soc.* **1998**, *75*, 1227.
7. Sebedio, J-L.; Grandgirard, A.; Septier, Ch.; Prevost, J. *Rev. Frang. Corps. Gras.* **1987**, *34*, 15-18.
8. Lavillonnièrre, F.; Martin, J. C.; Bougnoux, P.; Sébédio, J.-L. *J. Am. Oil Chem. Soc.* **1998**, *75*, 343-352.
9. Jung, M. Y.; Ha, Y. L. *J. Agric. Food Chem.* **1999**, *47*, 704-708.
10. Kramer, J. K. G.; Fellner, V.; Dugan, M. E. R.; Sauer, F. D.; Mossoba, M. M.; Yurawecz, M. P. *Lipids* **1997**, *32*, 1219-1228.
11. Shantha, N.C.; Decker, E.A.; Hennig, B. *J A OAC Int.* **1993**, *76*, 644-649.
12. Ha, Y. L.; Grimm, N. K.; Pariza, M. W. *J. Agric. Food Chem.* **1989**,*37*, 75-81.
13. Choi, B-D.; Kang, S-J.; Ha Y-L.; Ackman, R. G. In Quality Attributes of Muscle Foods, Xiong, Y. L.; Ho, C. T.; Shahidi, F., Eds; Klwer Academics/Plenum Publishers, New York, 1999; pp. 61-71.
14. Sehat, N.; Kramer, J. K. G.; Mossoba, M. M.; Yurawecz, M. P.; Roach, J. A. G.; Eulitz, K.; Morehouse, K. M.; Ku, Y. *Lipids* **1998**, *33*, 963-971.
15. O'Shea, M.; Lawless, F.; Stanton, D.; Devery, R. *Trends Food Sci. Technol.* **1998**, *9*, 192-196.
16. Chin, S. F.; Liu, W.; Storkson, J. M.; Ha, Y. L.; Pariza, M. W. *J. Food Comp. Anal.* **1992**, *5*, 185-197.
17. Ackman, R. G. In Marine Biogenic Lipids, Fats, and Oils; Ackman, R.G., Ed.; CRC Press: Boca Raton, FL, 1989; Vol. 1, pp 103-137.
18. Ackman, R. G. In Fish and Fishery Products: Composition, Nutritive Properties and Stah ility; Ruiter, A., Ed.. CAB International: Wallingford, Oxon, 1995; pp 117-156.
19. Ackman, R. G.; Ratnayake, W. M. N. In The Role of Fats in Human Nzitrition; Crawford, M.; Vergroesen, A.J., Eds; Academic Press: London, 1989; 2nd edn, pp 441-514.
20. Ackman, R. G. *Acta Med. Scand.* **1987**, *222*, 99-103.
21. Ackman, R. G.; Eaton, C. A.; Sipos, J. C.; Crewe, N. F. *Can. Inst. Food Sci. Technol. J.* **1981**, *14*, 103-107.
22. Sagredos, A. N. *Fat Sci Technol.* **1991**, *93*, 184-191. (in German)
23. Bjerkeng, B.; Refstie, S; Fjalestad, K. T.; Storebakken, T.; Rødbotten, M.; Roem, A. J. *Aquaculture* **1997**, *157*, 297-309.

24. Polvi, S. M.; Ackman, R. G. *J. Agric. Food Chem.* **1992**, *40*, 1001-1007.
25. Koshio, S.; Ackman, R. G.; Lall, S. P. *J. Agric. Food Chem.* **1994**, *42*, 1164-1169.
26. Parrish, C. C.; McLeod, C. A.; Ackman, R. G. *J. Sci. Food Agric.* **1995**, *68*, 325-329.
27. Farmer, L. J.; McConnell, J. M.; Graham, W. D. *In* Flavor and Lipid Chemistry of Seafoods; Shahidi, F.; Cadwallader, K. R.; ACS Symposium Series 674; American Chemical Society; Washington DC, 1997; pp 95-109.
28. Fish Info Service, April 16, 1999; http://www.fis-net.com/fis/hotnews/

Chapter 19

Impact of Emulsifiers on the Oxidative Stability of Lipid Dispersions High in Omega-3 Fatty Acids

Eric A. Decker[1], D. Julian McClements[1], Jennifer R. Mancuso[1], Larry Tong[1], Longyuan Mei[1], Shigefumi Sasaki[2], Sam G. Zeller[3], and James H. Flatt[3]

[1]Department of Food Science, Chenoweth Laboratory, University of Massachusetts, Amherst, MA 01003
[2]Hokkaido Food Processing Research Center, 589-4Bunkyod-l, Midori-Mach 1, Ebetsu 069–0836, Japan
[3]The Nutrasweet Kelco Company, 8225 Aero Drive, San Diego, CA 92123–1718

Lipid emulsions consist of three major regions that include the droplet interior, the droplet interfacial membrane and the continuous phase. Many compounds involved in lipid oxidation such as lipid peroxides and antioxidants are surface active meaning that they will partition at the interfacial membrane of the emulsion droplet surface. Algal or salmon oil emulsified with the cationic surfactant, dodecyltrimethyl ammonium bromide (DTAB) exhibited lower oxidation rates than emulsions stabilized with anionic (sodium dodecyl sufate, SDS) and nonionic (Brij 35) surfactants. The increased oxidation rates observed in anionic emulsion droplets is largely due to the ability of iron to interact with the emulsion droplet surface where it can catalyze the decomposition of lipid peroxides. Metal chelators are capable of removing iron from the surface of emulsion droplets and thus inhibiting the oxidation of highly unsaturated fatty acids (HUFA). Inhibition of iron catalyzed lipid oxidation can also be accomplished by manipulation of emulsion droplet charge using proteins. The ability of antioxidants to interact with emulsion droplets and thus protect highly unsaturated fatty acids from oxidation is also influenced by the interfacial properties of emulsion droplets. Thus, engineering the emulsion droplet interface to maximize antioxidant activity and minimize prooxidant-lipid interactions can be a useful technique for increasing the oxidative stability of emulsified lipids containing highly unsaturated fatty acids.

© 2001 American Chemical Society

Current dietary guidelines recommend increased consumption of unsaturated fats (*1*). In addition, recent research suggests that dietary bioactive lipids such as conjugated linoleic acid, ω-3 fatty acids and carotenoids are beneficial to health (*2, 3*). Incorporation of unsaturated and bioactive lipids into processed food would be beneficial to consumers, however, utilization of these lipids is difficult due to their susceptibility to oxidative degradation. If lipids oxidize during the processing and storage of food products, this will not only alter the nutritional composition of the products, but will also influence sensory quality since oxidation leads to rancid flavors and aromas and oxidation of carotenoids results in bleaching and thus color changes. Therefore, in order to produce foods with physiologically bioactive lipid components, methods must be developed to control oxidative reactions.

Lipid oxidation in foods can be controlled by a variety of antioxidant technologies including control of oxidation substrates (e.g. oxygen and lipid composition), control of prooxidants (e.g. reactive oxygen species and prooxidant metals) and addition of antioxidants which inactivate free radicals (*4*). Increasing the oxidative stability of foods by increasing saturated fatty acid composition is contrary to current nutritional recommendations. In addition, exclusion of oxygen from food products during both processing and subsequent storage is often not practical. Therefore, antioxidant applications and control of prooxidative metals are the most common methods used to inhibit lipid oxidation in processed foods. While synthetic food additives such as butylated hydroxyanisole (BHA), butylated hydroxytoluene (BHT) and ethylenediaminetetraacetic acid (EDTA) can be effective in controlling oxidation, consumer demand for all natural foods has limited the use of these "label unfriendly" additives. The synthetic chain breaking antioxidants (BHT, BHA etc.) can be replaced by several natural chain breaking antioxidants including tocopherols and extracts of herbs (e.g. rosemary). However, these natural antioxidants are often limited by their effectiveness (many are prooxidative at high concentrations), price and associated flavors and colors (*5*). If additional antioxidant technologies could be developed, this would increase the ability of food manufacturers to design oxidatively stable foods that contain high concentrations of unsaturated fatty acids and bioactive lipids.

The oxidation of bulk lipids has been studied extensively and there is now a fairly good understanding of the factors which affect oxidation in these systems (*6, 7*). Research in this area has elucidated many of the mechanisms by which lipid oxidation proceeds under various conditions, and the types of reaction products produced (*8*). The importance of the physical state and organization of lipids in foods on their susceptibility to oxidation was recognized many years ago (*9, 10*). Even so, it is only recently that systematic studies of lipid oxidation in food emulsions have been undertaken (*11-17*). This is surprising considering the large number of foods which consist either partly or wholly as emulsions, or which have been in an emulsified form sometime during their production (e.g. dairy products, mayonnaise, margarine, soups, sauces, baby foods, and beverages; *18*). There are significant differences between lipid oxidation in bulk fats and in emulsified fats due to the presence of a droplet membrane, the partitioning of ingredients between lipid and aqueous phases, the much larger interfacial area and the fact that the lipid is in contact with water rather than air. Therefore, there is a need to identify factors that influence lipid oxidation in food emulsions, since this information would be useful in the development of strategies to retard its progress. Such information could lead to improvements in the shelf-life and quality of existing and newly developed food emulsions, inhibit the oxidative destruction of bioactive lipids and decrease the production of potentially toxic lipid oxidation products (*19*). In particular, it would enable food manufacturers to create foods in which saturated fats were replaced by polyunsaturated fats, which are perceived to be more beneficial to health, but which are highly prone to lipid oxidation.

Lipid Oxidation in Emulsions

Emulsions are thermodynamically unstable because of the positive free energy needed to increase the surface area between oil and water phases (*18, 20*). For this reason emulsions tend to separate into a layer of oil (lower density) on top of a layer of water (higher density) with time. To form emulsions which are kinetically stable for a reasonable period (a few weeks, months or even years), chemical substances known as emulsifiers must be added prior to homogenization. Emulsifiers are surface-active molecules which are adsorbed to the surface of freshly formed droplets during homogenization, forming a protective membrane which prevents the droplets from coming close enough together to aggregate. The most common emulsifiers used in the food industry are surface active proteins (e.g. from casein, whey, soy and egg), phospholipids (e.g. egg or soy lecithin) and small molecule surfactants (e.g. Spans, Tweens, fatty acids).

An emulsion can be considered to consist of three regions: the interior of a droplet, the continuous phase, and the interfacial membrane. The interfacial membrane consists of a narrow region surrounding each emulsion droplet that consists of a mixture of oil, water and emulsifier molecules. Typically, the interfacial membrane has a thickness of a few nanometers, and can make up a significant proportion of the total number of molecules present in the droplet (*21*). The various molecules in an emulsion, including prooxidants and antioxidants, partition themselves between the three different regions according to their polarity and surface activity. Non-polar molecules are located predominantly in the oil phase, polar molecules in the aqueous phase and surface active molecules at the interface (Figure 1). The precise molecular environment of a molecule may have a significant effect on its chemical reactivity. Therefore the nature of the emulsion

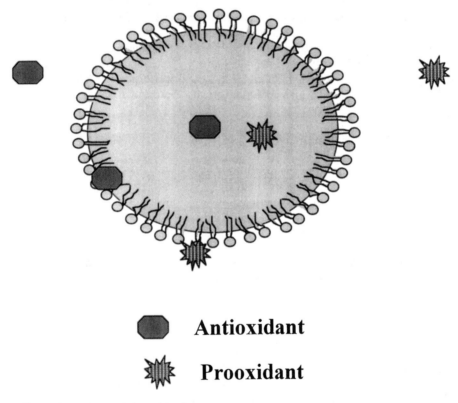

Figure 1. Potential partitioning patterns of antioxidants and prooxidants in oil-in-water emulsions.

droplet membrane would be expected to be extremely important in lipid oxidation reactions since it could dictate how lipids (e.g. unsaturated fatty acids and lipid peroxides) could interact with aqueous phase prooxidants (e.g. transition metals and reactive oxygen species). Lipid oxidation is a free radical chain reaction between unsaturated fats and oxygen that can proceed in an autocatalytic manner. However, in many foods lipid oxidation is not truly autocatalytic since it is accelerated by prooxidants such as UV light, photosensitizers, transition metal ions and certain enzymes (8, 22). Recent work in our laboratory has shown that the initial stages of lipid oxidation of Tween 20 stabilized salmon oil emulsions is more rapid than bulk oil (Figure 2).

Autoxidation of these emulsions is inhibited by metal binding agents including EDTA and the plasma iron binding protein, transferrin (Figure 3). Since these salmon oil emulsions do not contain added prooxidants, these results indicate that lipid oxidation in the emulsion is not truly due to autoxidation, but instead is promoted by endogenous transition metals. The most common prooxidant transition metals in foods are iron and copper. EDTA has been shown to inhibit both copper- and iron-promoted lipid oxidation. Transferrin also inhibits iron promoted lipid oxidation (23). However, the extremely high iron binding constant for transferfin (over 10^7 higher than other metals, 24) suggests that transferrin would mainly be effective at inhibiting the prooxidative activity of iron. In addition, chelators such as EDTA have limited lipid solubility while transferrin is a highly water-soluble protein. Therefore, the ability of transferrin to strongly inhibit the oxidation of Tween 20-stabilized salmon oil emulsions suggests that iron is the main lipid oxidation catalyst and that the prooxidative iron originates in the aqueous phase or is removed from the droplet interface or lipid phase by aqueous phase transferrin. Iron primarily accelerates lipid oxidation by promoting the breakdown of peroxides into free radicals. Most food grade lipids contain preexisting lipid peroxides and hydrogen peroxide can be produced in foods from the spontaneous dismutation of superoxide anion (22). The ferrous state of iron (Fe^{2+}) will decompose peroxides over 10^5 times faster than ferric iron (Fe^{3+}) (25). In addition, the reactivity of ferric iron is also limited by its low water solubility which is 10^{17} and 10^{13} times lower than ferrous iron at pH 7 and 3, respectively (26). While the ferrous state of iron is more reactive and more soluble, ferric ions can be more common in foods (27). Even with its low reactivity and solubility, ferric iron could be an important lipid oxidation catalyst during the long-term storage of emulsified lipids especially if it is able to interact with the interfacial membrane of emulsion droplets.

Impact of the Interfacial Properties of Emulsion Droplets on the Activity of Iron

Current work in our lab has shown that the interaction of iron with the emulsion droplet interface is an important factor in lipid oxidation rates. Using a

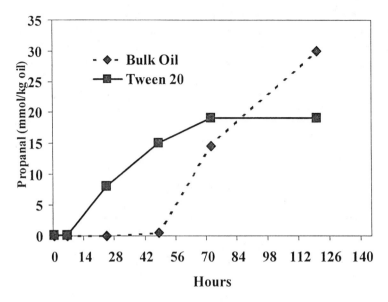

Figure 2. Oxidation of bulk and Tween 20 emulsified salmon oil (as determined by headspace propanal) in the presence of EDTA and transferrin.

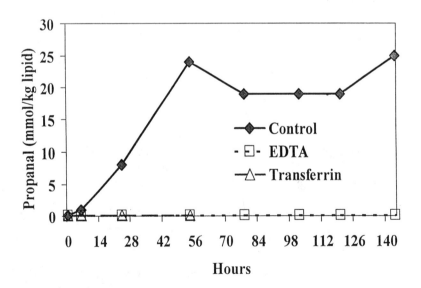

Figure 3. Oxidation of salmon oil emulsions (as determined by headspace propanal) in the presence of EDTA and transferrin.

model system consisting of corn oil-in-water emulsions stabilized with anionic (sodium dodecyl sulfate; SDS), cationic (dodecyltrimethylammonium bromide; DTAB) and nonionic (polyoxyethylene 10 lauryl ether; Brij) surfactants, it was found that iron-promoted lipid oxidation rates were highest with anionic (negatively charged) and lowest with cationic (positively charged) emulsion droplets (Figure 4; *15*). This seemed to be due to the positively charged iron ions being electrostatically attracted to the surface of the negatively charged emulsion droplets where they came into close proximity to the lipid substrate, whereas they were repelled from the surface of the positively charged droplets. Zeta potential was then used to show that both ferrous and ferric ions readily interacted with SDS- but not with DTAB- or Brij-stabilized hexadecane emulsion droplets (Table I, *16*). Factors which decreased iron-emulsion droplet interactions such as increasing pH, chelators (EDTA and phytate) and NaCl resulted in decreased lipid oxidation rates (*15, 16*). In a comparison of iron-promoted lipid oxidation rates in SDS-stabilized emulsions containing low (0. 12 [μmol peroxide/g oil) and high (17 [μmol oil) lipid peroxide concentrations, oxidation was much greater in the presence of high peroxide concentrations suggesting that lipid peroxides and not unsaturated fatty acids were actually the oxidation substrate (*16*). These data suggests that an important factor in the oxidation of emulsified lipids is the ability of iron to interact with lipids and/or lipid peroxides at the interfacial membrane of emulsion droplets.

Incorporation of oxygen into an unsaturated fatty acid during the formation of a lipid peroxide results in an increase in the polarity of the lipid molecule (*8*). Polar molecules are not thermodynamically favored in lipid environments therefore the peroxide would move towards the lipid interface of an emulsion droplet in an attempt to reach a more polar environment (*28*). Once the peroxide reaches the surface of the emulsion droplet it is free to interact with aqueous phase metals and/or metals associated with the emulsion interfacial membrane (Figure 5). This can be seen using a model consisting of Tween 20- or SDS-stabilized hexadecane emulsion containing cumene peroxide. Hexadecane was used in this model as a nonoxidizable lipid so that additional peroxides would not be formed from the free radicals originating from the breakdown of the cumene peroxide as would occur in an emulsion containing unsaturated fatty acids. In this model at pH 7.0, ferrous ions were able to rapidly breakdown peroxides in both SDS- (Figure 6) and Tween 20- (Figure 7) stabilized emulsions after which no change in peroxides occurred. At pH 3.0, the peroxide breakdown pattern was similar in the Tween-20 stabilized emulsions (Figure 7) while in the SDS-stabilized emulsions (Figure 6) peroxide decomposition continued throughout the incubation period. Ferric iron was only able to breakdown peroxides in the SDS-stabilized emulsions at pH 3.0 (Figure 8). Iron can decompose peroxides in a redox dependent manner:

$$Fe^{2+} + ROOH \rightarrow Fe^{3+} + RO^{\cdot} + OH^{-}$$

Table I. Zeta potential of SDS-, Brij-, and DTAB-stabilized hexadecane emulsions in the presence of iron salts at pH 3.0. Data represent means ± standard deviations. A decrease in zeta potential in iron-containing samples compared to the controls indicates iron binding. Adapted from Ref. *16*

Treatment	Conc. (μm)	SDS	Brijj	DTAB
		Zeta Potential (mV)		
$FeSO_4$	0	-105.7±2.1	-2.8±1.2	74.7±1.6
	50	-105.4±1.7	-4.4±0.6	74.2±1.4
	150	-103.1±1.8	-4.5±0.9	74.0±0.9
	500	-95.9±1.2	-5.2±0.7	71.6±1.2
$FeCl_2$	0	-105.3±1.8	-2.7±0.8	75.8±0.4
	5	-104.8±0.1	nd	nd
	50	-103.6±4.1	-2.6±0.8	76.0±0.3
	150	-97.9±1.0	-2.8±0.6	75.6±0.9
	500	-94.6±0.9	-2.9±1.0	74.6±0.7
$FeCl_3$	0	-105.8±0.6	-3.2±0.7	75.5±0.3
	50	-101.3±0.2	-3.6±0.7	76.0±0.2
	150	-93.9±0.4	-3.0±0.6	75.5±0.8
	500	-82.9±0.7	-3.3±1.2	74.3±0.7

Abbreviations are: SDS, sodium dodecyl sulfate; Brij, polyoxyethylene 10 lauryl ether; DTAB, dodecyltrimethylammonian bromide; and nd, not determined.

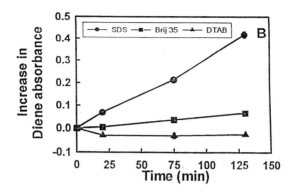

Figure 4. Oxidation of tocopherol stripped corn oil-in-water emulsions stabilized by SDS, DTAB and Brij. Oxidation was accelerated by 50 [μM $FeCl_3$ and 100 [μM ascorbate at pH 4.0. Lipid oxidation was determined by monitoring the formation of conjugated dienes. Adapted from ref. *15*.

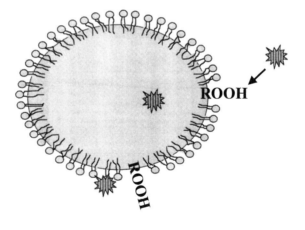

ROOH = Fatty Acid Peroxide

Prooxidant

Figure 5. Potential orientation of lipid peroxides in oil-in-water emulsions.

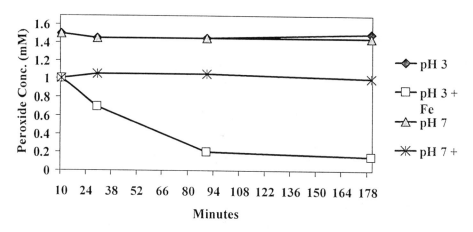

Figure 6. Fe 21 (500 ppm FeCl$_2$)promoted breakdown of cumene peroxide in hexadecane emulsions stabilized by SDS at pH 7.0 or 3.0.

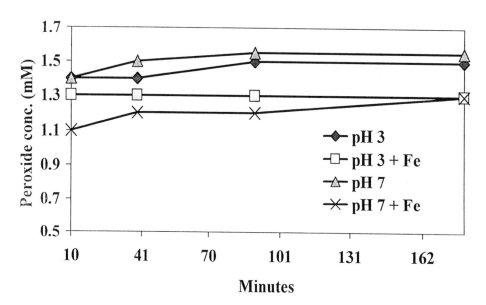

Figure 7. Fe 2+ (500 ppm FeCl$_2$)promoted breakdown of cumene peroxide in hexadecane emulsions stabilized by Tween 20 at pH 7.0 or 3.0.

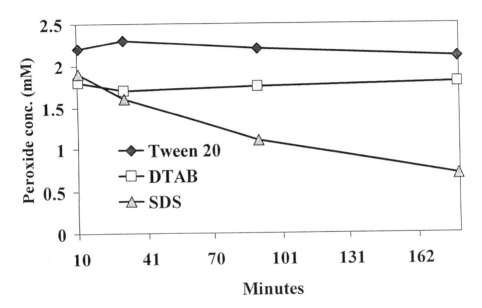

Figure 8. Fe 3+ (500 ppm FeCl$_2$) promoted breakdown of cumene peroxide in hexadecane emulsions stabilized by Tween 20 at pH 7.0 or 3.0.

$$Fe^{3+} + ROOH \rightarrow Fe^{2+} + ROO^{\cdot} + H^{+}$$

Ferric ions breakdown peroxides at a much slower rate than ferrous ions. However, the reactivity of ferric ions increases with decreasing pH due to increases in Fe^{3+} solubility. Since Fe^{3+} was not able to promote peroxide breakdown in the Tween stabilized emulsions and was only able to decompose peroxide in the SDS-stabilized emulsions at pH 3.0 (Figure 8), this suggests that both high Fe^{3+} solubility and attraction to the emulsion interface is required for its activity. This data also suggests that the ability of Fe^{2+} to promote prolonged decomposition of peroxides at pH 3.0 in the SDS-stabilized emulsions was due to the ability of Fe^{3+} to also participate in the reaction through the redox cycling pathway outlined above.

Synthetic surfactants serve as useful tools to study how the interfacial region of an emulsion droplet impacts lipid oxidation rates. However, some of these surfactants are not approved for food use. Utilization of proteins as emulsifiers represents an excellent opportunity to engineer the properties of the emulsion droplet interface to decrease lipid oxidation. Protein-stabilized emulsion droplets can be manipulated to produce emulsion droplets that differ in membrane charge and thickness, both of which could impact lipid oxidation rates. The charge of protein-stabilized emulsions is dictated by the pI of the protein with pH's below the pI producing positively charged emulsion droplets and pH's above the pI producing negatively charged droplets. These differences in charge influence lipid oxidation in a manner similar to small molecule surfactants with different charges (eg. SDS, DTAB and Tween ,*15, 16*). In an emulsion containing Menhaden oil stabilized with whey protein isolate, emulsion droplets were positive at pH < 5.0 and negative at pH > 5.0 (Figure 9). As expected, lipid oxidation rates were lower at pH 3.0 where the emulsion droplets were positively charged compared to pH 7.0 where the droplets were negatively charged (Figure 10). The opposite was observed for Tween 20-stabilized Menhaden emulsions where lipid oxidation rates increased with decreasing pH (because of increases in iron solubility with decreasing pH). When whey protein isolate was displaced from the emulsion droplet surface by Tween 20, the influence of pH on oxidation rate became similar to Tween 20 (e.g. high oxidation at low pH; *17*).

Conclusions

The oxidative stability of oils containing high amounts of polyunsaturated fatty acids is a limiting factor in their utilization in foods. The stability of these oils can be increased using traditional methods such as antioxidant addition. However, modification of the interfacial properties of lipid emulsions also represent a possible technique that could be utilized to inhibit lipid oxidation. Techniques that engineer the properties of the emulsion droplet interface could be developed to

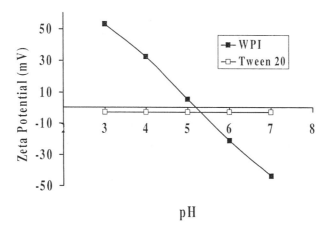

Figure 9. Changes in the charge whey protein isolate (WPI)- and Tween 20-stabilized hexadecane emulsions as a function of pH.

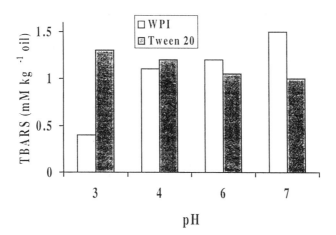

Figure 10. Iron-catalyzed formation of thiobarbituric acid reactive substances in 0.5% Tween 20 and 0.2% whey protein isolate stabilized emulsions over the pH range of 3-7. Oxidation was carried out in the dark at 25 C for 45 min. Adapted from ref. *17*.

decrease interactions between aqueous phase prooxidants and lipid peroxides. Alternatively, chelators can be added to the aqueous phase to prevent prooxidant metals from binding from the emulsions surface as well as decreasing metal reactivity. Such techniques could be used as an additional hurdle against oxidative reactions that would enhance the stability of foods containing high amounts of unsaturated fatty acids.

References

1. Kritchevsky, D. 1997. In Lipid Chemistry. Ed. C. C. Akoh and D.B. Min, Marcel Dekker, Inc. New York, NY, pp. 449-461.
2. Watkins, S.M.; German, J.B. 1998. In. Lipid Chemistry. Ed. C.C. Akoh and D.B. Min, Marcel Dekker, Inc. New York, NY, pp. 463-493.
3. Decker, E.A. *Nutr. Rev.* **1994**, *5*, 49-58.
4. Decker, E.A. 1998. In Lipid Chemistry. C.C. Akoh and D.B. Nfin, Ed.; Marcel Dekker, Inc. New York, NY, pp. 397-421.
5. Giese, J. *Food Technol.* **1996**, *50* (11), 73-77.
6. Fritsch, C.W. *Inform* **1994**, *5*, 423-436.
7. Halliwell, B.; Murcia, M.A.; Chirico, S.; Aruoma, O.I. *Crit. Rev. Food. Sci. Nutr.* **1995**, *3*, 5:7-20.
8. Nawar, W.W. (1996). In Food Chemistry. Third Edition; O.R. Fenema, Ed.; Marcel Dekker Inc. New York, NY, pp. 275-319.
9. Labuza, T. *Crit. Rev. Food Sci. Technol.* **1971**, *10*, 355-405.
10. Coupland, J.N.; Weiss, J.; Lovy, A.; McClements, D.J. *J. Food Sci.* **1996**, *61*, 1114-1117.
11. Frankel, E.N.; Huang, S.W.; Kanner, J.; German, J.B. *J. Agric. Food Chem.* **1994**, *42*, 1054-1059.
12. Roozen, J.P.; Frankel, E.N.; Kinsella, J.E. *Food Chem.* **1994**, *50*, 33-38.
13. Roozen, J.P.; Frankel, E.N.; Kinsella, J.E. *Food Chem.* **1994**, *50*, 39-43
14. Huang, S.-W.; Frankel, E.N.; Aeschbach, R.; German, J.B. *J. Agric. Food Chem.* **1997**, *45*, 1991-1994.
15. Mei, L.; McClements, D.J.; Wu, J.; Decker, E.A. *Food Chem.* **1998**, *61*, 307-312.
16. Mei, L.; Decker, E.A.; McClements, D.J. *J. Agric. Food Chem.* **1998**, *46*, 5072-5074.
17. Donnelly, J.L.; Decker, E.A.; McClements, D.J. *J. Food Sci.* **1998**, *63*, 997-1000.
18. Dickinson, E. 1992. An Introduction to Food Colloids, Oxford University Press, Oxford, UK.
19. Kubow, S. *Free Rad. Med. Biol.* **1992**, *12*, 63 -81.
20. McClements, D.J. 1999. Food Emulsions: Principles, Practice and Techniques, CRC Press, Boca Raton, FL.

21. Dickinson, E.; McClements, D.J. 1995. Advances in Food Colloids, Blackie Academic & Professional, Glasgow, UK.
22. Kanner, J.; German, J.B.; Kinsella, J.E. *Crit. Rev. Food Sci. Nutr.* **1987**, *25*, 319-364.
23. Baldwin, D.A.; Jenny, E.R.; Aisen, P. *J. Biol. Chem.* **1984**, *259*, 13391-13394.
24. Harris, W.R. *Biochem.* **1983**, *22*, 3920-3927.
25. Dunford, H.B. *Free Rad. Biol. Med.* **1987**, *3*, 405-421.
26. Zumdahl, S.S. 1989. Chemistry, Second Edition; D.C. Heath and Co., Lexington, KY.
27. Clydesdale, F.M. 1988. In. Nutrient Interactions. C.E. Bodwell and J.W. Erdman, Ed.; Marcel Dekker Inc, New York, NY, pp.
28. Buettner, G.R. *Arch. Biochem. Biophys.* **1993**, *2*, 535-543.

Chapter 20

Application of Natural Antioxidants in Stabilizing Polyunsaturated Fatty Acids in Model Systems and Foods

Leon C. Boyd

Department of Food Science, North Carolina State University,
Raleigh, NC 27695–7624

Fish lipids provide an ideal environment to determine the relative effectiveness of antioixidants. Seafood lipids, and fish in particular, contain a variety of highly unsaturated fatty acids. Even though synthetic antioxidants have been proven to be cost effective in prolonging shelf life and maintaining product quality, there continues to be increased consumer interest in obtaining more natural sources antioxidants and other natural additives. Plant extracts provide rich sources of primary and secondary antioxidants, many of which the mechanism of action has not been fully explored nor explained. This review will provide an update on research efforts on the application of natural antioxidants, their mode of actions in model systems, synergism between various types of natural and synthetic antioxidants, and their potential application in food systems. Of the natural antioxidants, phospholipids, amino acids, amine-containing compounds, and flavonoids are prevalent among the natural antioxidants.

Interest in Natural Antioxidants

Three major areas of interest have fueled the resurgence in the development and application of natural antioxidants to foods. These major fronts include consumer desires for "all natural" foods containing less synthetic antioxidants, clinical studies showing a relationship between the consumption of natural antioxidants and the risk associated with the occurrence of cardiovascular diseases and selected cancers, and the desire of food manufacturers to follow consumer trends. Recent interest in "functional foods" has caused a dramatic increase in consumer interest in finding quality foods that contain functional ingredients that can help prevent or delay the onset of natural aging process while containing fewer synthetic preservatives.

Several studies (1-4) show that increased consumption of antioxidant vitamins, primarily Vitamin E, C, and carotenoids, are associated with decreased risk of cardiovascular diseases and selected forms of cancer. Though vitamin studies predominate the discussions on functional foods, other epidemological studies (4) have shown positive association between the consumption of foods containing other natural antioxidants and a number of degenerative diseases. These plant materials include tocopherol, ascorbic acid, flavonoids, carotenes, polyphenols, tannins, and other antioxidants. As many segments of the food industry are charged with developing the technology associated with the use and application of natural antioxidants, interest from the industrial segment has focused on providing consumers with an array of food products and food supplements that contain selected functional ingredients. These functional ingredients including antioxidants have been shown to have the potential of preventing or delaying degenerative diseases and conditions associated with aging while providing antioxidant protection to lipid-containing foods.

The quest for more natural antioxidants appear to be fueled by decreasing consumer purchasing patterns foods containing synthetic antioxidants and reports that some of the more popular synthetic antioxidants such as butylated hydroxyanisole (BHA) and butylated hydroxytoluene (BHT) show possible carcinogenicity and toxicity in model studies (1,2). By contrast, several *in vitro* model studies of natural antioxidants show they can act as powerful biological antioxidants preventing the oxidation of low density lipoproteins (LDL), and guarding against the damages caused by other free radical initiators, possibly involved in carcinogenesis and mutagenesis (1).

Therefore, the objectives of the review were to examine some the natural antioxidants holding potential as natural ingredients for inclusion in functional foods, and whose presence in many model studies and epidemological studies show potential in performing multiple functions in foods. Since many nutritionists and medical clinicians continue to recommend increased consumption of omega-3 containing foods and a balanced consumption of polyunsaturated fatty acids, monounsaturated, and saturated fatty acids, one of the major challenges facing the food industry is the stabilization of lipids containing highly unsaturated fatty acids.

Function of Natural Antioxidants

The success of natural antioxidant depends not only their "perceived safety" but like synthetic antioxidants, they too must stabilize highly unsaturated fatty acids and provide a modicum of safety similar to the native material from which they are extracted. The antioxidant activity is generally related to their ability to scavenge free radical, decompose free radicals, quench the development of singlet-oxygen, and to act in synergism with other antioxidants present within the micro reaction environment. Even though many natural antioxidants are given the general classification as "antioxidants" many do not function as primary antioxidants having the ability to stop the free radical process or to prevent propagation phase of oxidation. The latter category of compounds is more specifically classified as synergist, metal chelators, and retarders of oxidation. Synergist type compounds generally enhance the antioxidant activity of primary antioxidants, whereas metal chelators will remove the presence of pro-oxidizing metals. Unlike many single-component synthetic antioxidants, many natural antioxidant extracts are composed of one or more compounds having multiple modes of action; thus allowing them to act as primary antioxidants, synergists, retarders, metal scavengers, or singlet oxygen quenchers depending on the origin and nature of the material extracted (*1,6*)

Sources of natural antioxidants

Plants are the major source of numerous natural antioxidants and many contain multiple types of antioxidants each possessing one or more modes of activity. These plant materials can be classified into several broad categories (Table I). These classes of plant material include spices, oilseeds, grains, legumes, vegetables, hydrolyzed proteins, and a variety of miscellaneous extracts that have been used as medical herbs and dietary supplements including barks, leaves, etc. Many of the natural plant extracts will contain multiple compounds, such as tocopherols, phospholipids, and flavonoids (*5,6*).

Spices

Of the many natural sources of antioxidants, spices have enjoyed one the longest history of safe usage as food colorants, seasoning, and stabilizers of highly sensitive ingredients, such as unsaturated lipids, protein, and vitamins. Many of our modern spices pre-date recorded history and have been used primarily in flavoring a variety of meats and vegetable dishes. These include such spices as thyme, clove, rosemary, sage, ginger, oregano, and garlic. Modern day usage of spices includes their application as flavorants and enhancers but also their use in combination with other spices and aromatic extracts. Most spices generally contain volatile oils and oleoresins that give them their characteristic flavor. However recent application of spices and their extracts have expanded to include their potential to mask the accumulation of off-flavors and to stabilize fat-containing foods against oxidation and the ensuing rancid flavors and aromas.

Table I. Major sources of natural antioxidants and their proposed mechanism(s) of action.

Major sources	Oxidative Inhibitor/Activity
Oils and oilseeds	Tocopherols, resins, phenolic, compounds phospholipids
Oat and rice bran	Lignin-derived compounds
Fruits and vegetables	Ascorbic acid, hydroxycarboxylic acids flavonoids, carotenoids
Spices, herbs	Phenolic and polyphenolic compounds flavonoids
Tea	Flavonoids: flavonols
Protein and protein hydrolysates	Amino acids, dihydropyridines
Legumes, cereals	Flavonoids, lignin

The occurrence of warned-over flavor (WOF) in cooked meats, often referred to as meat flavor deterioration (MFD), provides an ideal application for spices and herbs in that meats products are generally lower in natural antioxidants and because of membrane lipids, metal-containing heme compounds, they deteriorate very rapidly. The antioxidant properties of spices have been compared in a number of model oxidation studies (1,5,7,8) as well as product applications singly and in combination with synthetic antioxidants. One such study conducted by Shahidi et al. (7) compared the efficacy of rosemary, ground clove, ginger, thyme, and oregano to the commercial antioxidants, tert-butylhydroquinone (TBHQ) and BHT, using a comminuted pork model system. Spices were added at 200 to 2000 ppms whereas TBHQ and BHT were held at the maximum regulatory limit of 200 ppm. Using the accumulation of 2-thiobarbituric acid reactive substances (TBARS) as an indicator of oxidation, the most effective antioxidants were clove, sage, and rosemary with ginger and thyme being least and equally ineffective (Table II). Additions of spices at the 2000 ppm were equivalent to 200 ppm of TBHQ and BHT, respectively (Table II).

Table II. Ground spices stabilize pork emulsions at 200 (A) and 1000 (B) ppm when stored at 4°C for 21 days.*

Treatments	A: TBARS (µg/g)	B: TBARS (µg/g)
Control	10.6	10.6
Rosemary	10.6	1.6
Clove	4.7	0.4
Sage	9.1	0.7
Ginger	8.5	5.8
Thyme	9.4	3.8
Oregano	8.8	1.6
TBHQ (200 ppm)	0.4	0.4
BHT (200 ppm)	0.9	0.9

TBARS, thiobarbituric acid reactive substances; TBHQ, tert-butylhydroquinone; BHT, butylated hydroxytoluene.
SOURCE: Adapted with permission from reference 7. Copyright 1995.

Of the spices, rosemary has perhaps received the greatest amount of study in that commercial extracts of water miscible and oil miscible antioxidant compounds are available and have been standardized to a given antioxidant equivalency. Addition of the water-soluble extract to frozen cooked rainbow trout flakes at 2.5g /kg of fish was similar though not as effective as 0.5g/kg of fish TBHQ. However, addition of TBHQ and the rosemary extract increased the effectiveness of the TBHQ enabling a reduction in the amount of synthetic antioxidant necessary to stabilize one of the most highly unstable oil systems (9). This study is significant in that it demonstrated that the rosemary extract was effective in reducing the degree of oxidation as measured by TBARS as well as in reducing the intensity of fish odors and aromas typically associated with seafood products and the oxidation of their oils.

Several studies have well documented both the positive and negative consequences of irradiation on food quality. Even though irradiation of foods at low dosages of less than 10 kilogray (kGy) has been approved by many food regulatory agencies through out the world, studies show that foods containing polyunsaturated fatty acids may generate volatile oxidation compounds, most notably alkanes and alkenes. This is especially true of meat products noted to contain significant sources of unsaturated fatty acids such as linoleic acid (18:2, n-6) and arachidonic acid (20:4, n-6). In a model study *(10)* designed to determine the effects of low dosages of gamma-irradiation on radiolysis of linoleic acid and arachidonic, unextracted powdered rosemary and thyme were effective in reducing the formation of volatile hydrocarbons. When rosemary and thyme were applied at ratio of 1/0.1 (lipid/ plant), both spices were effective in reducing the total amount of C_{10}-C_{19} below the acceptable hydrocarbon levels allowed for 10 kGy applications used in commercial pasteurization operations.

The chemical composition of the active antioxidant components of rosemary appears to be related to the concentrations of carnosic acid, carnosol, carnosolic acid, rosmaridiphenol, and rosmarinic acid *(10-12)*. As these are all phenolic compounds, the major mode of antioxidant activity of rosemary is hydrogen donation; thus allowing it to be grouped as a primary antioxidant. The antioxidant activity of thyme has been shown to be associated with the presence of volatile components, thymol and carvacrol as well as with two very active flavonoid glycosides, identified as eriodictyol-7-rutinoside and luteolin-7-0-β-glucopyranoside *(13)*. All of the compounds appear to have radical scavenging capacity and the potential of reacting with singlet oxygen. It is interesting to note that while structural similarities exist between the thymol, carvacrol and BHT and BHA, neither of the latter synthetic antioxidants was effective in reducing radiation-induced hydrocarbons *(14)*.

Seed Oil Extracts

Second only to spices, the application of seed oils and their extracts as natural stabilizers of foods pre-date recorded history *(1)*. A variety of seed extracts including soybean, cottonseed, olive oils, and sesame, macadamia nut, peanuts, etc have all been used as sources of raw material from which antioxidant compounds have been extracted. Extracted materials have varied greatly from that of the hull of peanuts to the defatted flour, protein concentrate and isolate of cottonseeds, soy, and peanuts. Historically, sesame seed oil and olive oil extracts have been most often cited for their antioxidant capacity *(1,15,16)*. During the screw-press extraction of sesame seeds and olives, active antioxidant compounds are carried into the oils giving them a shelf stability unseen in many other oils due not only to their low content of polyunsaturated fatty acids but the presence of variety of antioxidants. The active compound in sesame seeds has been described as sesamolin, a lignin, which is hydrolyzed to form sesamol upon heating.

During the extraction of olive oil from olives, polyphenolic compounds pass into the raw oil and provide an increased level of stability to the oil. In addition to high levels of natural tocopherols, phospholipids and other polyphenolic compounds contribute to the stability of olive oil. The total hydrophilic phenols and the oleosidic forms of 3,4-hydroxyphenolethanol have been shown to have a higher correlation to the stability of virgin olive oil than total tocopherol content (*15*).

Antioxidant activity due to Flavonoids

Soy extracts including soy flour, protein, isolates, and concentrates are perhaps the most popular modern antioxidant extracts. Several epidemiological studies (*16-19*) have shown excellent correlation between soy protein consumption and reduced risk of cancer as well as the inhibition of tumor formation in animal studies (*16,17*). The major antioxidant compounds believed to be responsible for the antioxidant activity are the isoflavone aglycons, genistein, and daidzein. Their effectiveness has been amply demonstrated in model biological studies (*18,19*) as well as in stability food model studies (*20-24*). One such study compared the effectiveness of glandless cottonseed peanut, and soy protein extracts on the rancidity and quality of ground beef. At the 10% replacement level (Table III), cottonseed flour was most effective in protecting cooked ground beef patties from rancidity during refrigerated storage. Soy products and peanut extracts were equally effective in reducing TBARS. However, sensory evaluations revealed that only the soy products were equal in sensory quality to the "all beef" patties (*21*).

Table III. Seed oil proteins reduce rancidity in ground beef at 10% stored at 4°C over 6 days*.

Treatments	*TBARS[1]*	*Sensory Scores*
All Beef	3.0	4.7
+ Cottonseed	1.5	3.6
+ Peanut	2.2	3.5
+ Soy	1.7	4.6
+ Soy textured	1.7	4.9

[1]TBARS = 2-Thiobarbituric acid reactive substances
*Adapted from Ref. 21.

Comparison of different soy products reveals that differences exist between the efficacy of the various soy products as antioxidants and that this is a function of the concentration of individual flavonoids present. For example, a soy protein isolate antioxidant fraction obtained from soy protein was more effective at 900 ppm than at 300 ppm with the differences attributable to the presence of selected isoflavones compared to glucosides (23,24). The antioxidant properties of soy products have been primarily attributed to the isoflavones aglycons, genistein and daidzen (24).

Tea flavonoids

The polyphenols contained in tea, most notably green tea has received lots of research attention because of their effectiveness as antioxidant in lipid peroxidation model systems and *in vitro* model studies examining antioxidative and antimutagenic properties of teas. In model studies comparing the effectiveness green tea extracts to tocopherol and BHT, extracts of green tea added at 40 ppm to 1000 ppm to bulk oil oxidation model systems of lard, soybean oil, and fish oil showed antioxidative activity comparable to BHT and α-tocopherol. In fish oil models, 250 ppm of tea extracts showed stronger antioxidative activity than 500 ppm of BHT whereas tocopherol was ineffective. Similarly, in lard and soybean oil models, a dose dependent relationship was observed in that increased addition tea extracts at 40 to 60 ppms were as effective as 200 ppm of α-tocopherol (25). Other studies show that compared to rosemary extracts, tea extracts are effective in stabilizing model systems containing pork fat, chicken fat, and canola oil. Extracts obtained from jasmine and longjing teas when added at the 200 ppm showed stronger antioxidant activity than extracts of rosemary (26).

The major compounds associated with the antioxidative activity of tealeaves have been identified and ranked with regards to their potency. The compounds were identified as catechins and included epicatechin (EC), epigallocatechin (EGC), epicatechin gallate (ECG), and epigallocatechin gallate (EGCG). It appears that much of the antioxidant properties of individual catchins, and therefore the teas from which they are extracted, is dependent upon dosage and the relative concentrations of individual catechins in any given tea when extracted. For example, the addition of 200 mg/kg of the fore-mentioned catechins to a comminuted pork model system showed that the order of effectiveness was EGCG = ECG> EGC> EC with each of the catechins demonstrating greater antioxidant activity than comparable levels of α-tocopherol or BHT (27,28). It is significant to note also that even though investigators have used a variety of methods to show that flavonoids have strong antioxidant properties, there is consistency in their findings that crude forms of flavonoids and or refined synthetic forms of flavonoid are comparable to commercial antioxidants. A variety of very sensitive oil systems such as lard and pork fat, soybean oil, and fish oil have been used to demonstrate the potential application of flavonoids as natural antioxidants (29).

Tea flavonoids have also been tested for their ability to function as primary antioxidants whereby they scavenge free radicals to prevent the peroxidation of lipids and lipoproteins. As free radicals, including hydroxyl radicals, superoxides, and hydrogen peroxides may be produced by carcinogens, tea flavonoids are often regarded as natural anticarcinogens and antimutagens (30). In a simulated *in vitro* oxidation model for heart disease, tea flavonols exhibited a dose-response inhibition of copper-induced peroxidation of LDL similar to that shown by vitamins C, E and β-carotene (5). When tested against the vitamins, epigalloepicatechin gallate, a tea polyphenol, exhibited an antioxidant capacity 20 times that vitamin C.

The potency of the flavonols appears to be associated with the number of o-hydroxy groups as EGCG has four such groups. The type of tea being examined and the fermentation method(s) used to produce a given tea are major contributing factors to the potency of a given tea as an antioxidant. For example, the superoxide scavenging effects of four tea extracts decreased in the order of oolong tea > pouchong tea> green tea > black tea, whereas the length of processing and potency decreased in the order of semi-fermented > nonfermented > fermented (31). The antimutagenicity of the teas often correlated with the antioxidant capacity but also varied with mutagen and the antioxidative properties (5,30,31).

Role of Phospholipids as Antioxidants

Like many of the natural sources of antioxidants, phospholipids (PL) are found in a variety of plant sources and are believed to exert multiple modes of antioxidant activity. The most abundant sources of PL include egg yolk, soybean lecithin, and the many sources of vegetable oils from which crude lecithins are removed during the degumming phase of oil refinery. As a class of natural antioxidants, PL represent the most controversial group of antioxidants in that many *in vivo* assays, they have been shown to be pro-oxidants. By contrast, many lipid model studies show that PL, alone and in conjunction with other antioxidants, have the ability to stabilize lipids. Though the exact mechanism(s) are still not clearly defined, several factors, including the class of PL, their headgroup, associated fatty acids, and the presence or absence of other minor constituents associated with PL have all been demonstrated to affect their activity in model systems and in mixed food systems. The proposed mechanism(s) of action include their ability to decompose hydroperoxides, chelate metal ions, and act synergistically in the presence of other primary and secondary antioxidants, such as tocopherols, flavonoids, carotenoids, and selected vitamins such as vitamins C and E (32,33).

Studies of the antioxidant activity of natural and synthetic antioxidants in fish oil model systems show that when blue fish PL were isolated and separated into PL vs. neutral lipids, the addition of the more unsaturated PL fraction stabilized salmon oil compared to the neutral lipids that promoted oxidation (34). The addition of crude PL fractions at concentrations of 2.5% and 5.0% (wt/wt) were equivalent to the addition of BHT as measured by both TBARS and change in the polyene index (Figure 1 & 2).

Subsequent follow-up studies (35) using purified PL further clarified that the antioxidant activity of PL was associated with amine-containing PL, such as phosphatidyl choline (PC), phosphatidyl ethanolamine (PE), and sphingomyelin (SPH). The ability to stabilize lipids was also affected by the chain length and degree of unsaturation of the fatty acids on the PL. Those PL with longer chain length (i.e C: 18 to C22: 0 vs. C16: 0) and PL containing more saturated fatty acids were the most effective antioxidants. A dose response Curve was also observed for the addition of PC from as low as 0.01% (wt/wt) (Figures 3).

Synergism between PL and other natural antioxidants contained in seed oils and other plant materials has been observed between tocopherols in a number of model studies (36,37) as well as several product applications (37,38). Addition of PL to tocopherol-containing oils shows that PL tends to increase the efficacy of the tocopherols and that the classes of PL most frequently associated with this synergism are PC and PE. The four proposed mechanisms put forth to explain the synergist activity of PL with tocopherols include the chelation of trace metal ions; decomposition of hydroperoxides; delay or extension of the irreversible oxidation of tocopherol to tocopherylquinone (32,39); and the possible formation of Maillard reactive compounds resulting from the reaction of PL (i.e. PC and PE) with conjugated carbonyl compounds following the degradation unsaturated fatty acids (32,36,39).

Phospholipids have also been shown to exhibit a high degree of synergism with polyhydroxy flavonoids that also act as primary antioxidants. The addition of PE at levels upward of 0.1% along with flavonoids (i.e, quercetin) of 0.007% to 0.07% were effective in stabilizing a soybean oil model system (38). The proposed mechanisms appear to be a combination of the ability of PL to function as metal chelators and proton donors in that the amounts of PL required for metal chelation is greater than the trace quantities of metal ions typically present in oils. It is significant to note that in all the studies where PL have been shown to be effective in stabilizing lipids, the model systems contained oils with a high degree of polyunsaturated fatty acids and or fats noted for their instability. Studies of Perilla oil (40), fish oils (34,35), leaf tissue lipids (37), and soybean oil (32) show that these oils contain significant quantities of omega-3 fatty acids including 18:3, 20:5 and 22:6 and that an additive or synergistic effect can readily be observed with the addition of PL to tocopherols and or flavonoids.

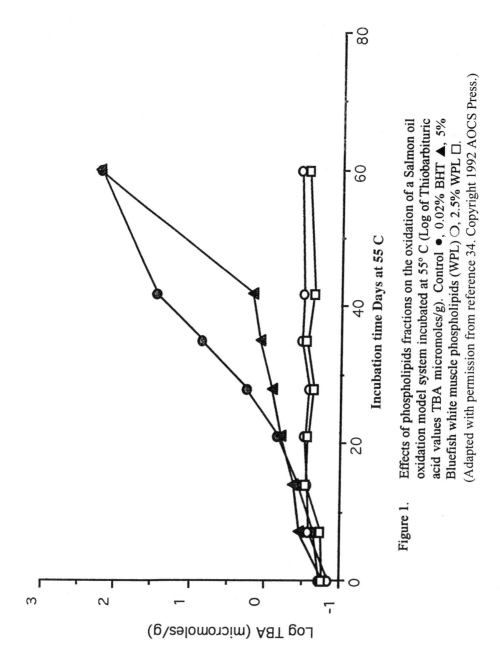

Figure 1. Effects of phospholipids fractions on the oxidation of a Salmon oil oxidation model system incubated at 55° C (Log of Thiobarbituric acid values TBA micromoles/g). Control ●, 0.02% BHT ▲, 5% Bluefish white muscle phospholipids (WPL) ○, 2.5% WPL □. (Adapted with permission from reference 34. Copyright 1992 AOCS Press.)

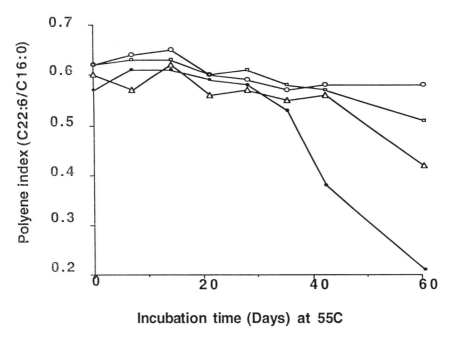

Figure 2. Effects of phospholipids and butylated hydroxytoluene (BHT) on change in fatty acid profile (Polyene Index) Control ●, BHT (.02%) Δ, 2.5% WPL. □, 5.0% WPL ○.

(Adapted with permission from reference 34. Copyright 1992 AOCS Press.)

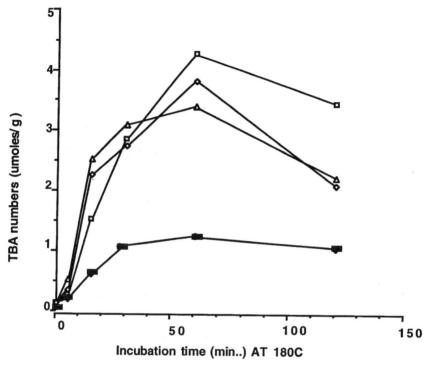

Figure 3. Effects of commercial phosphatidylcholine (PC) on the TBA values of a salmon oil oxidation model system heated at 180 C. Control □, 0.01% PC ◇, 0.1% PC Δ, 1% PC ■.

(Adapted with permission from reference 35. Copyright 1992 AOCS Press.)

Finally it is also important to note that during the typical refining of many oils, for all practical purposes, all of the PL are removed during the degumming process whereas significant reductions in the tocopherol levels can be seen in most refined vegetable oils. Studies show that the retention of minimum levels of PL in the range of 0.025% to 0.22% add significantly to the stability of highly unsaturated oils containing residual quantities of tocopherols (*41,42*). The retention of as little as 1.5% or 11 µg/g of the original 43 µg/g of γ-tocopherol has been shown to influence the stability of rapeseed oil. It should also be pointed out that γ-tocopherol levels typically present in rapeseed oil represent less than 6% of the total tocopherols and that the order of effectiveness in hydrogen-donating power is α > β > γ > δ. These two studies (*41,42*) as well as many others on the potential role of natural antioxidant show the need to seriously reevaluate the role of processing in maximizing the retention of health-promoting constituents, i.e. antioxidants.

Fruit, Vegetable and Wine Antioxidants

No discussion of natural antioxidants would be complete without some special attention given to the massive body of research evidence accumulating that fruits and vegetables and their by-products contain some of the most potent sources of antioxidants. While most traditional studies tend to relate the antioxidant properties of fruits and vegetables to their content of β-carotene and vitamins A, C, and E, more recent evidence shows that in many of these fruits and vegetables the predominant antioxidant activity appears to more related to the quantities of polyphenolic compounds. These compounds include flavonoids (i.e. flavonols, isoflavones, flavonones, anthocyanins, catechin, and isocatechin) and others. As many fruits, vegetables, and their by-products contain a number of antioxidants, some of which have been shown to be primary antioxidants, investigators have developed *in-vitro* techniques to determine and compare their total antioxidant capacity to each other and to synthetic and biological antioxidants. The ORAC (i.e. oxygen radical absorbance capacity) and the ability of antioxidants to inhibit the peroxidation of LDL are two frequently used assays (*43,44*). Using the ORAC assay, strawberries have been demonstrated to have one of the highest antioxidant activities, followed by plum, orange, red grape, kiwi fruit, pink grapefruit, white grape, banana, apple, tomato, pear, and honeydew melon (*44*). When *in vitro* oxidation of LDL is used, blackberries produced the highest antioxidative effects, followed by red raspberries, sweet cherries, blueberries, and strawberry (*43*). Even though the mechanism(s) by which flavonoids exert their antioxidant activity is still not fully established, application of these assays to fruits and vegetables have progressed the theoretical basis for which scientists believe wine and other plant phenolics are associated with reduced risk of chronic diseases such as heart attack and cancer. It is postulated that *in vivo*, they may down-regulate the enzymes associated with the production of cyclooxygenases, and lipoxygenases, two enzymes associated with atherogenesis, platelet aggregation, and immune suppression (*43*).

Antioxidant Activity of Miscellaneous Plant and Food Extracts

Numerous lipid model studies have demonstrated that significant antioxidant activity reside in the natural extracts of a numerous of plants including peanut hulls, died beans, sweet potatoes, leaves, bark of trees, natural Chinese and Japanese drugs, fresh peppers, etc. Many of the proposed antioxidant components and their proposed mechanism(s) of activity are similar to those mentioned in the above text. A few of this showing unusual antioxidant activity when compared to commercial and or synthetic antioxidants will be briefly discussed.

Peanut hulls

Methanol extracts of peanut hulls show antioxidative activity equal to BHA, greater than that of α-tocopherol, and varied with the cultivar and quantity of luteolin and total phenolic compounds present (*45, 46, 47*). The content of luteolin and total phenolic of cultivars examined decreased in the order of Spanish >Valencia > Runner > Virginia and increased with maturity (*37*).

Macadamia nuts

Cold pressed Macadamia nut oil has been shown to have a resistance to rancidity development greater than its refined oil and when oils were examined, the antioxidant properties were found to be due to the quantity of phenolic compounds coming from the shell and nut kernels. These compounds were identified as 2,6-dihydroxybenzoic acid, 2'-hydroxy-4'-methoxyacetophenone, 3'5'-dimethoxy-4'-hydroxyacetophenone, and 3,5-dimethoxy-4-hydroxycinnamic acid. Synergism between the 4 phenolic compounds and some unidentified compound(s) is believed to exist in that the lowest levels for which individual compounds were tested were not equivalent to the levels found in the nut itself, whereas both crude oil and crude oil with shells showed greater antioxidant than the refined oil (*48*).

Phytic acid

The presence of phytic acid at 1-5% by weight of many plant seeds, cereals, legumes, and nuts has in the traditional sense been viewed as negative factor in that phytates are known chelators of iron and thus considered to be antinutrients with regards to mineral bioavailability. However, more recent studies show that phytic acid can decrease the incidence of colon cancer (*49*), lower blood cholesterol (*50*) and are potent inhibitors of iron-driven hydroxyl radical formation (*51*). When used as a natural antioxidant, phytic acid (1mM) was able to extend the shelf life of oil-in-water emulsions by fourfold. Phytic acid also prevented the development of warmed-over flavor notes in refrigerated chicken as determined by decreased oxygen uptake and reduced TBARS. Additionally, phytic acid has been shown to be effective in inhibiting both iron and non-iron induced lipid peroxidation in beef homogenates and was more effective at equal concentrations than either ascorbic acid, BHT or ethylenediaminetraacetic acid (*52*).

Sweet potatoes

The antioxidative activity of sweet potatoes has been attributed to it content of phenolic compounds. These compounds have been identified as chlorogenic acid and isochlorogenic acid-1, -2 and –3 as well as some other minor phenolic compounds. Contrary to what one might expect, the antioxidant properties were not associated with the high content of β-carotene but with the concentration of phenolic components and the synergism between free amino acids and phenolic compounds (*53*).

Dried beans

The antioxidant capacity of dried beans has primarily been associated with the presence of polyphenolic compounds and their ability to chelate metal ions. When several types of beans, including pinto, kidney, white (Great Northern), pink, and black beans were tested in a soybean oil model oxidation system in the presence and absence of iron, all of the bean extracts inhibited iron-catalyzed oxidation but failed to show a protective effect in the absence of iron. At equal concentrations, bean extracts were more effective than BHA, propyl gallate, and ascorbic acid in preventing iron-catalyzed oxidation (*54*).

Leaf extracts

Since many extracts from leaves and other plant materials are used medicinally and contain essential oils that are also used medicinally, the development of natural antioxidants to protect these extracts would be expected. Extracts of *Ecalputus globulus* leaves were found to contain β-diketones with long alkyl side chains that were effective in protecting oil-in- alcohol oxidation systems but provided no protection for bulk oil systems (55).

Leaf extracts of Boldine (*Peumus boldus*) have also been shown to have antioxidative properties comparable to commercial antioxidants. In a fish oil oxidation model system, extracts of boldine showed antioxidant activity greater than that of dl-α- tocopherol, BHA or BHT, and was comparable to quercetin. The antioxidant properties appear to be associated with the concentration of an alkaloid [(S)-2,9-dihydroxy-1, 10-dimethoxyaporphine) and the presence of two phenolic groups in its structure and their capacities to delocalize electrons (56). The recurring and common themes occurring across the several classes of polyphenolic compounds found in a variety of plant extracts appear to be the dihydroxy substutients located in the orthro- position of the B ring and the conjugation between the A and B rings. Both properties have been shown to enhance the capacity of flavonoids and other polyphenolic types of compounds to delocalize free electrons and provide resonance stabilization (57).

Maillard Reaction Products (MRP)

The antioxidant properties of MRP have been demonstrated in a number of model lipid oxidation studies as well as in processed food systems. The formation of these products is characterized by the presence of glycine and glucose in the presence of nitrogenous compounds such as amino acids. In the presence of heat, a reducing sugar, and a source of nitrogen, MRP have been shown to provide antioxidative protection to lipids in a number of systems including cereal products, lard, cottonseed oil, soybean oil, and shortening (58). These products are also most often associated with increases in color and the occurrence of fluorescence products (59). When examined in foods, MRP have been found effective in reducing warmed-over flavors, odors, and chemical parameters associated with oxidation such as hexanal and TBARS. For example, the application of MRP resulting from the autoclaving of egg albumin acid hydrolysate and glucose were effective in stabilizing cooked beef, refrigerated at 4° C over several days. The development of off-flavors characterized as cardboardy and painty was reduced whereas TBARS values appeared to reflect the level of addition of MRP (60).

Factors affecting the potency of natural antioxidants

It is important when evaluating the potency of antioxidants, especially in model systems, that consideration be given to the variety of factors that can influence the outcome of the results. These can include such factors as the type of assay, reaction medium from which the reaction is performed, characteristics of the antioxidant, and the presence of minor and unknown components which may influence the activity of the antioxidant, and synergism between different types of antioxidants. For example, Frankel et al (*61*) demonstrated that the lipophilic and hydrophilic nature of the antioxidant influences its effectiveness. Lipophilic antioxidants such as α-tocopherol were more effective in an oil-in water emulsion system whereas hydrophilic antioxidants such as ascorbic acid performed better in a bulk oil system. The differences in performance of the antioxidants were attributed to an "interfacial phenomena" in which hydrophilic antioxidants exhibit greater affinities for the air-oil interfaces of bulk oil systems whereas the lipophilic antioxidants have a greater affinity for water-oil interfaces present within emulsions. This theory was subsequently demonstrated in a series of antioxidant tests in which a hydrophilic rosemary extract was effective in inhibiting oxidation in bulk oil systems containing soybean, corn, peanut or fish oil. However, when rosemary extracts were tested in an oil-water emulsion system, they were far less effective (*61*).

The importance of the reaction environment has also been demonstrated in studies that examined the antioxidant activities of β-carotene and α-tocopherol. In model peroxidation studies in which β-carotene and α-tocopherol were tested singly and together, it was shown that the rate of reaction for β-carotene reaction increases with it own concentration and that of oxygen and decreased with methyl linoleate concentration, whereas the rate of reaction for α-tocopherol was independent of its own concentration as well as methyl linoleate level (*63*). These findings are similar to more recent studies of β-carotene in model studies by Henry et al (*64*) in which it was shown that under high oxygen pressure and temperature, β-carotene became a pro-oxidant increasing the oxidation of unsaturated lipids. Synergism between antioxidants is also a critical issue in that many studies have demonstrated that high concentration (i.e. above 500 ppm); α-tocopherol may exhibit pro-oxidant activity. In the presence of 0.01% citric acid and 468 ppm of α-tocopherol, 566% synergism has been demonstrated where as in the presence of 1500 ppm of tocopherol, only 285% synergism was observed (*43*). Similarly, when γ-Tocopherol is used in the presence of ascorbic acid in a fish oil model system, 100% synergism can be demonstrated by increasing γ-tocopherol at a fixed concentration of ascorbic. However when γ-tocopherol levels are fixed and ascorbic acid levels varied, synergism efficiency rose sigmoidally with increasing ascorbic acid levels (*65*). Both studies demonstrate the need to examine multiple sources of antioxidants, and in many foods systems, multiple isomers of tocopherol have been shown to be more effective than α-tocopherol alone.

The results obtained with antioxidant(s) may also be highly dependent on the type of assay applied. Though many investigators have adopted the Oxidative Stability Instrument (OSI) as a means of measuring oxidation, Frankel (66) and others (67) caution the application of OSI, Active Oxygen Method (AOM), the oxygen bomb and other high temperature assays (i.e. 100^0 C) to measure the antioxidant properties of antioxidants. Their basic premise is that at high temperatures, the mechanism of oxidation differs from that observed at lower temperature and that resulting peroxides values (PV) derived at such temperatures are not in focus with lower PV obtained where food products are actually consumed or rejected (68).

Summary

Renewed interest in the application of natural antioxidants has caused many research investigators and many segments of the food industry to examine their application in stabilizing lipids and as potential sources of biological antioxidants. The major sources of antioxidants include spices and herbs, seed oils, cereals, protein hydrolysates and others. Unlike many synthetic antioxidants, natural extracts contain a variety of compounds whose modes of antioxidant activity may be varied, exhibit multiple sources of activity, and in many instances show a great deal of synergism with natural and synthetic antioxidants. In assessing the activity of natural antioxidants, it is important to develop and apply relevant assays to measure the activity of the antioxidant under conditions similar to planned product applications. As many natural antioxidants may be used under good manufacturing conditions with no maximum limits set on usage levels, it is important that their safety and efficacy for use in human foods be determined with the same vigor as synthetic antioxidants. This is especially true for compounds isolated in chemically pure forms and where they possess structures and modes of action similar to some of the synthetic antioxidants already under investigations as potential carcinogens.

References

1. Namiki, Mitsuo. *Crit. Rev. Food Sci Nutr.* **1990**, *29*, 273-300.
2. Jacob, R. A. *INFORM* **1994**, *5*, 1271-1275.
3. Gaziano, J. M.; Manson, J. E.; Ridker, P. W.; Hennekens, C. H. *Circulation* 1990. 82 supp III, 201.

4. Enstrom, J. E.; Kanim, L. W.; Klein, M. A. *Epidemiology* **1992**, *3*, 194.
5. Vinson, J. A.; Dabbagh, Y. A.; Serry, M. M.; Jang, J. *J. Agric. Food Chem.* **1995**, *43*, 2800-2802.
6. Pokorny, J. *Trends Food Sci. Technol.* **1991**, *2*, 223-227.
7. Shahidi, F.; Pegg, R.; Saleemi, Z. O. *J. Food Lipids* **1995**, *2*, 145-153.
8. Olson, J. A.; Kobayashi. P. *Soc. Exp. Biol. Med.* **1992**, *200*, 245-281.
9. Boyd, L. C.; Green, D. P.; Giesbrecht, F. B.; King, M. F. *J. Sci. Food Agric.* **1993**, *61*, 87-93.
10. Lacroix, M.; Smoragiewicz, W.; Pazdernik, L.; Kone, M. I.; Krzstyniak, K. *Food Res. Int.* **1997**, *30*, 457-462.
11. Wu, J. W.; Lee, M.; Ho. C.; Chang, S. S. *J.Am. Oil Chem. Soc.* **1982**, *59*, 339-345.
12. Hall, III, C. A.; Cuppett, S. L.; Dussault, P. *J. Am. Oil Chem. Soc.* **1998**, *75*, 1147-1154.
13. Wang, M.; Li, J.; Ho, G. S.; Peng, X.; Ho, C. T. *J. Food Lipids* **1998**, *5*, 313-321.
14. Deighton, M.; Glidewell, N.; Deans, S. M.; Goodman, S. G. *J. Food Sci.* **1997**, *63*, 221-225.
15. Baldioli, M.; Servili, M.; Perretti, G.; Montedoro, G. F. *J. Am. Oil Chem. Soc.* **1996**, *73*, 1589-1593.
16. Caragay, A. B. *Food Technol.* **1992**, *46*, 65-68.
17. Messina, M.; Barnes, S. J. *Natl. Cancer Inst.* **1991**, *83*, 541-546.
18. Bartholomew, R. M.; Ryan, D. S. *Mutat. Res.* **1980**, *78*, 317-320.
19. Henderson, B. R.; Ross, R. K.; Pike, M. C.; Casagrande, J. T. *Cancer Res.* **1982**, *42*, 3232-3239.
20. Hammerschmitdt, P. A.; Pratt, D. E. *J. Food Sci.* **1978**, *43*, 556-557.
21. Ziprin, Y. A.; Rhee, K. S.; Carpenter, Z. L.; Hostetler, R. N.; Terrell, R. N.; Rhee, K. *J. Food Sci.* **1981**, *46*, 58-61.
22. Rhee, K. S.; Ziprin, Y. A.; Rhee, K. C. *J. Food Sci.* **1981**, *46*, 75-77.
23. Wu, S. Y.; Brewer, M. S. *J. Food Sci.* **1994**, *59*, 702-706.
24. Wang, H.; Murphy, P. A. *J. Agric. Food Chem.* **1994**, *42*, 1666-1673.
25. Koketsu, M.; Satoh, Y. J., *J. Food Lipids* **1997**, *4*, 1-9.
26. Chen, Z. Y.; Wang, L. Y.; Chan, P. T.; Zhang, Z.; Chung, H. Y.; Liang, C. *J.Am. Oil Chem. Soc.* **1998**, *75*, 1141-1145.
27. Shahidi, F.; Alexander, D. M. *J. Food Lipids* **1998**, *5*, 125-133.
28. Wanasundara, U. N.; Hardin, R. T.; Sims, J. S. *J. Food Sci.* **1993**, 43-46.
29. Nieto, S.; Garrido, A.; Sanhueza, J.; Loyola, L. A.; Morales, G.; Leighton, F.; Valenzuela, A. *J.Am.Oil Chem. Soc.* *70*, 773-778.
30. Yuting, C.; Ronggliang, Z.; Zhongijamm, J.; Yong, J. *Free Rad. Biol. Med.* **1990**, *9*, 19-21.
31. Yen, G.; Chen, H. *J. Agric. Food Chem.* **1995**, *43*, 27-31.

32. Hilderbrand, D. H.; Terao, J.; Kito, M. *J. Am. Oil Chem. Soc.* **1984**, *61*, 552-555.
33. Bertram, J.; Hudson, F.; Lewis, J. I. *Food Chem.* **1983**, *10*, 111-120.
34. King, M. F.; Boyd, L. C.; Sheldon, B.W. *J. Am. Oil Chem. Soc.* **1992**, *69*, 237-242.
35. Nwosu, C. V.; Boyd, L. C.; Sheldon, B. *J. Am. Oil Chem. Soc.* **1997**, *74*, 293-297.
36. Bandarra, N. M.; Campos, R. M.; Batista, I.; Nunes, M. LL.; Empis, J. M. *J. Am. Oil Chem. Soc.* **1999**, *76*, 905-913.
37. Bertram, J. F.; Hudson, J. F.; Mahgoub, S. E. O. *Soc. Chem. Ind.* **1981**, *32*, 208-210.
38. Bertram, J. F.; Hudson, F.; Lewis, J. I. *Food Chem.* **1983**, *10*, 111-120.
39. Ohshima, T.; Fujita, Y.; Koizumi, C. *J. A. Oil Elec. Soc.* **1993**, 269-276.
40. Kashima, M.; Cha, G.; Isoda, Y.; Hirano, J.; Miyazawa, T. *J. Am. Oil Chem. Soc.* **1991**, *68*, 119-122.
41. Prior, E. M.; Vadke, V. S.; Sosulski, F. W. *J.Am. Oil Chem. Soc.* **1991**, *68*, 407-411.
42. Lampi, A.; Hopia, A.; Piironen, V. *J. Am. Oil Chem. Soc.* **1997**, *74*, 549-555.
43. Frankel, E. N. *INFORM* **1999**, *10*, 889-896.
44. Wang, H.; Cao, G.; Prior, R. L. *J. Agric. Food Chem.* **1996**, *44*, 701-705.
45. Duh, P.; Yeh, D.; Yen, G. *J. Am. Oil Chem. Soc.* **1992**, *69*, 814-818.
46. Yen, G.; Duh, P. *J. Am. Oil Chem. Soc.* **1993**, *70*, 383-386.
47. Yen, G.; Duh, P. *J. Am. Oil Chem. Soc.* **1995**, *72*, 1065-1067.
48. Quinn, L. A.; Tang, H. H. *J. Am. Oil Chem. Soc.* **1996**, 1585-1588.
49. Graf, E.; Eaton, J. W. *Cancer* **1985**, *56*, 717-7618.
50. Jariwalla, R. J.; Sabin R.; Lawson, S.; Bloach, D. A.; Prender, M.; Andrew, V.; Herman, Z. S. *Nutr. Res.* **1988**, *8*, 813-827.
51. Empson, K. L.; Labuz, T. P.; Graf. E. *J. Food Sci.* **1991**, *56*, 560-563.
52. Lee, B. J.; Hendricks, D. G. *J. Food Sci.* **1995**, *60*, 241-244.
53. Hayase, F.; Kato, H. *J. Nutr. Sci Vitaminol.* **1984**, *10*, 37-46.
54. Ganthavorn, C.; Hughes, J. S. *J. Am. Oil Chem. Soc.* **1997**, *74*, 1025-1030.
55. Osawa, T.; Namiki, M. *J. Agric. Food Chem.* **1985**, *33*, 777-779.
56. Valenzuela, A.; Nieto, S.; Cassels, B. K.; Speisky, H. *J. Am. Oil Chem. Soc.* December **1991**, *68*, 935-937.
57. Rice-Evans, C.; Miller, N. J.; Bolwell, P. G.; Bramley, P. M.; Pridham, J. B. *Free Rad. Res.* **1995**, *22*, 375-383.
58. Park, C. K.; Kim, D. H. *J. Am. Oil Chem. Soc.* **1983**, *60*, 98-102.
59. Severini, C.; Lerici, C. R. *Ital. J. Food Sci.* **1995**. N. 2. 189-196.
60. Smith, J. S.; Alfawaz, M. *J. Food Sci.* **1995**, *60*, 234-235.
61. Frankel, E. N.; Huang, S.; Kanner, J.; German, J. B. *J. Agric. Food Chem.* **1994**, *42*, 1054-1059.

62. Prior, E.; Aeschbach, R. *J. Sci. Food Agric.* **1996**, *72*, 201-208.
63. Tsuchihashi, H.; Kigoshi, M.; Iwatsuki, M.; Niki, E. *Arch. Biochem. Biophys.* **1995**, 137-147.
64. Henry, L.; Catignani, G. L.; Schwartz, S. J. *J. Am. Oil Chem. Soc.* **1998**, *75*, 1399.
65. Yi, O.; Han, D.; Shin, H. *J. Am. Oil Chem Soc.* **1991**, *68*, 881-883.
66. Frankel, E. N. *Trends Food Sci. Technol.* **1993**, *4*, 220-226.
67. Warner, T.; Frantel, E. N.; Mounts, T. L. *J. Am. Oil Chem. Soc.* **1989**, *66*, 555-563.
68. Hill, S. E. *INFORM* **1994**, *5*, 104-109.

Chapter 21

The Effect of Additives on the Rancidity of Fish Oils

R. J. Hamilton, G. B. Simpson, and C. Kalu

Chemistry Department, School of Pharmacy and Chemistry, John Moores University, Liverpool L3 3AF, United Kingdom

The stabilization of Chilean fish oil by adding fractions from Maize (corn) oil has indicated the utility of additives. Chilean fish oil has been stabilized by binary and ternary mixtures containing tocopherols, lecithin, ascorbyl palmitate and cholesta-3,5-diene .

About 80-100 million tones of fish are caught each year .For such a large amount, they are obviously of major importance to the modern day diet. seventy five percent of the fish caught annually are classed as "oily" , which means that they contain between 5 and 25 % lipid.They are therefore a significant source of oil for the food industries of teday and contain plentiful supplies of n-3 polyunsaturated fatty acids, such as eicosapentaenoic acid (EPA, 20:5) and docosahexaenoic acid (DHA , 22:6). From a nutritional point of view these oils are attractive because they are thought to combat heart disease (*1-10*) and inhibit the growth of cancerous tissue (*11-18*).Imports of fish oil to the United Kingdom are compared to the values for other major vegetable fats and oils in Table 1from which it can be seen that the imports of marine oils has been falling as the impact of the trans fatty acid story reduces the use of partially hydrogenated fats.

Blending of fish oils with other dietary oils such as vegetable oils (e.g. sunflower, maize and corn) can be beneficial for two important reasons. Firstly, to influence texture during mastication (*19,20*) and secondly to aid the inhibition of oxidation of the fish oil with which it is blended ,in the presence of certain antioxidants (*21-24*). Due to their highly unsaturated nature , oils are susceptible to oxidation and therefore rancidity which in turn leads to spoilage of the oil or oil blends. Vegetable oils contain a higher percentage of oleic acid (OA , 18:1) and linoleic acid (LA , 18:2),which although susceptible to oxidation themselves, help to lower the overall amount of the more highly unsaturated EPA and DHA in the blend. Antioxidants and in particular synergistic mixtures of antioxidants are of a major benefit to the stability of dietary fish oils and blends. Antioxidants inhibit the initiation and propagation steps of the free radical mechanism of autoxidation increasing the oxidative stability of the oil or oil blends. The most common antioxidants used in the food industry today are tocopherols (α, β/γ, and δ), butylated hydroxytoluene (BHT), tert. butyl hydroquinone (TBHQ) and butylated hydroxyanisole (BHA) together with other additives such as ascorbic acid, ascorbyl palmitate and lecithin which can be used in synergistic mixtures to limit the degree of oxidation (*25-28*). Synergists are combinations of two or more antioxidants which inhibit oxidation to a greater extent than the individual components.

Stability studies on oils and oil blends can be followed in a number of ways. Organoleptic assessment of the oil or oil blends takes the form of gustation and olfaction , in other words the senses of smell and taste respectively. Tastes and aromas of oils taken over a period of time will change as the oxidation processes within the oils take place. Each taste and each aroma is particular to a specific oxidation-or breakdown-product of the oil and sometimes a combination of products.For taste perception, the tongue can assess four qualities: sweet, sour,bitter and salt. Metallic qualities can also be perceived when 1-penten-3-one and 1-octen-3-one are present (*29*). Olfaction, on the other hand gives a much wider perception of flavour although it is particular to each individual. A major analytical procedure for the assessment of oxidative stability of oils or oil blends is peroxide value (PV). This is measured by an iodometric titration and shows the level of peroxides (primary oxidation products) present in the oil or oil blend. Anisidine value (AV) and Thiobarbituric Acid (TBA) Assay can also used in conjunction with PV since they measure the secondary oxidation and breakdown products of the oil or oil blend. Chromatographic analytical techniques such as column chromatography (CC) and gas chromatography-mass spectrometry (GC-MS) are important in separating components of oils and confirming their identity. These separate components may in fact be used in blending of oils (*30*) In this study, the effect on oxidative stability of the blending of vegetable and fish oils, together with synergistic studies of antioxidants on Chilean fish oil (anchovy) was investigated. Based upon the observations, the possibility of a novel antioxidant in a synergistic mixture or its activity in vivo is discussed.

Experimental Procedures

Materials

Chilean fish oil (anchovy) was obtained as a refined oil from Unilever Research (Sharnbrook, U K) (*31*). Cholesta-3,5-diene was purchased from Sigma Chemical Co. (Poole, U K). Delta-tocopherol was a gift from Eisai Pharma-Chem Europe Ltd (London, U K). Soy lecithin (approx. 56% phospholipid in soybean oil) was a gift from Unilever Research (Sharnbrook, U K).

Methods

Chilean fish oil samples were placed in wide-necked, loosely fitted screw capped amber glass bottles (free access to air) at 20-22°C. The single antioxidant oil sample contained cholesta-3,5-diene (0.02%w/w). The binary mixture contained cholesta-3,5-diene (0.02%w/w) and soy lecithin (0.5%w/w). The ternary mixture contained cholesta-3,5-diene (0.02%w/w), δ-tocopherol (2%w/w) and soy lecithin (0.5%w/w). A blank reaction was set up which contained Chilean fish oil only. Oxidation was measured by peroxide value over 10 weeks using the method given by Hamilton et al. (*31*).

References to other articles (*25,30*) in this publication will be made for comparative purposes in the Results and Discussion section.

Alumina/silica/citric acid fractionation of maize (MZO), sunflower (SFO) and Chilean fish oils (CFO) was performed and sensory evaluations, PVs, tocopherol level, trace metal level and GC-MS were taken on single and blended oils (*30*). Antioxidant studies on Chilean fish oil were performed as previously discussed (*25*).

Results and Discussion

Sensory evaluations of maize, sunflower and Chilean fish oils are shown by the flavor wheels in **Figure 1**. These are a combination of the two organoleptic senses, namely gustation and olfaction. Maize oil (Table II) has a high level of tocopherols (1662 ppm) compared to fish oil (85 ppm). It is significant that the level of beta-/gamma- tocopherols and delta- tocopherol is only 10 ppm and 0 ppm, respectively in fish oil. It is known that maize oil can be used in admixture with fish oil to extend the induction period of the fish oil from 6 weeks to greater than 10 weeks (Table III).

283

MAIZE OIL

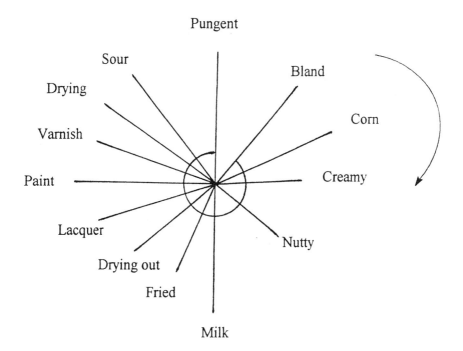

SUNFLOWER OIL

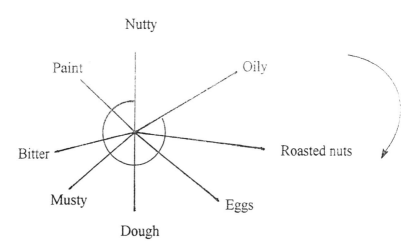

Figure 1. Flavour wheels of maize, sunflower and Chilean fish oil.

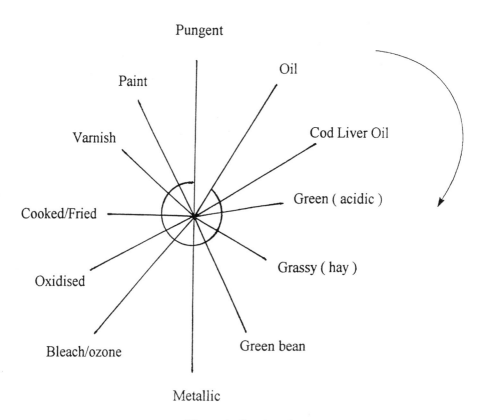

Figure 1. *Continued.*

Table I. Oil imports to the UK.

Oil	1995	1996	1997
Marine	117,429	90,615	71,436
Soya	84,228	83,186	107,990
Palm	376,908	505,973	492,360
Rape	61,530	111,589	110,128
Sunflower + Safflower	60,517	116,355	102,588
Coconut	74,791	59,829	42,493
Palm Kernel + Babassu	64,390	101,920	82,919

Table II. Composition of fish and maize oils (ppm).

Oil	α-tocopherol	β-/γ-tocopherol	δ-tocopherol	Copper	Iron
Fish	75	10	0	0.001	0.25
Maize	192	1390	80	0.03	0.15

Table III. Peroxide values (meq O2/kg) for fish oil (FHO), maize oil (MZO) and blend (MZO/FHO 75:25).

Week No.	0	2	4	6	8	10
Fish Oil	1.7	3.5	9.9	3.4	15.7	18.6
Maize Oil	0.3	0.6	1.2	1.2	1.2	1.5
MZO/FHO (75:25)	0.2	0.9	1.5	1.7	2.0	1.5

We attempted to add fractions from maize oil separated by chromatography to fish oil with a view to localising the active component. Chromatography of fish and vegetable oils over alumina had the expected result of lowering the tocopherol levels e.g. the level of alpha- tocopherol goes from 192 ppm to 0 ppm. However, there was a significant amount of copper left after chromatography (Table IV).

The chromatography of fish and maize oils was performed under nitrogen. The oils were unstable after chromatography e.g. fish oil had a peroxide value of 172 meq/kg after 3 weeks compared to the untreated oil of 4.3 meq/kg. (Table V). Chromatography in air leaves the fish oil even more unstable --- the PV after 3 weeks being 275 meq/kg. It was decided to perform all chromatography under nitrogen.

When maize oil was chromatographed over silica gel, it was possible to minimize the loss of tocopherols. Thus the levels after silica chromatography were 1/3 to 2/3 of the tocopherol levels in maize oil (Table VI). Maize oil chromatographed over silica gel is more stable than the corresponding oil over alumina (Table VII). After 5 weeks the peroxide value has reached 29 meq/kg after treatment with silica gel compared to 170 meq/kg after alumina treatment. However, it is evident (Table VIII) that there is still substantial quantities of copper entering the oil.

A novel silica column with a citric acid (CA) plug at the foot to retain any iron or copper was first reported at ISF in the Hague (*31*) was thus possible to fractionate maize oil without causing the oil to become less stable than it was originally. One of the fractions from such chromatography shown in Table 8 MZO was eluted from a silica column without the citric acid plug using light petroleum: diethyl ether (95:5 v/v) whereas MZO/CA did have the citric acid plug. It can be seen (Table IX) that the fraction eluted from silica is much less stable than the material eluted from the novel silica / citric acid column. The material.from the silica/citric acid has a similar PV of 1.7 meq/kg at week 8 compared to untreated maize oil with a PV of 1.2 meq/kg whilst the oil chromatographed on silica alone had reached a PV of 330 meq/kg by that time. Blends of the chromatographed maize oil with fish (1:1 v/v) and (3:1 v/v) showed some improvement in the blend compared to untreated fish oil up to 4 weeks. Thereafter the blends increased in PV much more quickly than either maize oil or fish oil alone (Table X). This maize oil fraction was eluted from silica gel with light petroleum ether and is referred to as the non-polar fraction. It appeared that this non-polar fraction contained some components which stabilized (at least temporarily) the fish oil.

The non-polar fraction of maize oil was analysed by GC-MS which showed that the majority of the peaks were n-alkanes (hydrocarbons). Three peaks stood out as non-straight chain hydrocarbons. They were elucidated by GC-MS to be phytosteradienes (methyl cholesta-3,5- dienes) (ergosta -3,5-diene) and beta sitosterol dienes (ethyl cholesta-3,7-diene). The GC trace is shown in Figure 2.
The mass spectra of ergostadiene and stigmastadiene are shown in Figure 3. Since these components appeared to be the only components other than n-alkanes in this fraction of the oils, it was thought that these phytosteradienes might help to stabilise the fish oil /vegetable oil blends.

These hydrocarbons are believed to come from the sterols and the sterol esters during the processing of the oils and have been used to detect the processing of adulterated oils such as olive oil where no processing other than pressing should be used.

As part of a set of experiments designed to study the role of additives in the keeping qualities of fish oil, Chilean fish oil (mainly anchovy) was blended with alpha- tocopherol and delta- tocopherol. Figure 4 shows that delta- tocopherol is better than alpha in keeping the PV low. Indeed it is believed that even at 0.2 % w/w alpha- tocopherol is pro-oxidant

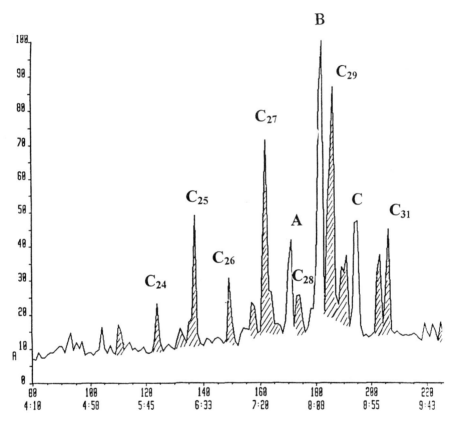

Figure 2. GC chromatogram of the non polar fraction of maize oil.

288

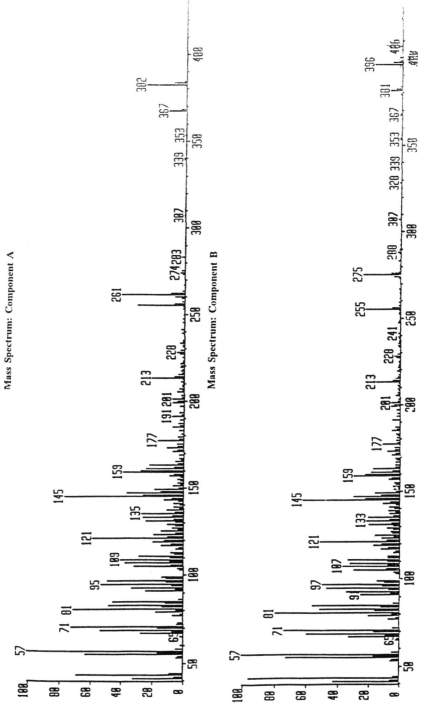

Figure 3. Mass spectra of [A] ergostadiene and [B] stigmastadiene.

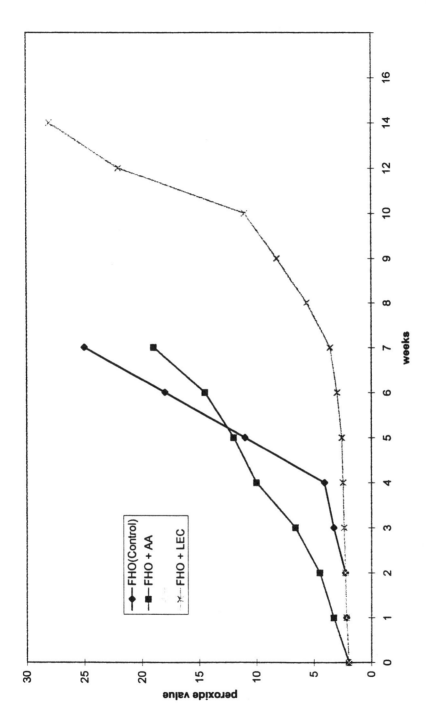

Figure 4. Induction Curves of Chilean Fish Oil (FHO) containing 0.2 % gamma/delta-tocopherols alone (Control) with added 0.1 % ascorbyl palmitate (AA) and with added 0.5 % lecithin (Lec).

It is generally accepted that fish oil cannot be stabilised by one antioxidant alone. Figure 4 shows that Chilean fish oil containing 0.2 % gamma-/delta-tocopherol is slightly pro-oxidant in the first five weeks when 0.1 % ascorbyl palmitate is present. This binary mixture is not a success. When lecithin is added to the fish oil containing 0.2 % gamma-/delta- tocopherol the oil is more stable than the control which has only one antioxidant present. It is clear that these binary mixtures lead to an improvement in the keeping quality of fish oil compared to one antioxidant alone. When ternary mixtures of antioxidants are used (Figure 5), a significant improvement in the keeping quality of the fish oil is achieved. The induction period for the oil plus 0.2 % gamma-/delta- tocopherol is approximately 3 weeks whilst the binary mixture of fish oil plus 0.2 % gamma-/delta- tocopherol plus 0.5 % lecithin has an induction period of 21 weeks.

The preliminary results (Table X) had shown that the non-polar fraction of maize ooil had some antioxidant properties when mixed with fish oil. These results lead to further investigation into the steradiene antioxidant/synergistic properties since sufficient amounts of steradienes were unobtainable ,cholesta -3,5-diene was purchased for the study.

It was considered that the major unusual feature of the phytosteradienes was the conjugated double bond system found in rings A and B. Interest in these conjugated double bond systems is high with the finding of the significance of conjugated linoleic acid in the diet (*32*).

Table IV. Concentration of tocopherols and metal ions (ppm) in maize oil (MZO) and fish oil (FHO)

	α-tocopherol	β-/γ tocopherol	δ-tocopherol	Copper	Iron
FHO (untreated)	115	10	0	<0.01	0.25
*FHO + N_2	0	0	0	0.02	0.08
MZO (untreated)	192	1390	80	0.01	0.06
*MZO + N_2	0	46	5	0.01	<0.05

* = chromatographed on alumina.

Table V. Peroxide values (meq O_2/kg) of post alumina treated oils under N_2

Week No.	0	1	2	3
FHO (untreated)	0.1	0.4	2.2	4.3
*FHO + N_2	1.0	56.8	118.7	172.1
MZO (untreated)	0.1	0.3	0.4	0.6
*MZO + N_2	2.3	18.5	27.3	55.8

* = chromatographed on silica

Table VI. Concentration of tocopherols and metal ions (ppm) of maize oil (MZO).

Oil	α-tocopherol	β/γ-tocopherol	δ-tocopherol	Copper	Iron
MZO (untreated)	192-260	1390-1855	80-105	<0.01	0.06
MZO (a)	0	46	5	0.01	0.05
MZO (b)	137-175	856-1056	16-34	0.6	0.01

(a) alumina chromatography and (b) silica chromatography.

Table VII. Peroxide values (meq O_2/kg) of maize oil (MZO) eluted on alumina and silica columns under nitrogen.

Week No.	0	1	2	3	4	5
MZO (untreated)	<0.1	0.3	0.4	0.6	0.6	1.2
MZO (alumina)	2.3	18.5	27.3	55.8	100.0	170.0
MZO (silica)	0.3	2.1	9.8	16.7	22.0	29.0

Table VIII. Levels of tocopherols and metals (ppm) in maize oil (MZO*).

Oil	α-tocopherol	γ/β-tocopherol	δ-tocopherol	Copper	Iron
MZO*	45	650	15	<0.01	<0.05
MZO/CA	125	700	10	<0.01	<0.05

Maize oil (MZO*) eluted from silica gel, and maize oil (MZO/CA) eluted from silica/citric acid column. Both eluted with light petroleum ether:diethyl ether (95:5 v/v).

Table IX. Peroxide values (meq O_2/kg) for maize oil (MZO).

Week No.	0	4	8	12
MZO (untreated)	<0.2	0.9	1.2	1.8
MZO (a)	0.4	16.8	33.0	36.0
MZO (b)	1.0	1.3	1.7	6.7

(a) silica column chromatography, (b) silica + citric acid column chromatography.

293

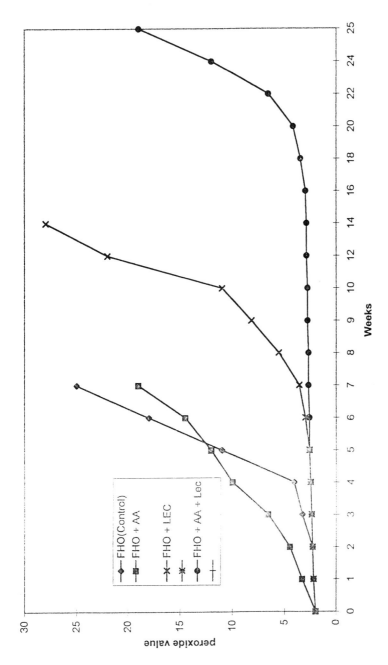

Figure 5. Induction curves of Chilean fish oil (FHO) containing 0.2% gamma/delta-tocopherol alone (Control) with added 0.1% ascorbyl palmitate (AA) and with added 0.5% lecithin and with a combination of 0.1% ascorbyl palmitate and 0.5% lecithin.

Table X. Peroxide values (meq O2/kg) of a non-polar maize oil fraction from a silica column.

Week No.	0	1	2	3	4
MZO (untreated)	<0.2	0.4	1.1	0.9	0.9
FHO (untreated)	<0.2	0.2	5.5	1.7	9.5
Non Polar-MZO / FHO (1:1)	0.1	4.2	6.6	3.4	6.7
Non Polar-MZO / FHO (3:1)	1.8	3.3	6.7	30.9	50.0

Maize oil (MZO), Chilean fish oil (FHO), maize oil : Chilean fish oil (1:1) and maize oil: Chilean fish oil (3:1).

Single Component Antioxidant

Combination of cholesta-3,5-diene (0.02 %, w/w) with Chilean fish oil was shown to be pro-oxidant (Figure 6) as was the case with ascorbyl palmitate (Figure 4).

Binary Mixtures of Antioxidants

The combination of fish oil, cholesta-3,5-diene, lecithin (Table XI) is better than cholesta-3,5-diene itself up to week 5. However, it becomes pro-oxidant and by week 10 it has a similar PV (322 meq/kg) to that of the mixture with the diene alone (Figure 6).

Ternary Mixture of Antioxidants

As with our earlier results where ascorbyl palmitate, tocopherol and lecithin are needed to stabilise fish oil, so in the present investigations. A mixture of cholesta-3,5-diene, lecithin and delta - tocopherol were able to keep the oil in good condition for at least 10 weeks (Table XI).

Conclusions

It has already been shown that silica treatment of an oil will help to stablise it, and the addition of a citric acid plug maintains the tocopherol level, and inhibits trace metals from interfering with the oil stability (*30*). Blending of the non-polar fraction of maize oil with Chilean fish oil led to a stabilisation of the blend and opened up the question of phytosteradienes having antioxidant/ synergistic properties. Previous work on the stabilisation of Chilean fish oil with tocopherols, lecithin and ascorbyl palmitate in binary and ternary synergistic mixtures led to the study of synergistic mixtures of commercial cholesta-3,5-diene (a derivative of natural sterols) with lecithin and δ-tocopherol. Here it was shown that a ternary mixture of cholesta-3,5-diene (0.02%w/w) and δ-tocopherol (2%w/w) and lecithin (0.5%w/w) showed the best stabilisation of the fish oil over the 10 week trials. The cholesta-3,5-diene alone was unsuccessful and showed pro-oxidant activity, as was the binary mixture of cholesta-3,5-diene and lecithin.

It may be that the steradienes which are present in many processed vegetable oils have some synergistic antioxidant effects. It could be that the unexpected efficacy of avenasterol as an antioxidant in frying oils (*33*) is the result of its dehydration to produce a phytosteradiene. In addition the conjugated double system in conjugated linoleic acid seems to have many biologically important effects. It would be interesting if the conjugated double bond system in the phytosteradienes had an equally significant bioactivity.

Table XI. Peroxide values (meq O2/kg) of fish oil.

Week No.	Fish Oil	Fish Oil (I)	Fish Oil (II)	Fish Oil (III)
0	2.1_1.6	2.1_1.6	2.1_1.6	2.1_1.6
1	2.5_1.7	2.4_0.3	2.8_3.6	2.1_1.8
2	3.4_0.1	3.5_2.1	2.6_0.0	1.5_1.9
3	6.3_5.1	8.8_3.5	3.5_1.7	1.4_0.0
4	12.2_3.4	25.5_7.2	7.8_0.7	1.6_0.0
5	18.1_1.5	47.4_7.6	20.6_2.0	1.7_1.8
8	44.1_0.8	185.4_1.2	91.9_5.7	2.6_3.7
10	68.8_2.1	331.5_11.7	322.4_9.4	2.8_7.3

(I) = fish oil + cholesta-3,5-diene, (II) = fish oil + cholesta-3,5-diene + lecithin, (III) = fish oil + cholesta-3,5-diene + lecithin + _-tocopherol.

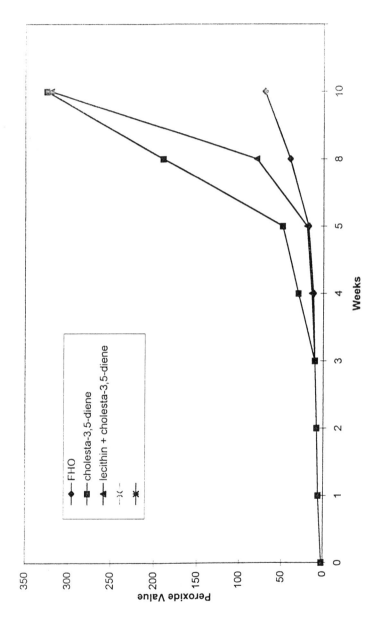

Figure 6. Induction curves of Chilean fish oil (FHO) and FHO with 0.02% Cholestadiene and FHO with 0.02% cholestadiene + 0.5% lecithin.

References

1. Simopoulos, A.P., *Am. J. Clin. Nutr.***1991**, *54*, 438-440.
2. Charnock, J.S., *Prog. Lipid Res.***1994**, *33* (4), 355-385.
3. Saynor, R.; Gillot, T. *Lipids* **1992**, *27*, 533-539.
4. Fujisawa, K.; Yagasaki, K.; Funabiki, R. *Lipids* **1994**, *29*, 779-783.
5. Herold, P.M;Kinsella, J.E. *Am. J. Clin. Nutr.* **1986**, *43*, 566-569.
6. Bang, H.O.; Dysberg, J.; Hjorne, N. *Acta. Med. Scand.* **1976**, *200*, 69-82.
7. Kato, H.; Tillotson, J.; Nichamal, M.Z.; Rhoads, G.G.;Hamilton, H.B. *Am. J. Epidemiol.* **1975**, *97*, 372-381.
8. Yotakis, L.D.O. *Thromb. Hemostas.* **1981**, *46*, 65-69.
9. Worth, R.M;, Kato, H.;Rhoads. G.G.; Kagan, A.;Syme, S.L. *Am. J. Epidemiol.* **1975**, *102*, 481-487.
10. Hirai, A.; Hamazaki, T.; Terano, T.; Nishikawa, T.; Tamura, T. Kumagai, A. *Lancet* **1980**, *ii*, 1132-1134.
11. Gonzalez, M.J.; Schemmel, R.A.; Dugan Jr. L.; Gray, J.I.;Welsch, C.W. *Lipids* **1993**, *28*, 827-832.
12. Dagnelie, P.C.; Bell, J.D.; Williams, S.C.R.; Bates, T.E.; Abel, P.D.; Foster, C.S. *Lipids* **1994**, *29*, 195-203.
13. Welsch, C.W. *Cancer Res.* **1992**, *52*, 2040-2043.
14. Karmali, R.A.; Marsh, J.;Fuchs, C. *J. Nat. Cancer Inst.* **1984**, *73*, 457-459.
15. Jurkowski, J.J.; Cave Jr., W.T. *J. Nat. Cancer Inst.* **1985**, *74*, 1145-1149.
16. Gabor, H. ;Abraham, S. *J. Nat. Cancer Inst.* **1986**, *75*, 1223-1228.
17. Gonzalez, M.J.; Schemmel, R.A.; Gray, J.I.; Dugan Jr., L.; Sheffield, L.G.;Welsch, C.W. *Carcinogenesis* **1991**, *12*, 1231-1236.
18. Gabor,H.; Blank, E.W.; Ceriani, R.L. *Cancer Lett.* **1990**, *52*, 173-176.
19. Yamada, J. *Bull. Tokai. Reg. Fish. Res. Lab.* **1975**, *39*, 21-29.
20. Childs, E. *J. Fish Res. Bd. Can.* **1974**, *31*, 1142-1148.
21. Neff, W.E.; El-Agaimy, M.A.; Mounts, T.L. *J. Am. Oil Chem. Soc.* **1994**, *71*, 1111-1116.
22. Frankel, E.N.;Huang, S.W. *J. Am. Oil Chem. Soc.* **1994**, *71*, 255-259.
23. Wandasundara U.N.;Shahidi, F. *J. Am. Oil Chem. Soc.* **1996**, *73*, 1183-1190.
24. Chapman, K.W.; Sagi, I.; Regenstein, J.M.; Bimbo, T.; Crowther, J.B.;Stauffer, C.E. *J. Am. Oil Chem. Soc.* **1996**, *73*, 167-172.
25. Hamilton, R.J.; Kalu, C.; McNeill, G.P.; Padley, F.B.;Pierce, J.H. *J. Am. Oil Chemists' Soc.* **1998**, *75*, 813-822.
26. Kaitaranta, J.K. *J. Am. Oil Chem. Soc.* **1992**, *69*, 810-814.
27. Koschinski, I. ;Mcfarlane, N. **1993**, International Patent WO93/ 10207.
28. Yi, O-S.; Han, D ; Shin, H-K. *J. Am. Oil Chem. Soc.* **1991**, *68*, 881-883.
29. Kochar, S.P.; Meara, M.L. Leatherhead Food RA, Scientific and Technical Surveys No. 87.
30. Kalu, C.U.U. Ph.D. Thesis. 1995, John Moores University: Liverpool, U.K.
31. Hamilton, R.J.; Kalu, C.; Prisk, E.; Padley, F.B.; Pierce, J.H. *Oils-Fats-Lipids* **1995**, *2*, 285-289.
32. Brodie, A.E.; Manning, V.A.; Ferguson, K.R.; Jewell, D.E.; Hu, C.Y. *J.Nutr.* **1999**,*129*, 602-606.

Indexes

Author Index

Abedin, Lavinia, 79
Abril, Ruben, 108
Ackman, Robert G., 191, 235
Akoh, Casimir C., 151
Barclay, William, 108
Bierenbaum, M. L., 66
Billman, G. E., 28
Boyd, Leon C., 258
Cadwallader, Keith R., 221
Decker, Eric A., 243
Drapeau, Christian, 125
Finley, John W., 2
Flatt, James H., 243
Hamilton, R. J., 280
Hayes, K. C., 37
Holub, Bruce J., 54
Kaewsrithong, Janthira, 208
Kalu, C., 280
Kang, J. X., 28
Koizumi, Chiaki, 208
Kushak, Rafail I., 125
Kyle, David J., 92
Leaf, A., 28
Mancuso, Jennifer R., 243

McClements, D. Julian, 243
McLeod, Catherine A.191
Mei, Longyuan, 243
Nagai, Takashi, 176
Newton, Ian S., 14
Ohshima, Toshiaki, 208
Sasaki, Shigefumi, 243
Senanayake, S. P. J. Namal, 162
Shahidi, Fereidoon, 2, 142, 162, 221
Simpson, G. B., 280
Sinclair, Andrew J., 79
Terao, Junji, 176
Tong, Larry, 243
Ushio, Hideki, 208
Utsunomiya, Hiroyuki, 208
Van Cott, Elizabeth M., 125
Voskuyl, R. A., 28
Wanasundara, Udaya N., 142
Watkins, T. R., 66
Winter, Harland H., 125
Xiao, Y-F., 28
Xu, Xueliang, 191
Zeller, Sam G., 108, 243

Subject Index

A

Acidolysis, production of structured lipids, 153
Aldehyde products, autoxidation of selected unsaturated fatty acids, 169t
Algae
  species affecting cholesterol, 138
  worldwide as food source, 125
  See also Blue-green alga *Aphanizomenon flos-aquae*
Alkaline refining process, refining crude fish oil, 7
Alzheimer's disease, docosahexaenoic acid (DHA), 60
Amberjack
  hydroperoxide levels in blood and liver, 215, 216t
  See also Fish aroma
Anchovy (Chilean fish oil). *See* Rancidity of fish oils
Animal products, docosahexaenoic acid (DHA) enriched, 118
Anisidine value (AV), measuring secondary oxidation products of oil or oil blend, 281
Antioxidants
  inhibition of oxidation of seal blubber oil, 149t
  seal blubber oil and its products, 148–149
  See also Oxidative stability of lipid dispersions; Rancidity of fish oils
Antioxidants, natural
  activity due to flavonoids, 264–265
  activity of miscellaneous plant and food extracts, 272–274
  catechins, 265

chemical composition of active antioxidant components of rosemary, 263
consequences of irradiation on food quality, 263
decreasing consumer purchasing patterns foods containing synthetic, 259
dried beans, 273
effects of commercial phosphatidylcholine (PC) on thiobarbituric acid (TBA) values of salmon oil oxidation model system, 270f
effects of phospholipids (PL) and butylated hydroxytoluene (BHT) on change in fatty acid profile, 269f
effects of PL fractions on oxidation of salmon oil oxidation model system, 268f
factors affecting potency, 275–276
fruit, vegetable, and wine antioxidants, 271–272
function of, 260
ground spices stabilizing pork emulsions at 200 and 1000 ppm at 4°C for 21 days, 262t
importance of reaction environment, 275
leaf extracts, 274
lipophilic and hydrophilic nature influencing effectiveness, 275
Maillard reaction products (MRP), 274
major sources and proposed mechanisms of action, 261t
meat flavor deterioration (MFD), 261–262

occurrence of warmed-over flavor (WOF), 261–262
peanut hulls, 272
phytic acid, 273
PL exhibiting high degree of synergism with polyhydroxy flavonoids, 267
potency of flavonols, 266
potential for inclusion in functional foods, 259–260
retention of minimum levels of PL adding to stability, 271
role of PL as antioxidants, 266–267, 271
rosemary for greatest amount of study, 262
seed oil extracts, 263–264
seed oil proteins reducing rancidity in ground beef at 10% at 4°C for 6 days, 264t
sources of, 260
spices, 260–263
sweet potatoes, 273
synergism between PL and other, 267
tea flavonoids, 265–266
three major areas of interest, 259
vitamins for decreased risk of cardiovascular disease and cancer, 259
Aphanizomenon flos-aquae. See Blue-green alga *Aphanizomenon flos-aquae*
Aquaculture, farmed fish, 191–192
Arachidonic acid (AA)
 essential fatty acid, 17
 infant development, 22
 nomenclature, 6t
Aroma
 characteristic by certain species of fish, 209
 factor in food selection, 192
 See also Fish aroma; Seal blubber oil (SBO) odorants

Arrhythmias
 action of polyunsaturated fatty acids (PUFA) modulating conductance of ion channels in plasma membranes of heart cells, 32
 canine model of sudden cardiac death, 29
 effects of PUFAs on electrophysiology of cardiac myocytes, 30
 electrical stabilizing effect of omega-3 PUFAs on cardiomyocyte, 30, 32
 inducing by toxin and using PUFAs, 30
 inhibition of Na^+ and Ca^{2+} currents by PUFAs, 33
 in vitro demonstration of electrical stabilizing effect of omega-3 PUFAs on isolated cultured neonatal rat hearts, 31f
 omega-3 fatty acid consumption, 128
 potent inhibitory action of PUFAs on L-type Ca^{2+} currents, 33
 PUFAs shifting steady state potential for inactivation of Na^+ channels to more hyperpolarized potentials, 32–33
 secondary prevention trials showing prevention of ischemia-induced sudden cardiac death, 34
 studying mechanism of PUFA producing antiarrhythmic effect, 29–30
Artery reclogging, omega-3 fatty acids preventing, 21
Atlantic salmon, farmed. *See* Salmon, farmed Atlantic
ATP charge and release, decreasing with increased dietary unsaturation, 72–73
Attention deficient disorder (ADD), omega-3 fatty acid consumption, 128

Attention Deficit Hyperactivity
Disorder (ADHD)
clinical trials with docosahexaenoic
acid single cell oil (DHASCO),
103–104
docosahexaenoic acid (DHA), 60
Autoimmune disorders, omega-3 fatty
acids reducing, 22
Autoxidation
aldehyde products from, of
unsaturated fatty acids, 169*t*
free-radical chain mechanism, 162
increasing unsaturation in dietary
fatty acids, 73, 75
rate of, of fatty acids, 163
Autoxidized lipids
accumulation representing serious
risk of cardiovascular disease
(CVD), 69
self-perpetuating chain reaction, 69

B

Beans, dried, natural antioxidant, 273
Beef cattle, docosahexaenoic acid
(DHA) enriched, 118
Behavior disorders, docosahexaenoic
acid (DHA), 60–61
Bleaching, mixing oil in tank with,
clay, 7
Blood lipid and lipoprotein levels,
effect of omega-3 fatty acids, 18
Blue-green alga *Aphanizomenon flos-aquae*
ability to provide essential lipids,
128
cholesterol and triacylglycerol, 134,
137
complete plasma lipid profile for
animals fed diets containing
soybean oil, coconut oil, and
alga, 133*f*, 135*f*
conditions with beneficial effects of
Aph. flos-aquae, 128

correlation between dietary and
serum levels of fatty acids, 137–
138
dietary fatty acids, 130–131
dynamics of algal blooms in Upper
Klamath Lake, 126–127
effect of inducing PUFA deficiency,
131, 132*f*, 137
experimental methods, 128–130
fatty acid analysis method, 130
fatty acid profiles of soybean oil,
coconut oil, and alga, 129*t*
harvesting and processing of *Aph.
flos-aquae*, 127–128
historic production of Klamath
Lake, 126
hypocholesterolemic properties, 138
levels of cholesterol, triglycerides,
and PUFA/SFA ratio for various
diets, 136*f*
lipid content for diets, 132*f*
market as health food supplement,
126
migration of natural killer cells
from blood to tissues, 126
plasma lipids, 131, 134
statistical analysis method, 130
test animals, 128–129
Upper Klamath Lake ecology and
Aph. flos-aquae bloom, 126–127
Body calorimeter, studying effect of
fatty acids in biological system, 67
Borage oil (BO)
conjugated dienes and TBARS
values during accelerated storage
at 60°C, 168*t*
degradation, 166, 170
enzymatic modification of, with
capric acid producing structured
lipids, 155
fatty acid profile before and after
enzymatic modification by
Pseudomonas sp. lipase, 167*t*
incorporating EPA and capric acid
with lipases, 165

lipase from *Pseudomonas sp.*
incorporating EPA and DHA, 165
main volatile aldehydes, 171*t*
modifying fatty acid composition with ethyl ester of EPA, 165–166
oxidative stability, 166, 170
Brain
important effects of polyunsaturated fatty acids (PUFA), 34
PUFAs patterns, 79–80
Breast cancer, docosahexaenoic acid (DHA), 61–62
Breast milk, human docosahexaenoic acid (DHA), 57–58
See also Infants
Brij 35 (nonionic surfactant). *See* Oxidative stability of lipid dispersions

C

Candida antarctica lipase
adding EPA and DHA to soybean and canola oils, 155–156
modifying soybean oil by incorporating omega-3 PUFA, 164
Canine, model of sudden cardiac death, 29
Capric acid
enzymatic modification of borage oil with, producing structured lipids, 154–155
enzymatically modified fish oil with, producing structured lipids, 153, 154*t*
Carcinogenesis, docosahexaenoic acid (DHA), 61–62
Cardiac arrhythmia
animal reduction by omega-3 fatty acids, 21

antiarrhythmic effect of polyunsaturated fatty acids (PUFA) in experimental animals, 29–30
See also Arrhythmias
Cardiac death
canine model, 29
docosahexaenoic acid (DHA), 58
Cardiovascular disease (CVD)
accumulation of autoxidized lipid in low density lipoprotein (LDL) particle representing serious risk, 69
antioxidant vitamins, 259
ATP charge and release and platelet aggregation decrease with increased dietary unsaturation, 72–73
autoxidation in vitro depleting antioxidant reserves, 73
clearing elevated cholesterol from circulatory system, 75
decreasing factors with unsaturation increase in diet, 75
diet composition for rat feeding trials, 70*t*
docosahexaenoic acid (DHA), 58–60
effect of degree of dietary lipid unsaturation upon platelet function and serum vitamin E in rat, 73*t*
effect of omega-3 fatty acids, 18, 21
functional response of human platelet from hypercholesterolemic human subject, 73
functional state of cellular, membrane associated enzymes, 74–75
hyperaggregation of platelets, 75
incidence, 67–68
incidence in Eskimos, 143
increased levels of unsaturation in

dietary fat decreasing serum lipid levels, 71, 72*t*
increased serum lipid hydroperoxides with unsaturated fatty acids in diet, 76
increasing autoxidation with increases in unsaturation sites, 75
influence of degree of unsaturation of dietary fat upon platelet response, serum fatty hydroperoxides, and serum vitamin E in hypercholesterolemic human subjects, 74*t*
influence of degree of unsaturation of dietary fat upon serum lipids, 72*t*
influence of increased unsaturation on platelets in rats and humans, 76
materials and methods, 70–71
risk factors, 67–68
Senate Select Committee on Nutrition and Human Needs, 68–69
serum cholesterol decrease with incorporation of unsaturated fatty acids for rat model and human subjects, 71–72
susceptibility of human platelet membranes to autoxidation, 73–74
unsaturated dietary fats facilitating lipid processing in tissue, 76
Catechins, antioxidant activity of tea leaves, 265–266
Chemistry
chemical structures and nomenclature of polyunsaturated fatty acids (PUFA), 6*t*
highly unsaturated fatty acids, 3–4
Chickens, docosahexaenoic acid (DHA) enriched broilers, 118
Children, clinical studies of docosahexaenoic acid single cell oil (DHASCO), 103–104
Chilean fish oil (anchovy). *See* Rancidity of fish oils
Chlorella vulgaris, decreasing blood cholesterol, 138
Cholesterol
clearing burden from circulatory system, 75
effect of plasma cholesterol by fats in Benedictine nuns experiencing normal or elevated lipoprotein profiles, 43*f*
effects after supplementation with docosahexaenoic acid single cell oil (DHASCO), 103–104
levels for animals fed diets containing soybean oil, coconut oil, and algae, 134, 136*f,* 137
lowering effect of dietary polyunsaturated fatty acids, 37
relative potency of dietary fatty acids on, 40, 44
See also Blue-green alga *Aphanizomenon flos-aquae*; Lipoprotein metabolism
Chromatography, fatty acids separation, 148
Clove, stabilizing pork emulsions, 262*t*
Coconut oil
fatty acid profile, 129*t*
plasma lipid profiles for animals fed, 131, 133*f,* 134, 135*f*
See also Blue-green alga *Aphanizomenon flos-aquae*
Conjugated linoleic acids (CLA)
analysis of methyl esters of fish lipid and of mixture of CLA prepared with linoleic acid with alkali, 238*f*
analysis using gas-liquid chromatography (GLC), 236
equivalent chain length (ECL) of

CLA isomers (methyl esters) on fused-silica capillary column, 237*t*
expected GC elution order of positional and geometric CLA fatty acid methyl ester isomers on CP-Sil 88 capillary column, 239*t*
fish research, 236–237
formation, 236
origin of interference problem with GLC, 237, 239
projects using GLC for identification, 240
salmon farming industry, 240
Consumer purchasing patterns, antioxidants, 259
Corn (maize) oil. *See* Rancidity of fish oils
Coronary heart disease (CHD)
docosahexaenoic acid (DHA), 58–59
polyunsaturated fatty acids (PUFAs) reducing CHD, 38, 40
relationship between fat intake and incidence of CHD as percentage of total mortality, 15*f*
See also Lipoprotein metabolism
Cottonseed, seed oil proteins reducing rancidity in ground beef, 264*t*
Crypthecodinium cohnii
asexual cell division, 98
cultivation in laboratory, 97
docosahexaenoic acid single cell oil (DHASCO) processing procedure, 99–100
fatty acid composition of DHASCO from *C. cohnii*, 101*t*
fermentation process for production of *C. cohnii* biomass, 98*f*
medium for growth from shake flask to production scale, 99
member of Dinophyta, 97
oil processing flow shear for production of DHASCO from, 99*f*
shear sensitivity, 98
strain for production of DHA-rich oil, 98
See also Single cell oil, docosahexaenoic acid (DHASCO)
Cumene peroxide. *See* Oxidative stability of lipid dispersions

D

Deodorization process
molecular distillation process, 7
two-stage process, 7
Depression, docosahexaenoic acid (DHA), 60
Diet. *See* Human diet
Dietary fatty acids
relationship between, and lipoprotein metabolism, 37–38
unsaturation in, facilitating lipid processing in tissue, 76
See also Blue-green alga *Aphanizomenon flos-aquae*; Cardiovascular disease (CVD); Lipoprotein metabolism
Dihomo-γ-linolenic, nomenclature, 6*t*
Direct thermal desorption (DTD)
alternative to static and dynamic headspace techniques, 222
apparatus, 223*f*
combining with gas chromatography–olfactometry (GCO) for evaluating volatiles, 224
Direct thermal desorption–gas chromatography–mass spectrometry (DTD–GC–MS)
method, 226–227
See also Seal blubber oil (SBO) odorants
Direct thermal desorption–gas chromatography–olfactometry (DTD–GCO)

apparatus for DTD, 223*f*
DTD–GCO and sample dilution analysis (SDA) methods, 224, 226
See also Seal blubber oil (SBO) odorants
Disease prevention/management, docosahexaenoic acid (DHA), 58–60
Distillation, method for removing fatty acids or their alkyl esters from admixture, 147–148
Docosahexaenoic acid (DHA)
antiarrhythmic in canine, 29
bioavailability in algae and algae-like organisms, 113
biosynthetic pathway involving elongation, desaturation, and peroxisomal beta-oxidation step, 93
breast-fed versus formula-fed infants, 94
component of phospholipids of neuronal cells, 92
critical functions in development and metabolism of neuronal cells, 93–94
dried DHA-rich microalgae from *Schizochytrium sp.*, 109–110
egg enrichment, 113, 116
enriched animal products, 118
enrichment, 6–7
essential fatty acid, 17
fatty acid profiles of different microorganisms and higher plants, 97*t*
fish as major dietary source, 108–109
importance in human infant nutrition, 94
incorporating into trilinolein using lipases, 156, 158*t*
infant formula lacking, 16
IQ differences between breast-fed and formula-fed children, 95*f*

large-scale production of photosynthetic microalgae, 96–97
microalgae as primary producers of DHA, 95–97
Mucor miehei and *Candida antarctica* lipases adding, to soybean and canola oils, 155–156
nomenclature, 6*t*
photosynthetic plankton as primary source, 95–97
prevalence and sources, 92–93
structure, 93*f*
supply for infant development, 22
See also Refined docosahexaenoic acid (DHA)-rich oil; Seal blubber oil (SBO); Single cell oil, docosahexaenoic acid (DHASCO)
Docosahexaenoic acid (DHA) in human health
Alzheimer's disease, 60
Attention Deficit Hyperactivity Disorder (ADHD), 60
behavioral problems, 61
breast cancer, 61–62
carcinogenesis inhibition, 61
cardiovascular health and disease prevention/management, 58–60
coronary heart disease (CHD), 58–59
depression, 60
desaturation, elongation, and retroconversion of omega-3 fatty acids, 56*f*
human breast milk, 57–58
Infantile Refsum's disease, 60
influencing selected risk factors for cardiovascular disease, 59
intake by pregnant and lactating women, 57–58
intake during pre-conception and pregnancy, 57
Mental Development Index in term infants, 58
mental stress-induced extra-

aggressive behavior, 61
neonatal adrenaleukodystrophy, 60
neurological/behavioral disorders, 60–61
pathophysiological conditions, 61–62
physiologically important nutrient for neuronal and visual functioning, 57–58
retinitis pigmentosa, 60, 61
schizophrenia, 60
seminal plasma, 62
serum triglyceride-lowering effects, 59–60
sleep problems, 61
sources and metabolism of DHA, 55, 57
spermatozoa, 62
supplementation in healthy men, 62
temper tantrums, 61
Zellweger syndrome, 60–61
Docosahexaenoic acid (DHA) production from microalgae
bioavailability, 13
composition of dried microalgae, 110, 112
DHA-enriched broiler chickens, 118
DHA ingredient options form microalgae and DHA-enriched eggs, 123*t*
egg enrichment in white leghorn hens, 113, 116
egg yolk and defatted egg yolk powders, 119
enriched animal products, 118
enriched beef cattle, 118
enriched swine, 118
fatty acid composition of DHA-enriched eggs compared to typical supermarket eggs, 117*f*
fatty acid composition of liver tissue from female rats fed control diet and admixtures of *Schizochytrium sp.* microalgae, 115*f*
fatty acid composition of sera from female rats fed control diet and admixtures of *Schizochytrium sp.* microalgae, 114*f*
fatty acid profile of crude oil and lipid fractions isolated from crude oil from *Schizochytrium sp.* microalgae, 116*t*
fatty acid profile of refined oil derived from dried microalgae and egg yolk powder and defatted egg powder from DHA-enriched eggs, 122*t*
fermentation, 110
flow diagram for fermentation and recovery of whole cell microalgae rich in DHA, 111*f*
nutritional supplements and food ingredients, 119
oil extraction and purification process producing DHA-rich oil, 120*f*
process flow diagram for production of egg yolk and defatted egg yolk powder, 121*f*
proximate composition of *Schizochytrium sp.* whole cell microalgae produced via fermentation process, 112*t*
refined oil, 119
Schizochytrium sp. of kingdom Chromista, 109–110
Docosapentaenoic acid (DPA)
attention deficit hyperactivity disorder and others, 96
docosahexaenoic acid (DHA) and DPA proportions in retinal phospholipids in guinea pigs fed diets containing different levels of omega03 fatty acids, 86*f*
increase with decrease of docosahexaenoic acid (DHA) in mammals, 80–81
nomenclature, 6*t*
See also Retinal function; Seal

blubber oil (SBO)
Dodecyltrimethyl ammonium bromide (DTAB). *See* Oxidative stability of lipid dispersions
Dogs, canine model of sudden cardiac death, 29
Dried beans, natural antioxidant, 273

E

Eczema, omega-3 fatty acid consumption, 128
Egg enrichment
  egg yolk and defatted egg yolk powders, 119
  fatty acid composition of docosahexaenoic acid (DHA)-enriched eggs compared to typical supermarket, 117f
  fatty acid profile, 122t
  potential applications, 123t
  process flow diagram for production of egg yolks and defatted egg yolk powder, 121f
  white leghorn hens, 113, 116
  See also Docosahexaenoic acid (DHA) production from microalgae
Eicosapentaenoic acid (EPA)
  antiarrhythmic in canine, 29
  blood levels in rats fed *Aphanizomenon flos-aquae* diet, 137–138
  enrichment, 7, 9
  enzymatic modification of borage oil with, producing structured lipids, 154–155
  epileptic seizures, 128
  fish as major dietary source, 108–109
  incorporating into trilinolein using lipases, 156, 158t
  microalgae as primary producer of EPA, 96f

Mucor miehei and *Candida antarctica* lipases adding, to soybean and canola oils, 155–156
  nomenclature, 6t
  structured lipids from fish oil, 153, 154t
  See also Seal blubber oil (SBO)
Electroretinographic (ERG) amplitudes
  dependence on omega-3 PUFA supply, 80
  ERG differences due to omega-3 deficiency, 81
  previous studies of α-linolenic acid (LNA) deficiency and ERG function, 87–88
  studies of omega-3 deficiency and ERG function in rat, monkey, and guinea pigs, 83t
  See also Retinal function
Emulsifiers. *See* Oxidative stability of lipid dispersions
Encapsulation, seal blubber oil and its products, 148–149
Enzymatic esterification with glycerol, producing concentrates in acylglycerol form, 147
Enzymatic hydrolysis, concentrating omega-3 fatty acids of seal blubber oil, 147
Epileptic seizures, eicosapentaenoic acid (EPA), 128
Ethylenediaminetetraacetic acid (EDTA). *See* Oxidative stability of lipid dispersions
Evening primrose oil (EPO)
  conjugated dienes and TBARS values during accelerated storage at 60°C, 168t
  degradation, 166, 170
  fatty acid profile before and after enzymatic modification by *Pseudomonas sp.* lipase, 167t
  lipase from *Pseudomonas sp.* incorporating eicosapentaenoic

acid (EPA) and docosahexaenoic acid (DHA), 165
main volatile aldehydes, 171*t*
oxidative stability, 166, 170
Excitable tissues. *See* Arrhythmias; Brain; Heart

F

Farmed Atlantic salmon. *See* Salmon, farmed Atlantic
Fat intake
  decreasing low density lipoprotein/high density lipoprotein (LDL/HDL) ratio along with total cholesterol without altering total, 46, 48*f*
  evolution of human diet, 4, 14, 16
  myocardial infarction and dietary, in Eskimos and Danes, 16*t*
  relationship to incidence of coronary heart disease as percentage of total mortality, 15*f*
Fatty acids
  balance between, becoming critical attempting to generate lowest total cholesterol (TC) and low density lipoprotein/high density lipoprotein (LDL/HDL) ratio, 46
  balancing dietary, improving LDL/HDL ratio, 46, 48*f*
  coronary heart disease (CHD) mortality as function of dietary polyunsaturated fatty acids (PUFA) fatty acid ratio, 44, 47*f*
  evolution of human diet, 14, 16
  exploring the fine-tuning of dietary acid balance enhancing lipoprotein profile, 46, 50
  fine-tuning, to enhance lipoprotein profile, 46, 49*f*, 50

  relationship between fat intake and incidence of coronary heart disease, 15*f*
  seal blubber oil, 144*t*
Fermentation. *See Crypthecodinium cohnii*; *Schizochytrium sp.*
Fetal development, docosahexaenoic acid (DHA), 57–58
Fish
  annual catch, 280
  antiarrhythmic effect from ingestion, 34
  characteristic aroma at live state by certain species, 209
  imports to United Kingdom, 280
  major dietary sources of long chain omega-3 polyunsaturated fatty acids, 108–109
  rich in highly unsaturated fatty acids, 209
  source of omega-3 polyunsaturated fatty acids, 55
Fish aroma
  analysis of volatiles, 211
  characteristic aroma of biogenerated volatile compounds, 217
  characteristic by certain species, 209
  compositions of volatiles compounds from total wet tissues of oyster, 213
  contents of phosphatidylcholine hydroperoxide in plasma, red blood cell, and liver of several fish, 216*t*
  determination of fatty acid compositions of total lipids (TL), phospholipids (PL), and neutral lipids (NL), 210
  determination of hydroperoxides, 211–212
  determination of lipid class compositions, 211

factors affecting formation and
  decomposition of hydroperoxides
  in tissues, 217
fatty acid compositions of oyster
  wet tissues and fish blood and
  liver, 212
fish and seafood samples, 210
generation of cucumber-like aroma
  in aromatic fish, 218f
hydroperoxide levels in fish blood
  and liver, 215
lipid extraction and fractionation
  method, 210
mechanisms for biogeneration of
  volatile aroma compounds, 217
prominent fatty acid compositions
  of TL in Pacific and European
  oysters, 213t
prominent fatty acid compositions
  of TL in sweet smelt plasma and
  red blood cell, 214t
prominent volatile compounds in
  Pacific and European oysters,
  215t
relationship between accumulated
  hydroperoxides and aroma
  development, 217, 219
Fish intake, coronary heart disease
  (CHD), 18, 19f
Fish oil
  benefits of blending with other
    dietary oils, 281
  human crossover studies of fish oil
    versus control oil, 50, 51t
  production process, 25f
  refining and quality assurance, 23
  role of phospholipids as
    antioxidants, 267
  source of omega-3 polyunsaturated
    fatty acids, 55
  structured lipids from, 153, 154t
  superior potential for reducing low
    density lipoprotein cholesterol
    (LDL-C) and total plasma
    cholesterol, 38, 40
  See also Rancidity of fish oils
Fish oil emulsion
  effect of lipase-hydrolysis on lipid
    peroxidation with different
    initiators, 179, 185
  effect of nonesterified EPA on lipid
    peroxidation, 185
  hydrolysis method, 177–178
  oxidation of lipase-hydrolyzed, 178
  See also Lipase hydrolysis
Flavonoids, polyphenolic compounds
  in fruits and vegetables, 271–272
Flavonols, potency, 266
Flounder
  hydroperoxide levels in blood and
    liver, 215, 216t
  See also Fish aroma
Food
  application of seal blubber oil, 145–
    149
  consequences of irradiation on
    quality, 263
  See also Antioxidants, natural
Formula for infants. *See* Infants
Fruit, antioxidants, 271–272

G

Gas chromatography–olfactometry
  (GCO)
  characterizing odor-active
    compounds, 222
  See also Seal blubber oil (SBO)
    odorants
Gas-liquid chromatography (GLC).
  See Conjugated linoleic acids
  (CLA)
Ginger, stabilizing pork emulsions,
  262t
Good Laboratory Practice (GLP),
  guidelines for safety studies of
  docosahexaenoic acid single cell oil
  (DHASCO), 102

Green algae, decreasing blood cholesterol, 138
Greenland Eskimos
diet and incidence of cardiovascular disease, 143
See also Seal blubber oil (SBO)

H

Harp seal
diet of Eskimos, 142–143
potential source of food and nutraceutical products, 142
See also Seal blubber oil (SBO)
Health benefits
highly unsaturated fatty acids, 4–5
Senate Select Committee on Nutrition and Human Needs, 68–69
See also Docosahexaenoic acid (DHA) in human health
Heart
canine model of cardiac death, 29
cardiac arrhythmia reduction by omega-3 fatty acids in animals, 21
See also Arrhythmias; Coronary heart disease (CHD); Coronary vascular disease (CVD)
Hexadecane emulsions. *See* Oxidative stability of lipid dispersions
High density lipoprotein (HDL)
increases after supplementation with docosahexaenoic acid single cell oil (DHASCO), 103–104
See also Lipoprotein metabolism
Highly unsaturated fatty acids (HUFA)
chemistry, 3–4
commercial products receiving marine oil processing, 5
encapsulation for supplements, 9
enriched omega-3 oils, 7, 9
evolution of human diet, 4

health benefits, 2–3
marine oil processing, 5–7
microencapsulation, 9
nomenclature of polyunsaturated fatty acids (PUFA), 6*t*
nutritional benefits, 4–5
Oxidative Stability Index (OSI), 158–159
oxygenation by lipoxygenases, 209
production of omega-3 concentrates from marine oils, 10*f*
production of refined, bleached, and deodorized (RBD) marine oils, 8*f*
stability, 156–159
stabilization of lipids containing, 259–260
two-stage deodorization process, 7
See also Fish aroma; Modified oils with highly unsaturated fatty acids; Structured lipids (SL)
Human development
clinical studies of docosahexaenoic acid single cell oil (DHASCO), 103–104
docosahexaenoic acid (DHA), 57–58
supplementation with DHASCO, 103–104
Human diet
balance between omega-3 and omega-6 oils, 16
coronary vascular disease, 18, 21
dependence of coronary heart disease (CHD) risk and death on consumption of long chain polyunsaturated fatty acids (PUFA), 20*f*
dependence of fish intake and mortality from coronary heart disease (CHD) indicating long term positive effects, 19*f*
dietary intake of fatty acids and risk of myocardial infarction (MI), 24*f*
effect of omega-3 long chain PUFA

on postprandial lipemia, 21*t*
essential fatty acids, 17
evolution, 4, 14, 16
infant development, 22
inflammatory and autoimmune disorders, 22
intake recommendations of omega-3 long chain PUFA, 24*f*
myocardial infarction and dietary fat intake in Eskimos and Danes, 16*t*
production process for refined omega-3 oils, 25*f*
PUFA containing two or more double bonds, 17
PUFA metabolism, 17–18
recommended intakes of omega-3 and omega-6 PUFA, 22–23
relationship between fat intake and incidence of CHD as percentage of total mortality, 15*f*
restenosis, 21
roles in human health, 18–22
saturated and unsaturated fatty acids, 17
special refining and quality assurance of fish oils, 23
Human health
Senate Select Committee on Nutrition and Human Needs, 68–69
See also Docosahexaenoic acid (DHA) in human health
Hydroperoxides
factors affecting formation and decomposition, 217
levels in fish blood and liver, 215, 216*t*
relationship between accumulated, and characteristic aroma development, 217, 218*f*
See also Fish aroma
Hydroperoxides, serum fatty, influence of degree of unsaturation of dietary fat, 74*t*

Hydrophilic antioxidant, nature influencing effectiveness, 275
Hyperlipidemia, clinical trials with docosahexaenoic acid single cell oil (DHASCO), 103–104
Hypocholesterolemic properties, blue-green alga *Aphanizomenon flos-aquae*, 137–138

I

Impact, medical product containing modified lipids, 164
Infantile Refsum's disease, docosahexaenoic acid (DHA), 60
Infants
clinical studies of docosahexaenoic acid single cell oil (DHASCO), 102–103
docosahexaenoic acid (DHA), 57–58
formula lacking DHA, 16
importance of DHA in human, nutrition, 94
IQ differences between children who were breast-fed versus formula-fed, 95*f*
neurological outcomes of breast-fed versus formula-fed, 94
supplementation of DHA, 152
supply of DHA for development, 22
Inflammatory disorders, omega-3 fatty acids reducing, 22
Iron catalysis. *See* Oxidative stability of lipid dispersions

L

Lactation, docosahexaenoic acid (DHA), 57–58
Leaf extracts, natural antioxidants, 274
Linoleic acid (LA)

correlation between dietary and
  serum levels, 137
essential fatty acid, 17
nomenclature, 6*t*
See also Conjugated linoleic acids
  (CLA)
α-Linolenic acid (LNA)
  animal studies showing effects of
    omega-3 PUFA on retinal
    function, 83*t*
  antiarrhythmic in canine, 29
  correlation between dietary and
    serum levels, 137
  essential fatty acid, 17
  materials and methods for retinal
    function study, 81–82
  metabolic conversion to
    docosahexaenoic acid (DHA), 55,
    56*f*
  microalgae as primary producers of
    LNA, 96*f*
  nomenclature, 6*t*
  role in retinal function, 80
  secondary prevention trials showing
    prevention of ischemia-induced
    sudden cardiac death, 34
  strategies to increase retinal DHA
    levels, 84, 88–89
  studies on LNA deficiency and
    electroretinographic (ERG)
    function, 87–88
  See also Retinal function
γ-Linolenic acid (GLA)
  nomenclature, 6*t*
  sources, 154
  structured lipids containing, and
    omega-3 highly unsaturated fatty
    acids (HUFA), 154–155
Lipase hydrolysis
  analysis of fatty acid composition of
    hydrolysis products, 178
  effect of intact lipase or heat-
    inactivated lipase on
    metmyoglobin-induced lipid
    peroxidation of fish oil emulsion,
    182*f*
  effect of nonesterified
    eicosapentaenoic acid (EPA) on
    lipid peroxidation of fish oil
    emulsion, 185
  effect of nonesterified EPA on
    metmyoglobin-induced lipid
    peroxidation of fish oil emulsion,
    188*f*
  effect of nonesterified fatty acid on
    metmyoglobin-induced lipid
    peroxidation of fish oil emulsion,
    189*f*
  effect on increase of thiobarbituric
    acid reactive substances
    (TBARS) of fish oil emulsion
    induced by different initiators,
    181*t*
  effect on lipid peroxidation of fish
    oil emulsions with different
    initiators, 179, 185
  effect on metmyoglobin-induced
    lipid oxidation of fish oil
    emulsion, 179
  fatty acid composition of lipase-
    hydrolyzate of fish oil, 181*t*
  hydrolysis of fish oil emulsion
    procedure, 177–178
  lipid peroxidation of nonesterified
    EPA and EPA methyl ester in
    emulsion system, 187*f*
  materials, 177
  measurement of peroxide value
    (PV) and TBARS, 178
  metmyoglobin-induced lipid
    peroxidation of fish oil and
    nonesterified EPA in emulsion
    system, 186*f*
  oxidation of lipase-hydrolyzed fish
    oil in emulsion, 178
  possible mechanism of lipid
    peroxidation in food and
    biological systems, 184*f*

relationship between, and oxidative stability of fish oil emulsion, 183*f*
thin layer chromatography (TLC) of extract of fish oil emulsion after incubation without lipase, with intact lipase, or with heat-inactivated lipase, 180*f*
Lipases, catalyzing reactions, 153
Lipid and lipoprotein levels, blood
effect of omega-3 fatty acids, 18
plasma lipids for animals fed diets containing soybean oil, coconut oil, and algae, 131, 133*f,* 134, 135*f*
See also Blue-green alga *Aphanizomenon flos-aquae*
Lipid dispersions, oxidative stability. *See* Oxidative stability of lipid dispersions
Lipid peroxidation
critical in edible oils, 176–177
effect of lipase hydrolysis on, of fish oil emulsions, 179, 185
non-enzymatic, via radical chain reaction, 177
See also Lipase hydrolysis
Lipids. *See* Structured lipids (SL)
Lipids, membrane, functional state with unsaturation in fatty acids, 74–75
Lipophilic antioxidant, nature influencing effectiveness, 275
Lipoprotein metabolism
balance between dietary fatty acids critical when generating lowest total cholesterol (TC) and low density lipoprotein/high density lipoprotein (LDL/HDL) ratio, 46
cebus monkeys exploring possibility of fine-tuning dietary fatty acid balance to enhance lipoprotein profile, 46, 50
cholesterol-lowering effect of dietary polyunsaturated fatty acids (PUFA), 37
decreasing LDL/HDL ratio along with TC without altering total fat intake, 46, 48*f*
effect of different lipoprotein setpoints on response to 18:2 relative to other fatty acids in Benedictine nuns, 43*f*
exchanging 2% energy from omega-6 to omega-3 high unsaturated fatty acids (HUFAs) and lack of effect on TC or LDL/HDL ratio in humans, 50, 51*f*
general summary of relative potency of dietary fatty acids on TC, 40, 42*f*
human crossover studies of fish oil versus control oil, 50, 51*t*
omega-6 and omega-3 balance and lipoproteins, 50–52
percent CHD mortality data from Seven Countries Study and relationship with fatty acid ratio in original diets, 47*f*
PUFAs reducing coronary heart disease (CHD), 38, 40
putative cholesterolemic response to each 1% energy from individual dietary fatty acids during low-cholesterol diets, 42*f*
putative relationship between dietary 18:2 threshold and low density lipoprotein cholesterol (LDL-C) in humans, 44, 45*f*
re-balancing dietary saturates:monounsaturates:polyunsaturates (S:M:P) ratio close to 10:10:10 improving LDL/HDL ratio, 46, 48*f*
re-balancing S:M:P ratio at 37% energy lowering TC, 46, 48*f*
relationship between classes of fatty acids consumed and relative risk for CHD, 39*f*

relationship between dietary fatty acids and, 37–38
replacing one-third of 4% energy from 18:2 omega-6 with 18:3 omega-3 and detrimental impact on HDL-C in cebus monkeys, 51*f*
replacing one-third of 4% energy from omega-6 to omega-3 and its detrimental impact on HDL-C in cebus monkeys, 53*f*
Seven Countries Study, 44
stepwise reduction of dietary saturated fatty acids (SFA) in cebus monkeys and improvement in TC, 49*f*
stepwise removal of saturated fatty acids from 35% energy total fat to 25% energy decreasing TC by lowering both LDL and HDL fractions, 46, 47*f*
study of Benedictine nuns, 40, 44
superior potential of fish oil and safflower oil reducing both LDL-C and total plasma cholesterol, 38, 40, 41*f*
threshold requirement for 18:2, 52
tile setpoint or lipoprotein profile at time of intervention influencing ultimate response to dietary fat saturation and its overall fatty acid profile, 44, 45*f*
Long chain fatty acids (LCFA)
metabolism and transportation, 151
See also Modified oils with highly unsaturated fatty acids; Structured lipids (SL)
Long chain hydroxyacylCoA dehydrogenase deficiency, clinical trials with docosahexaenoic acid single cell oil (DHASCO), 103–104
Low density lipoprotein (LDL)
effects after supplementation with docosahexaenoic acid single cell oil (DHASCO), 103–104
See also Lipoprotein metabolism

Low temperature crystallization, method for separating highly unsaturated fatty acids (HUFA) from others, 146–147

M

Macadamia nuts, antioxidants, 272
Maillard reaction products (MRP), natural antioxidants, 274
Maize oil. *See* Rancidity of fish oils
Mammals. *See* Retinal function
Marine oil processing
alkaline refining process for crude fish oil, 7
bleaching process, 7
commercial products, 5
enriched omega-3 oils, 7, 9
extraction of oil from fish by grinding, 5
levels of oil in fish, 6
production of omega-3 concentrates from marine oils, 10*f*
production of refined, bleached, and deodorized (RBD) marine oils, 8*f*
two-stage deodorization process, 7
Meads, nomenclature, 6*t*
Medium chain fatty acids (MCFA)
metabolism and transportation, 151
See also Modified oils with highly unsaturated fatty acids; Structured lipids (SL)
Melon seed oil
incorporating EPA using lipases, 164–165
oxidative stability, 170
Membrane lipids, functional state with unsaturation in fatty acids, 74–75
Mental acuity, improvements in infants with docosahexaenoic acid single cell oil (DHASCO), 102–103
Mental Development Index, docosahexaenoic acid (DHA), 58
Mental stress-induced extra-

aggressive behavior, docosahexaenoic acid (DHA), 61
Metabolism. *See* Lipoprotein metabolism
Metmyoglobin-induced lipid peroxidation
  effect of lipase hydrolysis, 179
  effect of nonesterified eicosapentaenoic acid (EPA) on, of fish oil emulsion, 177f
  effect of nonesterified fatty acid on, of fish oil emulsion, 189f
  fish oil and nonesterified EPA in emulsion system, 186f
  See also Lipase hydrolysis
Microalgae
  fatty acid profiles, 97t
  large-scale production of photosynthetic, 96–97
  potential applications from dried, 123t
  primary producers of docosahexaenoic acid (DHA), 95–97
  primary producers of eicosapentaenoic acid (EPA) and DHA, 108–109
  primary producers of omega-3 fatty acids in marine food web, 96f
  See also Docosahexaenoic acid (DHA) production from microalgae; Single cell oil, docosahexaenoic acid (DHASCO)
Microencapsulation, seal blubber oil and its products, 148–149
Model systems. *See* Antioxidants, natural
Modified oils with highly unsaturated fatty acids
  aldehyde products from autoxidation of selected unsaturated fatty acids, 169t
  benefits, 164

  conjugated dienes and thiobarbituric acid reactive substances (TBARS) values of borage oil (BO) and evening primrose oils (EPO) during accelerated storage at 60°C, 168t
  decreasing tumor growth in rodents, 164
  degradation of borage and evening primrose oils, 166, 170
  eicosapentaenoic acid (EPA) and capric acid incorporation into BO, 165
  evaluating oxidative stability of oils, 166
  fatty acid profile of BO and EPO before and after enzymatic modification by *Pseudomonas sp.* lipase, 167t
  incorporation of EPA into crude melon seed oil using lipases, 164–165
  incorporation of omega-3 PUFA into soybean oil, 164
  lipase from *Pseudomonas sp.* incorporating EPA and DHA into borage and evening primrose oils, 165
  main volatile aldehydes of BO and EPO stored at 60°C, 171t
  modifying fatty acid composition of BO using ethyl ester of EPA, 165–166
  oxidative stability, 166, 170
  oxidative stability of melon seed oil, 170
  production of TBARS, 166
  synthesis, 164–166
  synthesis methods, 163
  triacylglycerols with mixtures of short-, or medium-, and/or long-chain fatty acids, 163
Mucor miehei lipase
  adding EPA and DHA to soybean

and canola oils, 155–156
modifying soybean oil by incorporating omega-3 PUFA, 164
Multiple Risk Factor Intervention Trial (MRFIT), docosahexaenoic acid (DHA), 58
Myocardial infarction
coronary vascular disease, 18, 21
dietary intake of fatty acids and risk of, 24f
Eskimos and Danes, 16t

N

Neonatal adrenaleukodystrophy, docosahexaenoic acid (DHA), 60
Neuronal cells
critical functions of docosahexaenoic acid (DHA) in normal development and metabolism, 93–94
DHA component of phospholipids of, 92
Nomenclature, highly unsaturated fatty acids, 5, 6t
Nostoc commune, decreasing blood cholesterol, 138
Nutraceuticals, application of seal blubber oil, 145–149
Nutrition
benefits of highly unsaturated fatty acids, 4–5
See also Human diet

O

Odorants. *See* Seal blubber oil (SBO) odorants
Olfaction, perception of flavor, 281
Olives, extraction of olive oil, 264
Omega-3 fatty acids
abundance in oil from body of fatty fish species, 143
beneficial health effects, 143
cardiac arrhythmia reduction in animals, 21
dependence of coronary heart disease (CHD) risk and death on consumption of long chain polyunsaturated fatty acids (PUFA), 20f
dependence of fish intake and mortality from CHD, 19f
desaturation, elongation, and retroconversion, 56f
dietary intake of fatty acids and risk of myocardial infarction, 24f
effect on coronary vascular disease (CVD), 18, 21
effect on postprandial lipemia, 21t
electroretinographic (ERG) α- and β-wave amplitudes dependence on supply, 80
enriched oils, 7, 9
preventing restenosis, 21
production of omega-3 concentrates from marine oils, 10f
recommended intakes, 22–23, 24f
See also Salmon, farmed Atlantic; Structured lipids (SL)
Omega-6 and omega-3 oils, balance in diet, 16
Omega-6 fatty acids, correlation between dietary intake and total levels, 138
Omega-6 versus omega-3. *See* Lipoprotein metabolism
Oregano, stabilizing pork emulsions, 262t
Organoleptic assessment, stability studies of oils and oil blends, 281
Oxidation
formation of off-odors, 221–222
seal blubber oil and its products, 148–149

stabilization of structured lipids (SL), 152
Oxidative Stability Index (OSI), structured lipids, 158–159
Oxidative stability of lipid dispersions
  antioxidant technologies controlling lipid oxidation in foods, 244
  autoxidation inhibition in emulsions by metal binding agents, 247
  changes in charge whey protein isolate (WPI)- and Tween 20-stabilized hexadecane emulsions as function of pH, 255f
  emulsion consisting of three regions, 245
  factors affecting oxidation in bulk lipid systems, 244–245
  factors decreasing iron-emulsion droplet interactions, 249
  Fe^{2+} promoted breakdown of cumene peroxide in hexadecane emulsions stabilized by sodium dodecyl sulfate (SDS) at pH 7.0 or 3.0, 252f
  Fe^{2+} promoted breakdown of cumene peroxide in hexadecane emulsions stabilized by Tween 20 at pH 7.0 or 3.0, 252f
  Fe^{3+} promoted breakdown of cumene peroxide in hexadecane emulsions stabilized by Tween 20 at pH 7.0 or 3.0, 253f
  impact of interfacial properties of emulsion droplets on activity of iron, 247, 249, 254
  incorporation of oxygen into unsaturated fatty acid increasing polarity of lipid molecule, 249
  iron-catalyzed formation of thiobarbituric acid reactive substances (TBARS) in Tween 20- and WPI-stabilized emulsions over pH 3–7 range, 255f
  lipid oxidation in emulsions, 245, 247
  model system of corn oil-in-water emulsions stabilized with anionic, cationic, and nonionic surfactants, 249
  oxidation of bulk and Tween 20-emulsified salmon oil in presence of ethylenediaminetetraacetic acid (EDTA) and transferrin, 248f
  oxidation of salmon oil emulsions in presence of EDTA and transferrin, 248f
  oxidation of tocopherol stripped corn oil-in-water emulsions stabilized by SDS, dodecyltrimethyl ammonium bromide (DTAB), and Brij (nonionic surfactant), 251f
  partitioning of antioxidants and prooxidants between three regions, 245, 247
  potential orientation of lipid peroxides in oil-in-water emulsions, 251f
  potential partitioning patterns of antioxidants and prooxidants in oil-in-water emulsions, 246f
  reactivity of ferric versus ferrous ions with peroxides, 249, 254
  synthetic food additives controlling oxidation, 244
  synthetic surfactants studying impact of interfacial region on lipid oxidation rates, 254
  zeta potential of SDS-, Brij-, and DTAB-stabilized hexadecane emulsions in presence of iron salts at pH 3.0, 250t
Oysters
  compositions of volatile compounds from total wet tissues of, 213
  fatty acid compositions, 212, 213t
  prominent volatile compounds, 215t
  samples, 210
  See also Fish aroma

P

Peanut
  hulls as antioxidants, 272
  seed oil proteins reducing rancidity in ground beef, 264t
Peroxide value (PV), assessing oxidative stability of oils and oil blends, 281
Peroxisomal beta-oxidation step, biosynthetic pathway of docosahexaenoic acid (DHA), 93
Peroxisomal disorders, docosahexaenoic acid (DHA), 60–61
Pharmaceuticals, omega-3 concentrates from seal blubber oil, 145–148
Phoca groenlandica (harp seal). *See* Seal blubber oil (SBO)
Phospholipids (PL)
  abundant sources, 266
  activity of natural and synthetic antioxidants in fish oil model systems, 267
  controversial group of antioxidants, 266
  effects of commercial phosphatidylcholine (PC) on thiobarbituric acid (TBA) values of salmon oil oxidation model system, 270f
  effects of PL fractions on oxidation of salmon oil oxidation model system, 268f
  effects on change in fatty acid profile, 269f
  exhibiting high degree of synergism with polyhydroxy flavonoids, 267
  proposed mechanisms of action, 266
  removal during degumming process, 271
  retention of minimum levels of PL adding to stability, 271
  synergism between PL and other natural antioxidants, 267
  See also Antioxidants, natural
Photosynthetic phytoplankton
  large-scale production, 96–97
  primary source of docosahexaenoic acid (DHA), 95–96
  See also Single cell oil, docosahexaenoic acid (DHASCO)
Phytic acid, natural antioxidant, 273
Plant and food extracts, antioxidant activity
  dried beans, 273
  leaf extracts, 274
  macadamia nuts, 272
  Maillard reaction products (MRP), 274
  peanut hulls, 272
  phytic acid, 273
  sweet potatoes, 273
Platelet aggregation, omega-3 fatty acid consumption, 128
Platelets
  decreasing aggregation with increased dietary unsaturation, 72–73
  effect of degree of dietary lipid unsaturation upon platelet function in rat, 73t
  hyperaggregation, 75
  influence of degree of unsaturation of dietary fat upon platelet response, 74t
  physical influence of increasing unsaturation in diet, 76
Polyphenols, natural antioxidants from tea, 265
Polyunsaturated fatty acids (PUFA)
  antiarrhythmic action from effects on electrophysiology of cardiac myocytes, 30

coronary heart disease (CHD) mortality as function of dietary PUFAs, 47f
cholesterol-lowering effect of dietary, 37
coronary vascular disease, 18, 21
effect of inducing deficiency, 131, 132f, 137
electrical stabilizing effect on cardiomyocyte, 30, 31f, 32
essential fatty acids, 17
health benefits, 2–3
important effects on brain, 34
infant development, 22
inflammatory and autoimmune disorders, 22
metabolism, 17–18
pattern in phospholipids of tissues, 79–80
precursors to omega-6 and omega-3, 17
proportion of PUFA in mammalian brain, 79
recommended intakes, 22–23
restenosis, 21
roles in human health, 18–22
secondary prevention trials showing prevention of ischemia-induced sudden cardiac death, 34
See also Arrhythmias; Blue-green alga *Aphanizomenon flos-aquae*
Pork emulsions, ground spices stabilizing, 262t
Postprandial lipemia, effect of omega-3 long chain PUFA, 21t
Potatoes, sweet, natural antioxidant, 273
Potency of natural antioxidants
assay type, 276
factors affecting, 275–276
importance of reaction environment, 275–276
lipophilic and hydrophilic nature influencing effectiveness, 275
Pregnancy, docosahexaenoic acid (DHA), 57–58

Prooxidants. *See* Oxidative stability of lipid dispersions

R

Rainbow trout
hydroperoxide levels in blood and liver, 215, 216t
See also Fish aroma
Rancidity of fish oils
adding fractions of maize oil to fish oil, 285
assessing stability of oils and oil blends, 281
binary mixtures of antioxidants, 295
chromatography of fish and maize oils, 285
composition of fish and maize oils, 285t
concentration of tocopherols and metal ions in maize oil and fish oil, 290t
concentration of tocopherols and metal ions of maize oil, 291t
conjugated double bond system of phytosteradienes, 290
experimental materials, 282
experimental methods, 282
flavor wheels of maize, sunflower, and Chilean fish oil (FHO), 283f, 284f
gas chromatogram of nonpolar fraction of maize oil, 287f
induction curves of FHO, FHO with cholestadiene, and FHO with cholestadiene + lecithin, 297f
induction curves of FHO with binary mixtures of antioxidants, 289f
induction curves of FHO with ternary mixtures of antioxidants, 293f
levels of tocopherols and metals in maize oil, 292t

mass spectra of ergostadiene and stigmastadiene, 288f
minimizing loss of tocopherols from maize oil chromatographed over silica gel, 286
non-polar fraction of maize oil by gas chromatography–mass spectrometry (GC–MS), 286
novel silica column with citric acid (CA) plug, 286
oil imports to United Kingdom, 285t
peroxide values for maize oil (MZO), 292t
peroxide values of fish oil, 296t
peroxide values of maize oil eluted on alumina and silica columns under nitrogen, 291t
peroxide values of nonpolar maize oil fraction from silica column, 294t
peroxide values of post alumina treated oils under nitrogen, 290t
peroxide values (PV) for FHO, MZO, and blend (MZO/FHO 75:25), 285t
role of additives for maintaining quality of fish oil, 286, 290
sensory evaluations of maize, sunflower, and Chilean fish oils, 282
single component antioxidant, 295
ternary mixture of antioxidants, 295
Reaction environment, potency of natural antioxidants, 275–276
Reclogging arteries, omega-3 fatty acids preventing, 21
Recommended dietary allowances (RDA), omega-3 and omega-6 fatty acids, 22–23, 24f
Refined docosahexaenoic acid (DHA)-rich oil
fatty acid profile, 122t
nutritional supplements and food ingredients, 119

oil extraction and purification process producing, 120f
potential application, 123t
Refining
fish oils, 23
production process for omega-3 oils, 23, 25f
Reginal function, PUFAs patterns, 79–80
Restenosis, omega-3 fatty acids preventing, 21
Retinal function
animals and diets, 81–82
animal studies demonstrating effects of omega-3 PUFA on, 83t
docosapentaenoic acid (DHA) and docosapentaenoic acid (DPA) proportions in retinal phospholipids in guinea pigs fed diets containing different levels of omega-3 fatty acids, 86f
diet introductions, 84
electroretinographic (ERG) differences due to omega-3 deficiency, 81
importance of lipid nutrition during pregnancy and early postnatal life, 87
lipid analyses methods, 82
loss of DHA from linoleic acid rich diets, 80
materials and methods, 81–82
patterns of PUFA, 79–80
physiological role of α-linolenic acid (LNA), 80
previous studies on LNA and ERG function, 87–88
PUFA composition of retinal phospholipids from guinea pig fed diets containing different levels of omega-3 PUFA, 85t
reduced DHA levels, 84
statistical analyses methods, 82
strategies to increase retinal DHA levels, 84, 88–89

See also α-Linolenic acid (LNA)
Retinitis pigmentosa
   clinical trials with docosahexaenoic acid single cell oil (DHASCO), 103–104
   docosahexaenoic acid (DHA), 60, 61
Rosemary
   chemical composition of active antioxidant components, 263
   spice of greatest amount of study, 262
   stabilizing pork emulsions, 262*t*
   See also Antioxidants, natural

S

Safety testing, docosahexaenoic acid single cell oil (DHASCO), 101–102
Safflower oil, superior potential for reducing low density lipoprotein cholesterol (LDL-C) and total plasma cholesterol, 38, 40
Sage, stabilizing pork emulsions, 262*t*
Salmon, farmed Atlantic
   actual weights of fish on three diets for 12 months, 195*f*
   distribution of omega-3 fatty acids in phospholipids and triacylglycerols in white muscle of experimental versus farmed, 202*t*
   examples of lipid contents, and polyunsaturated fatty acids in lipid, comparing, with wild salmon, 200*t*
   experimental fish, 192
   fat content of diets, 192, 193*f*
   fat content of whole steak, trimmed, white muscle, and other quality parameters, 198*t*
   fat distribution, 197
   limiting factor in growth, 203
   lipid extraction method, 194
   misleading information regarding wild versus farmed, 203
   omega-3 fatty acids, 200–203
   percentages of major components in Atlantic salmon diets, 193*f*
   proximate compositions of diets for salmon feeding studies, 197*t*
   role of marine phytoplankton fatty acids in fats of seafoods, 204*f*
   texture analysis method, 194
   texture evaluation, 194, 197, 200
   texture indices in loin area of experimental salmon based on dietary treatments, 199*f*
   unusual accumulation of docosapentaenoic acid (DPA), 201
   wild salmon providing baseline for comparison, 201
   See also Conjugated linoleic acids (CLA)
Salmon oil
   effects of commercial phosphatidylcholine (PC) on thiobarbituric acid (TBA) values of salmon oil oxidation model system, 270*f*
   effects of phospholipids fractions on oxidation of, oxidation model system, 268*f*
   role of phospholipids as antioxidants, 267
   See also Oxidative stability of lipid dispersions
Sample dilution analysis (SDA) method, 224, 226
   See also Seal blubber oil (SBO) odorants
Saturated fatty acids (SFA)
   correlation between dietary and serum levels, 137
   dietary sources, 17
Scenedesmus acutus, decreasing blood cholesterol, 138
Schizochytrium sp.

bioavailability, 113
composition, 110, 112
dried docosahexaenoic acid (DHA)-
  rich microalgae, 109–110
fatty acid composition of liver
  tissue from female rates fed
  control and admixtures, 115f
fatty acid composition of sera from
  female rates fed control and
  admixtures, 114f
fermentation, 110
flow diagram for fermentation and
  recovery of whole cell
  microalgae rich in DHA, 111f
proximate composition of, whole
  cell microalgae produced via
  fermentation, 112t
See also Docosahexaenoic acid
  (DHA) production from
  microalgae
Schizophrenia, docosahexaenoic acid
  (DHA), 60
Sea bream
  hydroperoxide levels in blood and
    liver, 215, 216t
  See also Fish aroma
Seafood. *See* Conjugated linoleic acids
  (CLA)
Seal blubber oil (SBO)
  application in food and
    nutraceuticals, 145–149
  chromatography method, 148
  compositional characteristics, 143–144
  content of omega-3 fatty acids in
    urea complexing and non-urea
    complexing fractions of SBO,
    146t
  current harvest, 142
  distillation method, 147–148
  enzymatic esterification with
    glycerol method, 147
  enzymatic hydrolysis method, 147
  fatty acid distribution in different
    position of triacylglycerols of
    SBO, 144t
  inhibition of oxidation of SBO by
    antioxidants under accelerated
    storage condition, 149t
  low temperature crystallization
    method, 146–147
  major fatty acids, 144t
  methodologies for obtaining
    concentrates, 145
  omega-3 concentrations from SBO
    as pharmaceuticals, 145–148
  protection of SBO and products
    from oxidation, 148–149
  rich source of long-chain omega-3
    polyunsaturated fatty acids
    (PUFA), 142–143
  supercritical fluid extraction (SFE)
    method, 148
  traditional uses, 143
  urea complexation methodology,
    146
Seal blubber oil (SBO) odorants
  direct thermal desorption–gas
    chromatography–mass
    spectrometry (DTD–GC–MS)
    method, 226–227
  direct thermal desorption–gas
    chromatography–olfactometry
    (DTD–GCO) and sample dilution
    analysis (SDA) methods, 224,
    226
  DTD apparatus, 223f
  DTD–GC–MS analysis of crude
    and refined-deodorized SBO,
    228f
  experimental materials, 224
  general scheme for refining, 225f
  odor-active components, 227, 230–232
  predominant odor-active
    components of crude and refined-
    deodorized SBO, 229t, 230t
  relative concentrations of selected

volatile components of crude and refined-deodorized SBO, 231*t*
volatile profiles, 227
Seed oil extracts, natural antioxidants, 263–264
Seminal plasma and spermatozoa, docosahexaenoic acid (DHA), 62
Senate Select Committee on Nutrition and Human Needs, improving quality of life of Americans, 68–69
Sensory evaluations. *See* Rancidity of fish oils
Serum cholesterol
  incorporation of unsaturated fatty acids, 71–72
  risk factor for cardiovascular disease (CVD), 67–68
Serum fatty hydroperoxides, influence of degree of unsaturation of dietary fat, 74*t*
Serum lipid hydroperoxides, increase with unsaturated fatty acids in diet, 76
Serum lipids
  decreasing levels with unsaturation in fatty acids, 71
  influence of degree of unsaturation of dietary fat, 72*t*
Serum triglyceride, docosahexaenoic acid (DHA), 59–60
Serum vitamin E
  effect of degree of dietary lipid unsaturation upon, in rat, 73*t*
  influence of degree of unsaturation of dietary fat, 74*t*
Sesame seeds, oil extract, 263–264
Shellfish
  rich in highly unsaturated fatty acids, 209
  See also Fish aroma
Short chain fatty acids (SCFA)
  metabolism and transportation, 151
  See also Modified oils with highly unsaturated fatty acids; Structured lipids (SL)

Single cell oil, docosahexaenoic acid (DHASCO)
Crypthecodinium cohnii strain for production of DHASCO, 98
characterization, 100–101
children and adult clinical studies, 103–104
clinical studies, 102–104
DHASCO supplementation, 103
effect on high density lipoprotein (HDL), low density lipoprotein (LDL), and total cholesterol, 103–104
fatty acid composition of DHASCO from *C. cohnii*, 101*t*
fermentation process for production of *C. cohnii* biomass, 98*f*
Good Laboratory Practices (GLP) compliance, 102
infant clinical studies, 102–103
medium for growing *C. cohnii* from shake flask to production scale, 99
mental acuity improvements, 102–103
nonsaponifiable fraction, 100–101
oil processing flow sheet for production of DHASCO from *C. cohnii*, 99*f*
oil processing procedure, 99–100
production, 97–100
safety testing, 101–102
serum triacylglycerols and high density lipoprotein (HDL) cholesterol, 104*f*
shear sensitivity of *C. cohnii* biflagellate, 98
triglyceride structure of DHASCO, 101
visual acuity improvements, 102–103
See also Docosahexaenoic acid (DHA)
Sleep problems, docosahexaenoic acid (DHA), 61

Smelt
  characteristic aromas, 209
  fatty acid compositions of blood and liver, 212
  hydroperoxide levels in blood and liver, 215, 216*t*
  prominent fatty acid compositions of total lipids, 214*t*
  See also Fish aroma
Sodium dodecyl sulfate (SDS). *See* Oxidative stability of lipid dispersions
Soybean oil
  fatty acid profile, 129*t*
  plasma lipid profiles for animals fed, 131, 133*f*, 134, 135*f*
  See also Blue-green alga *Aphanizomenon flos-aquae*
Soy extracts
  antioxidant activity due to flavonoids, 264–265
  proteins reducing rancidity in ground beef, 264*t*
  seed oil proteins reducing rancidity in ground beef, 264*t*
Spermatozoa, docosahexaenoic acid (DHA), 62
Spices
  chemical composition of active antioxidant components of rosemary, 263
  consequences of irradiation on food quality, 263
  ground spices stabilizing pork emulsions, 262*t*
  meat flavor deterioration (MFD), 261–262
  natural antioxidants, 260–263
  occurrence of warmed-over flavor (WOF), 261–262
  rosemary for greatest amount of study, 262
  See also Antioxidants, natural
Spirulina sp., affecting cholesterol metabolism, 138

Stability
  oxidative, of modified oils, 166, 170
  Oxidative Stability Index (OSI), 158–159
  structured lipids, 156–159
  See also Oxidative stability of lipid dispersions; Rancidity of fish oils
Stearidonic acid, nomenclature, 6*t*
Structured lipids (SL)
  acidolysis in production of SL, 153
  antioxidants for stabilization, 152
  definition, 151
  difference between unmodified oils and SL, 158–159
  effects of food grade antioxidants on Oxidative Stability Index values of enzymatically synthesized SL containing HUFA, 159*t*
  eicosapentaenoic acid (EPA)-rich fatty acids (EPAX 6000) for transesterifying tricaprylin, 153
  enriched food products, 152
  enrichment or supplementation of existing food products, 160
  enzymatically modified fish oil with capric acid producing SL, 154*t*
  enzymatically synthesized SL with capric acid, γ-linolenic acid (GLA), and EPA in same glycerol backbone using borage oil, 154–155
  enzymatic modification of borage oil with capric acid producing SL, 155*t*
  enzymatic process for desired structural configuration, 151–152
  fatty acid composition of enzymatically modified trilinolein to contain omega-3 highly unsaturated fatty acid (HUFA), 158*t*
  fatty acid composition of linolenic acid containing vegetable oils

modified by *Candida antarctica* lipase to contain omega-3 HUFA, 157t
fatty acid composition of linolenic acid containing vegetable oils modified by *Rhizomucor miehei* lipase to contain omega-3 HUFA, 156t
incorporating EPA, docosahexaenoic acid (DHA), or both into pure trilinolein using lipases, 156, 158t
modification of tricaprylin with EPAX 6000 free fatty acids producing SL, 154t
Mucor miehei and *Candida antarctica* lipases adding EPA and DHA to soybean and canola oils, 155–156
Oxidative Stability Index (OSI) of SL with antioxidants, 158
reaction types by lipases, 153
requiring stabilization and proper packaging, 152
SL containing GLA and omega-3 HUFA, 154–155
SL from fish oil, 153
stability of omega-3 HUFA, 156–159
supplementing infant formula, 152
synthesizing SL from canola and fish, 157
use of SL containing HUFA in food formulations, 159–160
Sunflower oil. *See* Rancidity of fish oils
Supercritical fluid extraction (SFE), separation process for concentration of polyunsaturated fatty acids, 148
Supplements
docosahexaenoic acid (DHA) in healthy men, 62
encapsulation, 9
market for *Aphanizomenon flos-aquae* as health food, 126

Surfactants. *See* Oxidative stability of lipid dispersions
Sweet potatoes, natural antioxidant, 273
Swine, docosahexaenoic acid (DHA) enriched, 118
Synthetic antioxidants, possible carcinogenicity and toxicity, 259

T

Taste, factor in food selection, 192
Tea flavonoids
natural antioxidants, 265–266
primary antioxidant activity, 266
Temper tantrums, docosahexaenoic acid (DHA), 61
Tertiarybutylhydroquinone (TBHQ)
antioxidant for seal blubber oil, 148–149
antioxidant for structured lipids, 152
stabilizing pork emulsions, 262t
See also Antioxidants
Texture
analysis method, 194
evaluation for farmed Atlantic salmon, 194, 197, 200
factor in food selection, 192
indices in loin area of experimental salmon based on dietary treatments, 199f
Thiobarbituric acid, measuring secondary oxidation products of oil or oil blend, 281
Thiobarbituric acid reactive substances (TBARS)
effect of lipase hydrolysis on increase of TBARS of fish oil emulsion induced by different initiators, 181t
measurement method, 178
production of modified and unmodified oils, 166, 168t

relationship between increase of TBARS and level of released free fatty acid by lipase-treatment of fish oil emulsions, 179, 183f
Thyme, stabilizing pork emulsions, 262t
Tocopherols. *See* Rancidity of fish oils
Triacylglycerols
  fatty acid distribution in different position of triacylglycerols of seal blubber oil, 144t
  lowering effect, 18
  modifying for structured lipids (SL), 151–152
  See also Modified oils with highly unsaturated fatty acids; Seal blubber oil (SBO); Structured lipids (SL)
Tricaprylin, modifying, producing structured lipids, 153, 154t
Triglycerides
  decreases after supplementation with docosahexaenoic acid single cell oil (DHASCO), 103–104
  levels for animals fed diets containing soybean oil, coconut oil, and alga, 134, 136f
  See also Blue-green alga *Aphanizomenon flos-aquae*
Trout, rainbow
  hydroperoxide levels in blood and liver, 215, 216t
  See also Fish aroma

U

Unsaturated fatty acids
  dietary sources, 17
  rate of autoxidation increasing with increase, 69
  recommendation of Senate Select Committee on Nutrition and Human Needs, 68–69
  studying effect in biological system, 67
  See also Cardiovascular disease (CVD)
Upper Klamath Lake, Oregon. *See* Blue-green alga *Aphanizomenon flos-aquae*
Urea complexation, method to obtain high concentration of desired long-chain omega-3 fatty acid, 146

V

Vegetables, antioxidants, 271–272
Ventricular fibrillation, omega-3 fatty acid consumption, 128
Visual acuity, improvements in infants with docosahexaenoic acid single cell oil (DHASCO), 102–103
Vitamin E, serum, influence of degree of unsaturation of dietary fat, 74t
Vitamins, antioxidants, 259

W

Wild salmon. *See* Salmon, farmed Atlantic
Wine, antioxidants, 271–272

Y

Yellowtail
  hydroperoxide levels in blood and liver, 215, 216t
  See also Fish aroma

Z

Zellweger syndrome, docosahexaenoic acid (DHA), 60

Bestsellers from ACS Books

The ACS Style Guide: A Manual for Authors and Editors (2nd Edition)
Edited by Janet S. Dodd
470 pp; clothbound ISBN 0–8412–3461–2; paperback ISBN 0–8412–3462–0

Writing the Laboratory Notebook
By Howard M. Kanare
145 pp; clothbound ISBN 0–8412–0906–5; paperback ISBN 0–8412–0933–2

Career Transitions for Chemists
By Dorothy P. Rodmann, Donald D. Bly, Frederick H. Owens, and Anne-Claire Anderson
240 pp; clothbound ISBN 0–8412–3052–8; paperback ISBN 0–8412–3038–2

Chemical Activities (student and teacher editions)
By Christie L. Borgford and Lee R. Summerlin
330 pp; spiralbound ISBN 0–8412–1417–4; teacher edition, ISBN 0–8412–1416–6

Chemical Demonstrations: A Sourcebook for Teachers, Volumes 1 and 2, Second Edition
Volume 1 by Lee R. Summerlin and James L. Ealy, Jr.
198 pp; spiralbound ISBN 0–8412–1481–6
Volume 2 by Lee R. Summerlin, Christie L. Borgford, and Julie B. Ealy
234 pp; spiralbound ISBN 0–8412–1535–9

The Internet: A Guide for Chemists
Edited by Steven M. Bachrach
360 pp; clothbound ISBN 0–8412–3223–7; paperback ISBN 0–8412–3224–5

Laboratory Waste Management: A Guidebook
ACS Task Force on Laboratory Waste Management
250 pp; clothbound ISBN 0–8412–2735–7; paperback ISBN 0–8412–2849–3

Reagent Chemicals, Ninth Edition
768 pp; clothbound ISBN 0–8412–3671–2

Good Laboratory Practice Standards: Applications for Field and Laboratory Studies
Edited by Willa Y. Garner, Maureen S. Barge, and James P. Ussary
571 pp; clothbound ISBN 0–8412–2192–8

For further information contact:
Order Department
Oxford University Press
2001 Evans Road
Cary, NC 27513
Phone: 1-800-445-9714 or 919-677-0977

More Best Sellers from ACS Books

Microwave-Enhanced Chemistry: Fundamentals, Sample Preparation, and Applications
Edited by H. M. (Skip) Kingston and Stephen J. Haswell
800 pp; clothbound ISBN 0-8412-3375-6

Designing Bioactive Molecules: Three-Dimensional Techniques and Applications
Edited by Yvonne Connolly Martin and Peter Willett
352 pp; clothbound ISBN 0-8412-3490-6

Principles of Environmental Toxicology, Second Edition
By Sigmund F. Zakrzewski
352 pp; clothbound ISBN 0-8412-3380-2

Controlled Radical Polymerization
Edited by Krzysztof Matyjaszewski
484 pp; clothbound ISBN 0-8412-3545-7

The Chemistry of Mind-Altering Drugs: History, Pharmacology, and Cultural Context
By Daniel M. Perrine
500 pp; casebound ISBN 0-8412-3253-9

Computational Thermochemistry: Prediction and Estimation of Molecular Thermodynamics
Edited by Karl K. Irikura and David J. Frurip
480 pp; clothbound ISBN 0-8412-3533-3

Organic Coatings for Corrosion Control
Edited by Gordon P. Bierwagen
468 pp; clothbound ISBN 0-8412-3549-X

Polymers in Sensors: Theory and Practice
Edited by Naim Akmal and Arthur M. Usmani
320 pp; clothbound ISBN 0-8412-3550-3

Phytomedicines of Europe: Chemistry and Biological Activity
Edited by Larry D. Lawson and Rudolph Bauer
336 pp; clothbound ISBN 0-8412-3559-7

For further information contact:
Order Department
Oxford University Press
2001 Evans Road
Cary, NC 27513
Phone: 1-800-445-9714 or 919-677-0977